Feedback Strategies for Wireless Communication

Berna Özbek • Didier Le Ruyet

Feedback Strategies for Wireless Communication

 Springer

Berna Özbek
İzmir Institute of Technology
İzmir, Turkey

Didier Le Ruyet
Conservatoire National
des Arts et Métiers
Paris Cedex 03, France

ISBN 978-1-4614-7740-2   ISBN 978-1-4614-7741-9 (eBook)
DOI 10.1007/978-1-4614-7741-9
Springer New York Heidelberg Dordrecht London

Library of Congress Control Number: 2013948453

Printed on acid-free paper

Springer is part of Springer Science+Business Media (www.springer.com)

Preface

During the past decade, many wireless communication techniques have been developed to achieve various goals such as higher data rate, more robust link quality, and higher number of users in a given bandwidth. For wireless communication systems, depending on the availability of a feedback link, two approaches can be considered: namely open and closed loop. Open loop communication system that does not exploit the channel knowledge at the transmitter is now well understood from both a theoretical and practical point of view. On the other hand, closed loop based communication systems have emerged as a promising solution to increase the data rate and the robustness of the communication links by using channel state information. The performance of single input single output (SISO) closed loop wireless communication systems can be improved with adaptive modulation and coding and power allocation by exploiting channel state information at the transmitter. The potential gains are tremendous in the case of multiuser multiple input multiple output (MIMO) systems where it can be possible to communicate to more than one user. Furthermore, cooperative multicell transmission is a recent solution to increase data rate and reduces outage in wireless networks by mitigating intercell interference caused by the other cells. The channel state information at the transmitter can be obtained through the feedback link. Since it is important to limit the amount of information to feed back, different strategies have been considered up to now. This book will address all these feedback strategies to design efficient wireless communication systems from both theoretical and practical issues including the study of the quantization of the channel state information on the performance.

This book is intended to provide a comprehensive review of feedback strategies in wireless communication systems from a signal processing perspective. While there are many manuscripts on MIMO systems and LTE standard, the literature is quite scarce when it comes to wireless communication with feedback. We came up with the conclusion that there is a need for a monograph that carefully explains the theory and implementation of feedback strategies in wireless communication systems. We believe that this book which summarizes the most useful ideas and state-of-the-art algorithms and results in this important area of research will help to understand this emerging field.

The book is written for graduate students and researchers at universities and research institutes, as well as researchers and engineers working in the telecommunication industry, who are already familiar with technical concepts such as probability, digital communication systems, and signal processing for communication. We hope that the book will contribute to a better understanding of the value of feedback strategies for wireless communication systems and may motivate further investigation in this exciting research area.

In the first part of the book, we overview different feedback strategies and analyze the impact of feedback information on the capacity for single user SISO/MIMO and multiuser SISO/MIMO wireless communication systems.

In the second part of the book, we focus on advanced topics on wireless communications and examine different feedback strategies for multicell networks and the usage of feedback information on LTE/LTE-A standard.

The authors would like to thank the anonymous reviewers for their constructive suggestions. We would like to thank Ali Dziri, Yahia Medjahdi, Ilhan Basturk, and Esra Aycan for their careful reading and useful advice for book chapters. We would like to thank Mylene Pischella, Rostom Zakaria, Mustapha Amara, and Hajer Khanfir for the discussions on multiuser MIMO and interference alignment.

Berna Özbek would like to thank her husband Erdal, her son Umut, and her parents for their encouragement and endless support for writing this book.

Didier Le Ruyet would like to thank his parents and his wife Christine for their support and encouragement during the preparation of this book.

İzmir, Turkey   Berna Özbek
Paris, France   Didier Le Ruyet

Contents

Acronyms

ACK	Acknowledgment
AoA	Angle of arrival
AoD	Angle of departure
AMC	Adaptive modulation and coding
APC	Anti-polar correlation
AR	Autoregressive
ARQ	Automatic request
AS	Angle spread
ATCM	Adaptive trellis coded modulation
AWGN	Additive white Gaussian noise
BC	Broadcast channel
BD	Block diagonalization
BER	Bit error rate
BF	Beamforming
BICM	Bit interleaved coded modulation
BLER	Block error rate
BPSK	Binary phase shift keying
BS	Base station
CB	Coordinated beamforming
CS	Coordinated scheduling
CDM	Code division multiplexing
CER	Codeword error rate
CGI	Channel gain information
CM	Cubic metric
CoMP	Coordinated multi-point
CP	Cyclic prefix
CPC	Co-polar correlation
CPR	Co-polar ratio
CQI	Channel quality indicator
CRC	Cyclic redundancy check
CRS	Common reference signal

CSI	Channel state information
CSI-RS	Channel state information reference signal
CSIR	Channel state information at receiver
CSIT	Channel state information at transmitter
DAST	Diagonal space time
DCS	Dynamic cell selection
DFT	Discrete Fourier transform
DMT	Discrete memoryless channel
DM-RS	Demodulation reference signal
DoA	Direction of arrival
DoD	Direction of departure
DP	Dirty paper
DPB	Dynamic point blanking
DPC	Dirty paper coding
DPS	Dynamic point selection
DRS	Dedicated reference signal
ECC	Error correcting code
EESM	Exponential effective SINR mapping
eICIC	enhanced Intercell interference coordination
eNB	Evolved node B
ESM	Effective SINR mapping
FDD	Frequency division duplex
FFR	Fractional frequency reuse
FFT	Fast Fourier transform
GSCM	Geometry-based stochastic channel models
GSO	Gram-Schmidt orthogonalization
HARQ	Hybrid automatic request
HSDPA	High speed downlink packet access
H-S/MRC	Hybrid selection/maximum ratio combining
H-S/MRT	Hybrid selection/maximum ratio transmission
IA	Interference alignment
ICI	Intercell interference
IDFT	Inverse discrete Fourier transform
IFFT	Inverse fast Fourier transform
IR	Incremental redundancy
ISI	Inter symbol interference
JT	Joint transmission
LOS	Line of sight
LTE	Long-term evolution
MAC	Multiple access channel
MCS	Modulation and code scheme
MCW	Multiple codewords
MI-ESM	Mutual information based effective SINR mapping
MISO	Multiple input single output
MIMO	Multiple input multiple output

ML	Maximum likelihood
MMSE	Minimum mean square error
MRC	Maximum ratio combining
MRS	Maximum rate scheduler
NACK	Negative acknowledgment
NC	Nulling and canceling
NLOS	Non line of sight
OBF	Orthogonal beamforming
OFDM	Orthogonal frequency division multiplexing
OFDMA	Orthogonal frequency division multiple access
PAPR	Peak to average power ratio
PAS	Power azimuth spectrum
PBCH	Physical broadcast channel
PCCC	Parallel concatenated convolutional coding
PCFICH	Physical control format indicator channel
PDCCH	Physical downlink control channel
PDP	Power delay profile
PDSCH	Physical downlink shared channel
PEP	Pairwise error probability
PFS	Proportional fair scheduler
PHICH	Physical hybrid ARQ indicator channel
PMCH	Physical multicast channel
PMI	Precoding matrix indicator
PRACH	Physical random access channel
PSAM	Pilot symbol assisted modulation
PSS	Primary synchronisation signals
PUCCH	Physical uplink control channel
PUSCH	Physical uplink shared channel
PU2RC	Per unitary basis stream user and rate control
QAM	Quadrature amplitude modulation
QO-STBC	Quasi orthogonal space-time block code
QPP	Quadratic permutation polynomial
QPSK	Quadrature phase shift keying
RBF	Random beamforming
RCPT	Rate compatible punctured turbo
RI	Rank indicator
RR	Round-robin
RS	Reference signal
RSC	Recursive convolutional encoder
SDMA	Space division multiple access
SER	Symbol error rate
SFR	Soft frequency reuse
SINR	Signal to interference noise ratio
SISO	Single input single output
SJNR	Signal to jamming plus noise ratio

SM	Spatial multiplexing
SNR	Signal to noise ratio
SRS	Sounding reference signal
SSS	Secondary synchronization signals
STBC	Space-time block code
STTC	Space-time trellis code
SUS	Semi-orthogonal user selection
SVD	Singular value decomposition
TAST	Threaded algebraic space time
TCM	Trellis coded modulation
TDD	Time division duplex
TDMA	Time division multiple access
TM	Transmission mode
TTI	Transmission time interval
UE	User equipment
ULA	Uniform linear array
VQ	Vector quantization
WIFI	Wireless fidelity
WIMAX	Worldwide interoperability for microwave access
WSSUS	Wide sense stationary uncorrelated scattering
XPD	Cross-polar discrimination
XPI	Cross-polar isolation
XPR	Cross-polar ratio
ZF	Zero forcing
ZFS	Zero forcing selection
3GPP	Third generation partnership project

List of Symbols

b	Beam index
B	Number of feedback bits
B_W	Bandwidth
C	Capacity
$\mathbf{F}$	precoding matrix
$\mathbf{F}_N$	Fourier matrix
γ	Signal to noise ratio
$\mathbf{H}$	Channel matrix
k	User index
K	Number of users
n	Subchannel index
N	Number of subchannels
N_t	Number of transmit antennas
N_r	Number of receive antennas
N_0	Noise power spectral density
$\mathbf{n}$	Noise vector
P	Power value
q	Cluster index
Q	Total number of clusters
$\mathbf{Q}$	Covariance matrix
R	Rate
$\mathbf{R_H}$	Correlation matrix
$\mathbf{R}_{tx}$	Correlation matrix at the transmitter
$\mathbf{R}_{tx}$	Correlation matrix at the receiver
$\mathbf{s}$	Data vector
σ	Noise standard deviation
t	Time
u	Cell index
U	Number of cells
$\mathbf{U}$	Unitary matrix

V	Unitary matrix
W	Precoding matrix
x	Input channel vector
y	Output channel vector
z	Combining vector

Notations

$\mathbb{E}[x]$	Mean of the random variable x
$\text{tr}[\mathbf{F}]$	Trace of the square matrix $\mathbf{F}$
$\mathbf{F}^H$	Transpose conjugate of matrix $\mathbf{F}$
$||\mathbf{F}||$	Matrix 2-norm of $\mathbf{F}$
$||\mathbf{F}||_F$	Matrix Frobenius norm of $\mathbf{F}$
$\mathbf{F}^{-1}$	Inverse of $\mathbf{F}$
$\mathbf{F}^\dagger$	Pseudo inverse of $\mathbf{F}$
$\det(F)$	Determinant of $\mathbf{F}$
$\text{vec}\{F\}$	Vectorization of the matrix $\mathbf{F}$
$F(.)$	Cumulated density function
$p(.)$	Probability density function
$\mathbb{R}$	Real field
$\mathbb{C}$	Complex field
$\odot$	Hadamard product
$\otimes$	Kronecker product
$\text{diag}(a_1, \ldots, a_n)$	Diagonal matrix whose diagonal entries starting in the upper left corner are $a_1 \ldots a_n$
$[F]_{k,k}$	Entry (k, k) of $\mathbf{F}$
$\xi_m[F]$	Maximum eigenvalue of the matrix $\mathbf{F}$
$Pr(A)$	Probability of the event A

Part I
Fundamentals of Feedback Strategies for Wireless Communication

Chapter 1
Introduction

Over the last years, the interest in high data rate for wireless transmission has significantly increased. During the past decade, significant contributions have been made in modulation coding and resource allocation and have led to the current systems. Future wireless communication systems should provide a wide range of services at a reasonable cost and sufficient quality of services, comparable to wireline technologies. An important direction to increase the rate and performance is to design precoding and post-coding schemes by exploiting the knowledge of the wireless channel conditions at the transmitter and receiver.

Feedback has been initially studied in the field of control theory. Since the 1950s, the interest of feedback has continued to grow in areas including control systems, information theory, source coding and channel decoding, and communication theory. In this book, we will mainly focus on feedback in wireless communication. In a wireless communication system, a feedback link is used to send the channel state information from the receiver to the transmitter.

A first solution to obtain the CSI at the transmitter is the usage of reciprocity of the channels. These systems use the fact that the forward and reverse links share the same fading distribution. An example of such a system is when the access technique is time division duplexing and the transmitter is able to estimate the reverse channel and use it as channel state information. However, due to practical difficulties, statistical adaptation is difficult to implement. When the wireless communication is accomplished using frequency division duplexing (FDD), it is no more possible to exploit the reciprocity since the transmit and receiver bands are well separated in frequency and uncorrelated. Another solution is to obtain CSI through the feedback link. In this book, we will only consider the case where the transmitter and receiver lack channel reciprocity and where adaptation techniques that use feedback to obtain the instantaneous channel realization are performed.

The capacity of point-to-point single user transmission scheme with single transmit and single receive antenna can be improved by exploiting feedback information in wireless communication systems. Since multiple antenna systems are playing an increasing role in wireless communications, it is possible to extend the feedback strategies to the case of multiple input multiple output (MIMO)

B. Özbek and D. Le Ruyet, *Feedback Strategies for Wireless Communication*, DOI 10.1007/978-1-4614-7741-9_1, © Springer Science+Business Media New York 2014

systems where the transmitter and the receiver are equipped with more than one antenna. The wireless networks go beyond the point-to-point transmission scheme by employing multiuser MIMO systems with feedback strategies. In addition to that, in order to further increase the system performance, the concept of cooperative communications has recently emerged as a solution to exploit the potential MIMO gains on a distributed scale. The cooperative multicell transmission increases data rate and reduces outage in wireless networks by mitigating intercell interference caused by the usage of unity frequency reuse factor.

The quality feedback information affects the performance of wireless communication systems significantly when the beamforming and interference management are performed. The amount and reliability of feedback information is crucial to perform adaptive resource allocation efficiently for next-generation wireless communication systems in order to improve the quality of service. This book will address all these feedback strategies to design efficient wireless communication systems. Theoretical analysis as well as practical algorithms for the required feedback information at the base stations will be examined to perform adaptive resource algorithms and mitigate interference coming from other cells at the transmitter efficiently. Different strategies about limited feedback information including the impact of quantization and the delay of CSI on the performance, various algorithms on reduced feedback information will be given for wireless systems.

Chapter 2 will give brief information about wireless communication systems including single input single output (SISO), multiple input single output (MISO) and MIMO channels, orthogonal frequency division multiplex (OFDM) and MIMO transmission strategies.

Chapter 3 will consider single user SISO transmission schemes by exploiting feedback information in flat fading and frequency selective channels. The capacity of the single user single antenna transmission schemes will be given firstly and the feedback information will be defined for different adaptive schemes. The adaptation of one or more parameters such as modulation, rate, codes and power will be reviewed in order to maximize the spectral efficiency of the radio link for a given quality. The impact of quantization and delay due to the feedback link will be evaluated on the system performance.

Chapter 4 will examine single user MISO and MIMO transmission schemes with feedback in flat fading and frequency selective channels. The capacity of these transmission schemes will be first reviewed assuming full feedback information is available at both transmitter and receive side. Then, the fundamentals of limited feedback strategies including quantization methods MISO and MIMO single-stream and multiple-stream cases will be given in detail. The imperfect channel conditions including correlated channel models as well as delay issues will be discussed. The spatial correlation is exploited by restricting the search space, time correlation can be considered by designing differential feedback strategies and frequency correlation can be used to further reduce the quantity of feedback information or improve the wideband channel estimation quality.

Chapter 5 will describe user scheduling and reduced feedback strategies for multiuser single antenna framework in flat fading and frequency selective wireless channels. Since the amount of feedback information increases with the number of users, it is important to perform a user selection based on threshold. For wideband channels, reduced feedback schemes based on cluster selection for OFDMA-based wireless systems will be given in detail.

Chapter 6 will consider precoding, user scheduling, limited and reduced feedback strategies for multiuser MIMO systems with one and multiple receive antennas in flat fading and frequency selective wireless channels. The capacity of the multiuser MIMO system with single receive antenna for uplink and downlink will be analyzed by assuming full feedback information is available at both transmitter and receive side perfectly. The effect of reduced and limited feedback information including user selection algorithms at the receiver side and quantization issues will be shown for both single carrier and OFDMA-based wireless systems.

Chapter 7 will examine various user scheduling, power allocation and reduced and limited feedback schemes for cooperative multicell framework. Since the amount of feedback increases with the number of base stations, it is important to perform user selection at each cell based on the level of interference. In addition to that, the cooperative networks are quite sensitive to the quality of the CSI of serving and interfering base stations to eliminate intercell interference. Therefore, bit partitioning strategies for multicell systems will be given in detail.

Chapter 8 will describe some practical issues such as codebook designs and algorithms used in wireless communication system standards including LTE and LTE-advanced.

In this book, we provide a complete state of the research and application of the feedback strategies for both SISO and MIMO systems. After having considered the single user case, we will extend our derivation to the multi user case and cooperative communication systems.

Chapter 2
Background on Wireless Communication

2.1 Introduction

In this chapter, we present a brief overview of basic wireless communication systems. The chapter starts with a description of wireless communication channel models including multiple input multiple output (MIMO) channel models and dual-polarized antennas channel models. We then discuss about digital modulation and orthogonal frequency division multiplex (OFDM). Finally, we will provide a short introduction on diversity and spatial multiplexing gain in wireless communication systems and MIMO systems. We will introduce different MIMO space-time block codes (STBC) that will be used in the rest of the book. For a detailed treatment of wireless communication systems we refer to the books of Goldsmith [15], Tse and Viswanath [51] or Molisch [32], Proakis [40].

2.2 Wireless Communication Channel Models

2.2.1 Introduction

Compared to wireline channels, the wireless channels vary over time and frequency depending on the objects surrounding the transmitter, the radiated area and the receiver. The variations of the channel strength can be divided into two classes [41]:

- Large-scale fading: coming from the path loss due to the distance between the transmitter and the receiver (typically higher than 100 m) and the shadowing due to the obstacles (typically few meters to 100 m).
- Small-scale fading: the transmitted signal could arrive at the receiver through multiple paths, which experience different attenuations, arrive at different time delays and phases. It results in constructive or destructive summation of the transmitted signal and causes rapid variations. Furthermore, the path lengths

B. Özbek and D. Le Ruyet, *Feedback Strategies for Wireless Communication*,
DOI 10.1007/978-1-4614-7741-9_2, © Springer Science+Business Media New York 2014

Fig. 2.1 Free space model

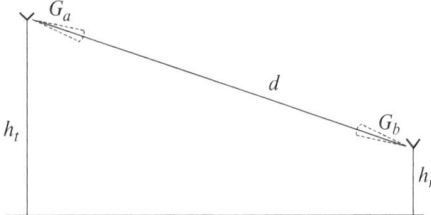

and the geometry alter due to the changes in the transmission environment or due to the relative motion of antennas, and the signal level might be subject to fluctuations.

While both classes are important, we will mainly focus on the second class since small-scale fading is more important for the design of feedback strategies in wireless communication systems.

2.2.2 Path Loss and Shadowing

2.2.2.1 Free Space

We first consider a free space transmission between a fixed transmitter and a receiver situated at a distance d from the transmitter as shown in Fig. 2.1.

Let's assume that the transmitted signal is $x(t) = \cos(2\pi f_0 t)$ where f_0 is the carrier frequency. In the far field, the received signal at time t is

$$r(t) = \frac{\lambda \sqrt{G_a G_b}}{4\pi d} \cos\left(2\pi f_0 \left(t - \frac{d}{c}\right)\right) \tag{2.1}$$

where G_a and G_b are, respectively, the gain of the transmit and receive antenna in the direction of interest ($G_a = 1$ if the transmit antenna is isotropic), $\lambda = \frac{c}{f_0}$ is the wavelength associated to the carrier frequency f_0 and $c = 3.10^8$ m/s is the speed of light. As expected, in the free space, since the electric field decreases as d^{-1}, the received power decreases as d^{-2}. In this model, the path loss $F_A(d)$ in dB is given by

$$F_A(d) = 10 \log_{10} \left(\frac{P_r}{P_t}\right)^{-1}$$

$$= 20 \log_{10} \left(\frac{4\pi d}{\lambda}\right) - 10 \log_{10} G_a G_b \tag{2.2}$$

$$= 32,44 + 20 \log_{10} f_0 + 20 \log_{10} d - 10 \log_{10} G_a G_b$$

A first simple model in mobile radio channel where the pathloss is proportional to d^α is given by

$$F_A(d) = -10 \log_{10} K_A + 10\alpha \log_{10} \left(\frac{d}{d_0} \right) \tag{2.3}$$

where d_0 is the distance from which the far field assumption is valid. $1 < d_0 < 10\text{m}$ for indoor systems and $10 < d_0 < 100\,\text{m}$ for outdoor systems. The exponent α range between 1,5 and 6,5. K_A and α can be obtained from measurement campaigns.

Different models are available in the literature and standards depending on the context (macrocell, microcell, picocell, urban, rural, indoor, . . .). As an example, for urban applications and considering a frequency range between 150 and 1,500 MHz, we can use the Okumura–Hata model where the empirical pathloss $F_A(d)$ in dB is given by

$$F_A(d) = 69.55 + 26.16 \log_{10} f_0 - 13.82 \log_{10} h_t - a(h_r) + (44.9 - 6.55 \log_{10} h_t) \log_{10} d \tag{2.4}$$

where f_0 is the carrier frequency in MHz (150 MHz $< f_0 <$ 1,500 MHz), h_t and h_r are the heights of the transmit and receive antenna, respectively, d is the distance between the transmitter and the receiver in km, and $a(h_r)$ is a correcting factor in dB to take into account the height of the receive antenna given by

$$a(h_r) = \begin{cases} (1.1 \log_{10} f_0 - 0.7)h_r - (1.56 \log_{10} f_0 - 0.8) & \text{small to medium size town} \\ 3.2(\log_{10}(11.75h_r))^2 - 4.97 & \text{large town} \end{cases} \tag{2.5}$$

The COST 231 models extend the Okumura–Hata model to cover a more elaborated range of frequencies (1.5–2 GHz).

2.2.2.2 Doppler Effect

Let's consider a car equipped with a receive antenna moving in the direction of the transmitter with a speed v as shown in Fig. 2.2.

During Δt, the distance between the transmitter and the receiver decreases by $\Delta d = v \Delta t \cos \theta$ corresponding to a phase shift of $\Delta \phi$

$$\Delta \phi = \frac{v \Delta t \cos \theta}{\lambda} 2\pi \tag{2.6}$$

The associated Doppler shift is given by

$$f_D = \frac{v f_0}{c} \cos \theta \tag{2.7}$$

Fig. 2.2 Doppler effect

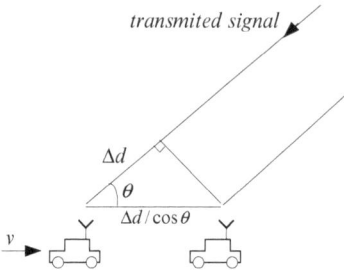

2.2.2.3 Shadowing

The shadowing effect is due to objects like trees and buildings, obstructing the propagation path between the transmit and the receive antennas. Since the properties of these objects (nature, location, etc.) is not known in advance, we model these properties as a random process. A good approximation of this effect consists in considering that the distribution of the shadow loss ψ is log-normal. The distribution of the logarithm of ψ in dB is Gaussian as follows:

$$p(\psi_{dB}) = \frac{1}{\sqrt{2\pi\sigma_{\psi_{dB}}^2}} \exp\left(-\frac{(\psi_{dB})^2}{2\sigma_{\psi_{dB}}^2} \right) \tag{2.8}$$

where $\psi_{dB} = 10\log_{10}\psi$ and $\sigma_{\psi_{dB}}$ is the standard deviation of ψ_{dB}. $\sigma_{\psi_{dB}}$ is generally chosen between 5 and 12 dB.

2.2.3 Multipath Channel Models

In wireless communications, multipath fading effect occurs in almost all environment. In urban area where the heights of the mobile antennas are below the height of the surrounding structures there is no line of sight (NLOS) path between the transmitter and the receiver. Thus, the transmitted signal arrives at the receiver through many different paths, reflection, refraction or diffraction over large objects. Multipath fading effects are difficult to predict and consequently they are commonly studied using statistic models. The modeling of multipath fading channels has been established in the early 1960s [3, 7]. We will start by introducing the most general description of the impulse response of the multipath fading channel.

2.2.3.1 General Description

Let $h(\tau, t)$ be the complex envelop of the time-varying channel response where t is the time-varying parameter and τ is the path delay parameter. The envelop $h(\tau, t)$

can be written as the sum of N_p elementary responses with amplitude $\alpha_n(t)$, phase $\phi_n(t)$ and excess time delay $\tau_n(t)$ (each response corresponds to an elementary physical path):

$$h(\tau, t) = \sum_{n=0}^{N_p-1} \alpha_n(t) e^{-j\phi_n(t)} \delta(t - \tau_n) \tag{2.9}$$

where $\delta(t)$ is the Dirac delta function. Hence, a path has zero amplitude for all time delays except $\tau = \tau_n$. The power $\mathbb{E}(\alpha_n^2(t))$ and delay τ_n of each path is determined with the power-delay profile (PDP) which is generally represented as plots of relative received power as a function of delay spread with respect to time.

2.2.3.2 Narrowband Model

In this model, we consider that the delay spread is small compared to the symbol period $x(t - \tau_n) \approx x(t)$. Under this hypothesis, the received signal can be approximated as follows:

$$r(t) = \sum_{n=0}^{N_p-1} \alpha_n(t) e^{-j\phi_n(t)} x(t) \tag{2.10}$$

where

$$\phi_n(t) = 2\pi f_0 \tau_n - 2\pi f_{D_n} t - \phi_0 \tag{2.11}$$

2.2.3.3 Rayleigh Distribution

As the multipath could arrive at the receiver in the same time delay with different phases, the amplitudes of these paths could add constructively or destructively. When the resulting amplitude is zero or near zero, we refer to it as a deep fade.

Moreover, the phase and delay of these paths can change significantly within a short period of time. Hence, the resulting amplitude of the channel at a particular time delay can vary within a short time interval. Since the number of paths is high, using the central limit theorem, the channel impulse response $h(t)$ can be modelled as a complex Gaussian random process. When there is no dominant signal path among all the paths, i.e., when there is NLOS, the amplitude of the channel $|h(t)|$ at any time instant is Rayleigh distributed and the distribution of the phase is uniform over the interval $[0, 2\pi]$.

The Rayleigh distribution is given by

$$p_R(r) = \frac{r}{\sigma^2} \exp\left(-\frac{r^2}{2\sigma^2}\right) \tag{2.12}$$

The square of the amplitude $|h(t)|^2$ is exponentially distributed with density

$$p_S(s) = \frac{1}{\sigma^2} \exp\left(-\frac{s}{\sigma^2}\right) \tag{2.13}$$

When there is a direct link between the transmit and receive antennas, i.e. when there is a line of sight (LOS), the received signal equals the superposition of a complex Gaussian component and an LOS component. The amplitude of the channel is modeled using Rician distribution and is correspondingly named the Rician fading channel. The Rician distribution is given by

$$p_R(r) = \frac{r}{\sigma^2} \exp\left(-\frac{r^2 + s^2}{2\sigma^2}\right) I_0\left(\frac{r \times s}{\sigma^2}\right) \tag{2.14}$$

where $I_0(.)$ is the modified Bessel function of the first kind with order zero. The Rician distribution is often described with the fading parameter $K_R = s^2/2\sigma^2$.

2.2.3.4 Autocorrelation and Uniform Scattering Environment Model

Let's assume that the path amplitudes $\alpha_n(t)$ are slowly varying such that we can consider them as constants during the observation time: $\alpha_n(t) \approx \alpha_n$. Under this assumption, the autocorrelation of $h^R(t)$, the real part of $h(t)$, can be written as:

$$R_{h^R,h^R}(\tau, t) = \mathbb{E}(h^R(t)h^R(t + \tau))$$

$$= 0.5 \sum_{n=0}^{N_p-1} \mathbb{E}(\alpha_n^2) \cos(2\pi f_{Dn}\tau) \tag{2.15}$$

We can see that $R_{h^R,h^R}(\tau, t)$ is independent of t: $R_{h^R,h^R}(\tau, t) = R_{h^R,h^R}(\tau)$

In Jakes' model [18], it is assumed that the reflectors are uniformly distributed around the receiver as shown in Fig. 2.3. The angle of arrival (AoA) of the N_p paths is $\theta_n = n\Delta\theta$ where $\Delta\theta = 2\pi/N_p$. We also assume that the average channel gain is constant $\mathbb{E}(\alpha_n^2) = 2/N_p$.

Since $f_{Dn} = v\cos\theta_n/\lambda$, we have

$$R_{h^R,h^R}(\tau) = \frac{1}{N_p} \sum_{n=0}^{N_p-1} \cos\left(2\pi \frac{v\tau}{\lambda} \cos\theta_n\right) \tag{2.16}$$

When $N_p \to \infty$, $\Delta\theta \to 0$, by replacing the summation by an integration we obtain

$$R_{h^R,h^R}(\tau) = \frac{1}{2\pi} \int_0^{2\pi} \cos\left(2\pi \frac{v\tau}{\lambda} \cos\theta\right) d\theta$$

$$= J_0(2\pi f_D\tau) \tag{2.17}$$

Fig. 2.3 Jakes' model

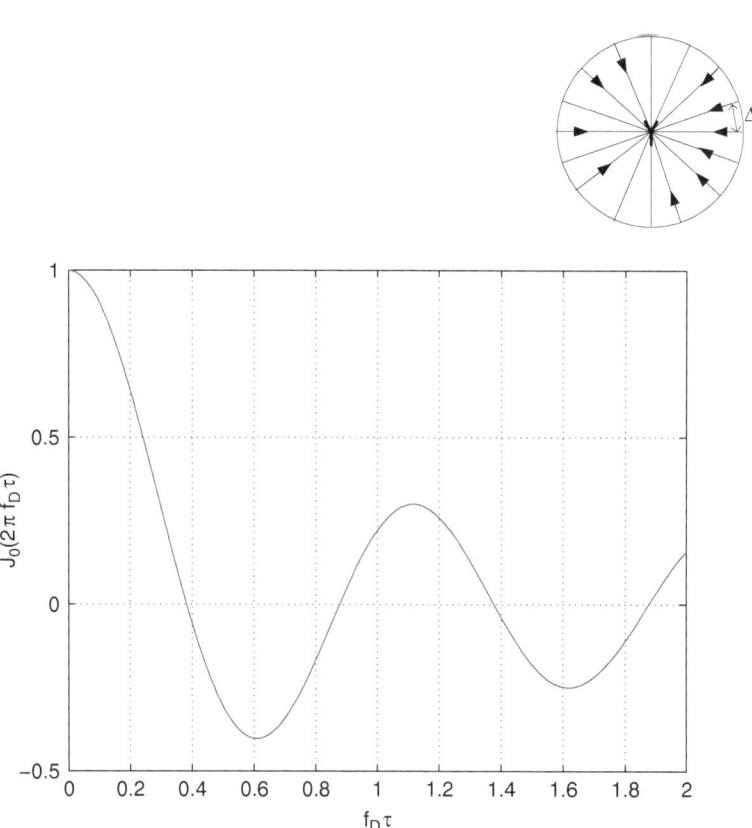

Fig. 2.4 Function $J_0(2\pi f_D \tau)$

where

$$J_0(x) = \frac{1}{\pi} \int_0^\pi \exp(-jx \cos \theta) d\theta \tag{2.18}$$

is the Bessel function of first kind and zeroth order as shown in Fig. 2.4 and f_D is the maximum Doppler spread $f_D = v/\lambda$.

The spectrum density of $h^R(t)$ and $h^I(t)$ is obtained by taking the Fourier transform of the autocorrelation

$$S_{h^R,h^R}(f) = S_{h^I,h^I}(f) = \begin{cases} \frac{1}{2\pi f_D} \frac{1}{\sqrt{1-(f/f_D)^2}} & \text{if } |f| \leq f_D \\ 0 & \text{else} \end{cases} \tag{2.19}$$

Fig. 2.5 Example of power
delay profile

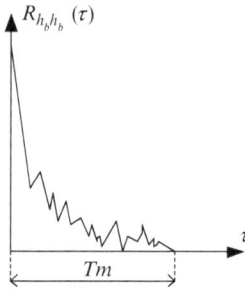

2.2.3.5 Wideband Model

In this section, we consider the general case where the signals are wideband. In this case, the maximum delay spread T_M is higher than the symbol period T_S. The approximation $x(t - \tau_n(t)) \approx x(t)$ is no longer valid. Therefore, the received signal is composed of a sum of copies of the transmitted signal where each copy is shifted in phase by $\phi_n(t)$ and delayed by $\tau_n(t)$.

Like previously, $h(\tau, t)$ is random due to the random variability of the phase, amplitude and delay of the different paths. The statistical properties of $h(\tau, t)$ are obtained from the autocorrelation function. Since the channel is usually wide-sense stationary (WSS), the autocorrelation is independent of t. Furthermore, the responses associated to the different delays are uncorrelated since they come from different scatterers. We say that this channel has wide-sense stationary uncorrelated scattering (WSSUS).

Then we have

$$R_{h,h}(\tau, \Delta t) = \mathbb{E}[h^*(\tau_1, t)h(\tau_2, t + \Delta t)] \tag{2.20}$$

with $\tau = \tau_1 - \tau_2$.

The scattering function is defined as the Fourier transform of $R_{h,h}(\tau, \Delta t)$ with respect to Δt:

$$S_{H,H}(\tau, \lambda) = \int_{-\infty}^{+\infty} R_{h,h}(\tau, \Delta t) \exp(-j2\pi\lambda \Delta t) d\Delta t \tag{2.21}$$

If we let $\Delta t = 0$ in $R_{h,h}(\tau, \Delta t)$, the resulting autocorrelation is the multipath intensity profile or the PDP $R_{h,h}(\tau)$ of the channel

$$R_{h,h}(\tau) \equiv R_{h,h}(\tau, 0) \tag{2.22}$$

The PDP is the average power output of the channel as a function of the delay τ. An example of PDP is given in Fig. 2.5.

From the delay profile, we can compute the maximum delay spread T_M, the mean delay μ_{T_M} and the root mean square delay spread σ_{T_M}.

$$\mu_{T_M} = \frac{\int_0^\infty \tau R_{h,h}(\tau) d\tau}{\int_0^\infty R_{h,h}(\tau) d\tau} \tag{2.23}$$

$$\sigma_{T_M} = \sqrt{\frac{\int_0^\infty (\tau - \mu_{T_M})^2 R_{h,h}(\tau) d\tau}{\int_0^\infty R_{h,h}(\tau) d\tau}} \tag{2.24}$$

2.2.3.6 Delay Spread and Coherence Bandwidth

The time-varying channel can also be characterized in the frequency domain by taking the Fourier transform of $h(\tau, t)$ with respect to τ denoted $H(f, t)$ and given as follows:

$$H(f, t) = \int_{-\infty}^{+\infty} h(\tau, t) \exp(-j2\pi f\tau) d\tau \tag{2.25}$$

As seen previously, for $h(\tau, t)$, $H(f, t)$ is the summation of complex Gaussian random variables and can be described using its autocorrelation function

$$R_{H,H}(\Delta f, \Delta t) = \mathbb{E}[H^*(f_1, t)H(f_2, t + \Delta t)] \tag{2.26}$$

with $\Delta f = f_2 - f_1$.

Again, if we let $\Delta t = 0$ in $R_{H,H}(\Delta f, \Delta t)$,

$$R_{H,H}(\Delta f) \equiv R_{H,H}(\Delta f, 0) \tag{2.27}$$

Since we have

$$R_{H,H}(\Delta f) = \int_{-\infty}^{+\infty} R_{h,h}(\tau) \exp(j2\pi \Delta f\tau) d\tau \tag{2.28}$$

$R_{H,H}(\Delta f)$ is the Fourier transform of the PDP.

The coherence bandwidth B_{coh} can be defined as the range of frequencies over which the amplitude of the spectral components of the channel response are correlated or equivalently for which $R_{H,H}(\Delta f)$ is different from 0. If we relax the constraint to ensure that the frequency correlation function is above 0.5, the coherence bandwidth can be defined as a function of the root mean squared delay spread σ_{T_M} as follows:

$$B_{coh} = \frac{0.2}{\sigma_{T_M}} \tag{2.29}$$

The delay spread of the multipath causes time dispersion of the transmitted signal. Depending on the delay spread and the symbol transmission time T_s, the channels can be categorized as flat fading or frequency selective fading channels.

Flat fading or non-frequency-selective assumption implies that the signal bandwidth B_W is much less than the channel coherence bandwidth B_{coh}. For perfect Nyquist pulses we have $B_W = 1/T_s$ and consequently the flat fading assumption implies that $T_s >> \sigma_{T_M}$.

When the bandwidth of constant amplitude and linear phase response of the channel is significantly less than the transmitted signal bandwidth ($B_{coh} << B_W$), the spectral characteristic of the signal cannot be maintained and the channel becomes frequency selective. In this case, the channel applies different gains or attenuations to different frequency components of the transmitted signal, causing spectral distortion in the signal. In the time domain, the time dispersion of the multipath channel is large enough such that some multipaths can be resolved at the receiver into symbol-spaced delay. In other words, a frequency selective fading channel creates inter symbol interference (ISI) onto the transmitted symbols.

2.2.3.7 Doppler Spread and Coherence Time

The Doppler effect can be seen by taking the Fourier transform of $R_{H,H}(\Delta f, \Delta t)$ relative to Δt:

$$S_{H,H}(\Delta f, \lambda) = \int_{-\infty}^{+\infty} R_{H,H}(\Delta f, \Delta t) \exp(-j 2\pi \lambda \Delta t) d\Delta t \qquad (2.30)$$

From $S_{H,H}(\Delta f, \lambda)$, if we set $\Delta f = 0$, it is possible to evaluate the Doppler power spectrum of the channel $S_{H,H}(\lambda) \equiv S_{H,H}(0, \lambda)$

$$S_{H,H}(\lambda) = \int_{-\infty}^{+\infty} R_{H,H}(\Delta t) \exp(-j 2\pi \lambda \Delta t) d\Delta t \qquad (2.31)$$

The function $S_{H,H}(\lambda)$ gives the signal intensity as a function of the Doppler frequency λ. The range of values of λ over which $S_{H,H}(\lambda)$ is nonzero is called the Doppler spread f_D. The Doppler spread measures the spectral broadening caused by the rate of change of the channel. As a time domain dual of the Doppler spread, coherence time measures the rate of change of the channel. We define the coherence time T_{coh} as the time during which $R_{H,H}(\Delta t)$ is above 0.5

$$T_{coh} = \sqrt{\frac{9}{16\pi f_D^2}} \qquad (2.32)$$

By using this coherence time, the multipath fading channels can be categorized into the fast or slow fading channels. While there is no consensus on those

Fig. 2.6 Relationship between the different functions

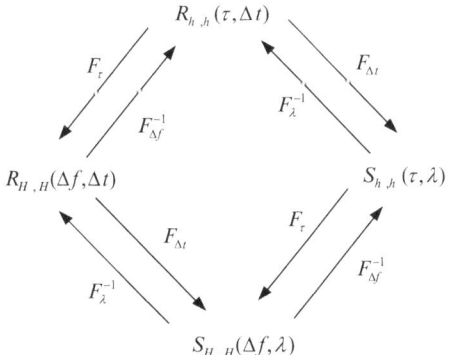

Table 2.1 ITU pedestrian A channel model for urban environment and 3 km/h mobile speed four multipaths with Rayleigh fading channel

Path	1	2	3	4
Delay (ns)	0	110	190	410
Attenuation (dB)	0	−9.7	−19.2	−22.8

Table 2.2 ITU vehicular A channel model for vehicular with low delay spread and 60 km/h mobile speed six multipaths with Rayleigh fading channel

Path	1	2	3	4	5	6
Delay (ns)	0	310	710	1090	1730	2510
Attenuation (dB)	0	−1	−9	−10	−15	−20

definitions, in this book we will call a channel slow fading if the condition $T_s \ll T_{coh}$ is satisfied. Since the time during which the channel remains correlated is long compared to the symbol duration, it is expected that the channel is unchanged while the symbol is transmitted. Conversely, if the condition $T_s \gg T_{coh}$ is satisfied, the channel is said to be fast fading since the fading characteristics of the channel change while transmitting the symbol. Depending on the channel variation, we will consider ergodic capacity (fast fading channel) or outage capacity (slow fading channel) which will be presented later. As a conclusion, delay spread, coherence bandwidth, Doppler spread and coherence time can be derived from four different functions as depicted in Fig. 2.6.

Two examples of multipath fading channel models are given in Tables 2.1 and 2.2.

2.2.4 MIMO Channel

MIMO systems are equipped with multiple antennas at both transmitter and receiver. Let N_t and N_r be the number of transmit and receive antennas. The MIMO channel can be described using a $N_r \times N_t$ matrix:

$$\mathbf{H}(t,\tau) = \begin{bmatrix} h_{11}(t,\tau) & \dots & h_{1N_t}(t,\tau) \\ \vdots & \ddots & \vdots \\ h_{N_r1}(t,\tau) & \dots & h_{N_rN_t}(t,\tau) \end{bmatrix} \qquad (2.33)$$

where $h_{ji}(t,\tau)$ is the time-varying impulse response between the ith transmit antenna and the jth receive antenna.

It is also possible to describe $h_{ji}(t,\tau)$ using the double-directional channel impulse model [45]:

$$h_{ji}(t,\tau) = \int \int \int h(\mathbf{r}_{Tx}^{(i)}, \mathbf{r}_{Rx}^{(j)}, t, \tau', \phi, \psi) G_{Tx}^{(i)}(\phi) G_{Rx}^{(j)}(\psi) f(\tau - \tau') d\tau' d\phi d\psi$$

$$(2.34)$$

where τ', ϕ and ψ denote, respectively, the excess delay, the direction of departure (DoD) and the direction of arrival (DoA), $\mathbf{r}_{Tx}^{(i)}$ and $\mathbf{r}_{Rx}^{(j)}$ are the coordinates of the ith transmit and jth receive antenna, and $h(\mathbf{r}_{Tx}^{(i)}, \mathbf{r}_{Rx}^{(j)}, t, \tau', \phi, \psi)$ is the double-directional channel impulse model and consists of the contributions of all individual multipath component. Furthermore, $G_{Tx}^{(i)}(\phi)$ and $G_{Rx}^{(j)}(\psi)$ represent, respectively, the transmit and receive antenna patterns and $f(\tau)$ is the overall impulse response.

Using the double-directional channel impulse model we have

$$h(\mathbf{r}_{Tx}^{(i)}, \mathbf{r}_{Rx}^{(j)}, t, \tau', \phi, \psi) = \sum_{l=0}^{N_p-1} h_l(\mathbf{r}_{Tx}^{(i)}, \mathbf{r}_{Rx}^{(j)}, t, \tau', \phi, \psi) \qquad (2.35)$$

where N_p is the total number of multipath components. For planar waves, the contribution of the lth multipath component is given from the single-input single-output (SISO) case by

$$h_l(\mathbf{r}_{Tx}, \mathbf{r}_{Rx}, t, \tau, \phi, \psi) = \alpha_l e^{j\beta_l} \delta(\tau - \tau_l)\delta(\phi - \phi_l)\delta(\psi - \psi_l) \qquad (2.36)$$

where α_l, β_l, τ_l, ϕ_l and ψ_l are, respectively, the amplitude, phase, delay, DoD and DoA associated with the lth multipath component. These parameters can also be time varying.

The input–output relation between the transmit signal vector $\mathbf{s}(t)$ and the received signal vector $\mathbf{y}(t)$ is given by

$$\mathbf{y}(t) = \int \mathbf{H}(t,\tau)\mathbf{s}(t-\tau)d\tau + \mathbf{n}(t) \qquad (2.37)$$

where $\mathbf{n}(t)$ is the noise vector.

If the channel can be treated as approximately constant over the total frequency bandwidth (frequency flat channel) and over the observation time, the corresponding input–output relationship (2.37) simplifies to

$$\mathbf{y}(t) = \mathbf{H}\mathbf{s}(t) + \mathbf{n}(t) \qquad (2.38)$$

where $\mathbf{H}$ is the $N_r \times N_t$ narrowband MIMO channel matrix given by

$$\mathbf{H} = \begin{bmatrix} h_{11} & \ldots & h_{1N_t} \\ \vdots & \ddots & \vdots \\ h_{N_r 1} & \ldots & h_{N_r N_t} \end{bmatrix} \tag{2.39}$$

2.2.4.1   MIMO Channel Model Classification

While we often assume that the elements of narrowband MIMO channel matrix are independent and identically distributed (i.i.d.), in reality, due to insufficient spacing between antenna elements and limited scattering in the environment, the fading is not independent. Therefore, the MIMO channel models should take this effect into account. The MIMO channel models can be divided into physical and nonphysical models.

Physical models choose some crucial physical parameters to describe the MIMO propagation channels. Some typical parameters include AoA, angle of departure (AoD), and time of arrival (TOA). However, under many propagation conditions, the MIMO channels are not well described by a small number of physical parameters and this makes it difficult, if not impossible, to identify the models. Although one attempts to separate the propagation channel from the measurement equipment (antenna responses, configuration, etc.) to allow extrapolation to other conditions, the model always contains some limitations related to the conditions under which the model was identified (point source assumption, etc.).

Physical models can be mainly split into deterministic models, geometry-based stochastic models, and nongeometric stochastic models:

- Deterministic channel models aim at reproducing the actual physical radio propagation process for a given environment. The characterization of the physical propagation parameters is performed in a completely deterministic manner like ray tracing and stored measurement data.
- In geometry-based stochastic channel models (GSCM), the impulse response is characterized by the laws of wave propagation applied to specific transmitter, receiver and scatterer geometries, which are chosen in a stochastic (random) manner.
- Nongeometric stochastic models describe and determine physical parameters (DoD, DoA, delay, etc.) in a completely stochastic way using underlying probability distribution functions without assuming an underlying geometry.

Nonphysical models characterize the impulse response or equivalently the transfer function of the channel between the individual transmit and receive antennas in an analytical way without explicitly accounting for wave propagation. The individual impulse responses are subsumed into the MIMO channel matrix. The main strengths of the nonphysical models is that they rely on a small set of parameters which fully characterize the communication scenario, namely the

power gain of the MIMO channel matrix, correlation matrices describing the
correlation properties at the transmitter and receiver side and the associated Doppler
spectrum of the channel paths. Nonphysical models can be further subdivided into
propagation-motivated models and correlation-based models:

- Propagation-motivated models use propagation parameters. Examples are the
  finite scatterer model and the maximum entropy model.
- Correlation-based models characterize the MIMO channel matrix statistically in
  terms of the correlations between the matrix entries. Examples are the Kronecker
  model [21, 39], the virtual channel representation model [43] and the eigenbeam
  model [55].

2.2.4.2 Nongeometric Stochastic Physical Channel Model

Nongeometrical stochastic physical models describe paths from transmitter and
receiver Rx by statistical parameters only, without reference to the geometry
of a physical environment. The most popular model is the so-called extended
Saleh–Valenzuela model [42, 54].

In [42] Saleh and Valenzuela have proposed a wideband SISO multipath channel
model for the indoor scenario based on the indoor measurements. They proposed
to model clusters of multipath components since they observed that the multipath
components arrive in groups and therefore the scatterers could be separated
into clusters. In the Saleh–Valenzuela model, the cluster amplitude is Rayleigh
distributed with an exponential decaying profile. The multipath components within
the individual clusters are zero-mean complex Gaussian also characterized by a
second exponential profile with a steeper slope.

Suppose there are L_C clusters and each cluster has K_C rays, the directional
channel response can be expressed as

$$h(\mathbf{r}_{Tx}, \mathbf{r}_{Rx}, t, \tau, \phi, \psi) = \frac{1}{\sqrt{L_C K_C}} \sum_{l=1}^{L_C} \sum_{k=1}^{K_C} \alpha_{kl} e^{j\beta_{kl}} \delta(\tau - \tau_{kl}) \delta(\phi - \phi_l - \phi_{kl}) \delta(\psi - \psi_l - \psi_{kl})$$

$$(2.40)$$

where α_{kl}, β_{kl}, and τ_{kl} are the amplitude, phase, and delay of the kth ray in the lth
cluster, ϕ_l and ψ_l are, respectively, the mean DoD and DoA associated with the lth
cluster, and ϕ_{kl} and ψ_{kl} are the relative transmit and receive angles for the kth ray
in the lth cluster.

Assuming that the AoA and AoD statistics are independent and identical, the
Saleh–Valenzuela model was extended to MIMO channels in [54].

In the extended Saleh–Valenzuela model the probability density function of the
ray AoA/AoD is Laplacian distributed:

$$p(\phi_{kl}) = \frac{1}{\sqrt{2}\sigma} \exp(-\sqrt{2}|\phi_{kl} - \phi_l|) \qquad (2.41)$$

$$p(\psi_{kl}) = \frac{1}{\sqrt{2}\sigma} \exp(-\sqrt{2}|\psi_{kl} - \psi_l|) \tag{2.42}$$

2.2.4.3 Correlation-Based Model

Following the MIMO modeling approach presented in [20, 44] that utilizes receive and transmit correlation matrices, the MIMO channel matrix $\mathbf{H}$ can be separated into a fixed (constant, LOS) matrix and a Rayleigh (variable, NLOS) matrix:

$$\mathbf{H} = \sqrt{\frac{1}{1 + K_R}}\mathbf{H}_s + \sqrt{\frac{K_R}{1 + K_R}}\mathbf{H}_d \tag{2.43}$$

where K_R is the Rician K-factor. The matrix $\mathbf{H}_d$ is the fixed LOS matrix and the matrix $\mathbf{H}_s$ is the NLOS (Rayleigh) matrix. The elements of $\mathbf{H}_s$ are correlated zero-mean, complex Gaussian random variables.

For simplicity, in the rest of this section we will assume $K_R = 0$ that is $\mathbf{H} = \mathbf{H}_s$ (Rayleigh fading).

Let $\mathbf{h_{vec}}$ be the $N_t N_r \times 1$ vector obtained by vectorization of the channel matrix $\mathbf{H}$ as follows:

$$\mathbf{h_{vec}} = [h_{11}\ h_{21}\ \ldots\ h_{N_t N_r}]^T$$
$$= \text{vec}\{\mathbf{H}\} \tag{2.44}$$

where $\text{vec}\{\mathbf{H}\}$ creates a column vector from the matrix $\mathbf{H}$. The zero-mean distribution multivariable complex Gaussian distribution of $\mathbf{h}_{vec}$ is given by

$$p(\mathbf{h_{vec}}) = \frac{1}{\det[\mathbf{R_H}]} \exp(-\mathbf{h_{vec}}^H \mathbf{R_H}^{-1}\mathbf{h_{vec}}) \tag{2.45}$$

where $\mathbf{R_H} = \mathbb{E}[\mathbf{h_{vec}}\mathbf{h_{vec}}^H]$ is the $N_t N_r \times N_t N_r$ semidefinite correlation matrix which describes the correlation between each pair of coefficient channels.

From (2.45), any channel realization is obtained by

$$\mathbf{h_{vec}} = \mathbf{R_H}^{1/2}\mathbf{g_{vec}} \tag{2.46}$$

where $\mathbf{g_{vec}}$ is a $N_t N_r \times 1$ vector of i.i.d unit variance elements.

2.2.4.4 Independent and Identically Distributed Model

The simplest analytical MIMO model is the independent and identically distributed (i.i.d.) model. The time varying multipath fading channel models can be used for each antenna pair from transmit antenna to receive antenna if MIMO channel coefficients are assumed to be i.i.d. This commonly used MIMO model is also called "rich scattering" model.

Since all elements of the MIMO channel matrix $\mathbf{H}$ are assumed uncorrelated we have

$$\mathbf{R_H} = \mathbf{I} \tag{2.47}$$

This model is often used for theoretical considerations like the information-theoretic analysis of MIMO systems.

2.2.4.5 Kronecker Model

The so-called Kronecker model, where it is assumed that the spatial correlation matrix of the MIMO radio channel is separable, was proposed in [21, 39].

Let's define the transmit and the receive correlation matrices, respectively:

$$\mathbf{R}_{tx} = \frac{1}{N_R}\mathbb{E}\left[\mathbf{H}^H\mathbf{H}\right] \qquad \mathbf{R}_{rx} = \frac{1}{N_T}\mathbb{E}\left[\mathbf{H}\mathbf{H}^H\right] \tag{2.48}$$

These symmetrical complex correlation matrices can also be written as

$$\mathbf{R}_{tx} = \begin{bmatrix} \rho_{11}^t & \rho_{12}^t & \cdots & \rho_{1N_t}^t \\ \rho_{21}^t & \rho_{22}^t & \cdots & \rho_{2N_t}^t \\ \vdots & \ddots & \cdots & \vdots \\ \rho_{N_t1}^t & \rho_{N_t2}^t & \cdots & \rho_{N_tN_t}^t \end{bmatrix} \quad \text{and} \quad \mathbf{R}_{rx} = \begin{bmatrix} \rho_{11}^r & \rho_{12}^r & \cdots & \rho_{1N_r}^r \\ \rho_{21}^r & \rho_{22}^r & \cdots & \rho_{2N_r}^r \\ \vdots & \ddots & \cdots & \vdots \\ \rho_{N_r1}^r & \rho_{N_r2}^r & \cdots & \rho_{N_rN_r}^r \end{bmatrix} . \tag{2.49}$$

The complex spatial correlation coefficients at the transmitter side between antenna i_1 and i_2 and at the receiver side between antenna j_1 and j_2 considering the channel transfer matrix in (2.39) and assuming all antenna elements in the two arrays have the same polarization and the same radiation pattern, are given by

$$\rho_{i_1i_2}^t = <h_{i_1j}, h_{i_2j}> \quad \text{and} \quad \rho_{j_1j_2}^r = <h_{ij_1}, h_{ij_2}> \tag{2.50}$$

respectively, where $<u, v>$ computes the correlation coefficient between u and v.

$$<u, v> = \frac{\mathbb{E}[uv^*] - \mathbb{E}[u]\mathbb{E}[v^*]}{\sqrt{(\mathbb{E}[|u|^2] - |\mathbb{E}[u]|^2)(\mathbb{E}[|v|^2] - |\mathbb{E}[v]|^2)}} \tag{2.51}$$

Provided that $\rho_{i_1i_2}^t$ and $\rho_{j_1j_2}^r$ are independent of j and i, respectively. The spatial correlation matrix of the MIMO radio channel is the Kronecker product of the transmit and receive correlation matrices as follows:

$$\mathbf{R_H} = \mathbf{R}_{tx}^T \otimes \mathbf{R}_{rx} \tag{2.52}$$

where $\otimes$ is the Kronecker product.[1]

From (2.46), the relation (2.52) simplifies to

$$\mathbf{h_{vec}} = (\mathbf{R}_{tx}^T \otimes \mathbf{R}_{rx})^{1/2} \mathbf{g_{vec}} \qquad (2.53)$$

An alternative approach is given by

$$\mathbf{H} = \mathbf{R}_{rx}^{1/2} \mathbf{G} \mathbf{R}_{tx}^{1/2} \qquad (2.54)$$

The matrix $\mathbf{G} = \text{unvec}\{\mathbf{g_{vec}}\}$ is an i.i.d. unit variance MIMO channel matrix of size $N_R \times N_T$ and $\text{unvec}\{\mathbf{g_{vec}}\}$ creates a matrix from the column vector $\mathbf{g_{vec}}$ with the same number of elements.

For the case $N_t = N_r = 2$, the two Hermitian matrices $\mathbf{R}_{tx}$ and $\mathbf{R}_{rx}$ have the form

$$\mathbf{R}_{tx} = \begin{bmatrix} 1 & t^* \\ t & 1 \end{bmatrix} \qquad \mathbf{R}_{rx} = \begin{bmatrix} 1 & r^* \\ r & 1 \end{bmatrix} \qquad (2.55)$$

where t and r are the transmit and received correlation. The 4×4 spatial correlation matrix is given by

$$\mathbf{R_H} = \begin{bmatrix} 1 & r^* & t & s_1 \\ r & 1 & s_2 & t \\ t^* & s_2^* & 1 & r^* \\ s_1^* & t^* & r & 1 \end{bmatrix} \qquad (2.56)$$

where $s_1 = \mathbb{E}[h_{1,1}h_{2,2}^*]$ and $s_2 = \mathbb{E}[h_{2,1}h_{1,2}^*]$ are the cross-channel correlations.

The Kronecker model is one of the most often-used channel model to model the second-order statistics of the channel. However, this separable model is not always accurate and discrepancies have been reported in measurement campaigns when the number of antennas is high [37]. To allow arbitrary coming between the transmit and receive parts, the virtual channel representation model has been proposed in [43]. The author utilizes a fixed and predefined virtual partitioning of the spatial domain to characterize the MIMO channel.

A more accurate model is the eigenbeam or Weichselberger model [55] that combines the advantages of both the Kronecker model and the virtual channel representation. The eigenbeam model treats the influence of the antennas and

---

[1]If $\mathbf{A}$ is a $M \times N$ matrix with elements a_{mn} and $\mathbf{B}$ is a matrix, the Kronecker product $\mathbf{A} \otimes \mathbf{B}$ is

$$\mathbf{A} \otimes \mathbf{B} = \begin{bmatrix} a_{11}\mathbf{B} & \dots & a_{1N}\mathbf{B} \\ \vdots & \ddots & \vdots \\ a_{M1}\mathbf{B} & \dots & a_{MN}\mathbf{B} \end{bmatrix}.$$

environment by means of eigenbases and a coupling matrix. Let $\mathbf{U}_{rx}$ and $\mathbf{U}_{tx}$ be the eigenbases of the unparameterized one-sided correlation matrices of sides A and B of the link (correlation as perceived from the other side of the link). A MIMO channel realization is generated as

$$\mathbf{H} = \mathbf{U}_{rx}(\Omega \odot \mathbf{H}_W)\mathbf{U}_{tx}^T \tag{2.57}$$

where $\odot$ is the Hadamard (entry wise) product and $\mathbf{H}_W$ is a matrix of i.i.d. random zero-mean complex-normal distributed values. The elements of the coupling matrix Ω specify the mean amount of energy that is coupled from the mth eigenvector of side Rx to the nth eigenvector of side Tx.

The spatial correlation function is the Fourier transform of the power azimuth spectrum (PAS). For MIMO channel model with uniform linear array (ULA) and with the omnidirectional antenna spacing, the PAS distribution in physical channels can be considered as uniform, Gaussian and Laplacian. Using these three distributions, the envelope of correlation coefficient is computed as a function of the relative antenna spacing at the receiver Δ_r and at the transmitter Δ_t, AoA and AS by considering multicluster model.

For ULA, the complex correlation coefficients for the receive antennas apart form Δ_r are expressed as [44]

$$\rho^r(\Delta_r) = R_{xx}^r(\Delta_r) + jR_{xy}^r(\Delta_r), \tag{2.58}$$

where $\Delta_r = d_r/\lambda$ and d_r stands for the distance between the receive antennas and $\lambda = c/f_c$ is the wavelength of a narrowband signal with center frequency f_c. R_{xx}^r and R_{xy}^r are, respectively, the correlation functions between the real parts and between the real and imaginary parts

$$R_{xx}^r = \int_{-\pi}^{\pi} \cos(2\pi\Delta_r \sin\phi_r)\text{PAS}(\phi_r)d\phi_r \quad \text{and} \quad R_{xy}^r = \int_{-\pi}^{\pi} \sin(2\pi\Delta_r \sin\phi_r)\text{PAS}(\phi_r)d\phi_r, \tag{2.59}$$

where ϕ_r is the AoA for the receiver. The Laplacian PAS distribution which gives better match than Gaussian or uniform distributions for typical rural environment can be expressed as

$$\text{PAS}(\phi_r) = \frac{1}{\sqrt{2}\sigma_r} \exp\left(-|\sqrt{2}\phi_r/\sigma_r|\right) \tag{2.60}$$

where σ_r is the standard deviation of PAS (which corresponds to the numerical rms value of AS). The same derivations can also be used to obtain correlation coefficients for the transmit antennas. For multipath fading channels, the derivation of correlation matrices is applied for each path individually.

When assuming that the AoD at the transmitter and AoA at the receiver are Gaussian distributed, respectively, around the mean AoD and AoA, i.e $\phi_r = \bar{\phi}_r + \hat{\phi}_r$

where $\hat{\phi}_r = N(0, \sigma_r^2)$ and $\phi_t = \bar{\phi}_t + \hat{\phi}_t$ where $\hat{\phi}_t = N(0, \sigma_t^2)$, it has been shown in [5] that for small angle spread, the spatial correlation coefficient using an ULA can be simplified as follows:

$$\rho_{i_1 i_2}^t = \exp(-j2\pi(i_1 - i_2)\Delta_t \cos(\bar{\phi}_t)) \exp(-0.5(2\pi(i_1 - i_2)\Delta_t \sin(\bar{\phi}_t)\sigma_t)^2) \quad (2.61)$$

$$\rho_{j_1 j_2}^r = \exp(-j2\pi(j_1 - j_2)\Delta_r \cos(\bar{\phi}_r)) \exp(-0.5(2\pi(j_1 - j_2)\Delta_r \sin(\bar{\phi}_r)\sigma_r)^2) \quad (2.62)$$

This model can be further simplified when the standard deviation of PAS $\sigma_t = 0$ and $\sigma_r = 0$. In this case, we have

$$\rho_{i_1 i_2}^t = \exp(-j2\pi(i_1 - i_2)\Delta_t \cos(\bar{\phi}_t)) \quad (2.63)$$

$$\rho_{j_1 j_2}^r = \exp(-j2\pi(j_1 - j_2)\Delta_r \cos(\bar{\phi}_r)) \quad (2.64)$$

2.2.5 Dual-Polarized Antennas

Since the vertical and horizontal polarizations are sufficient to characterize the far field, polarization can be taken into account by extending the impulse response to a polarimetric 2×2 matrix.

Handsets and small devices may not have adequate space for more than two elements linear array. For this reason, the use of dual-polarized antennas is a promising cost- and space-effective alternative, where two spatially separated uni-polarized antennas are replaced by a single antenna structure employing two orthogonal polarizations. The main advantage of exploiting polarization diversity is that the large antenna configurations of space diversity are not necessary to achieve performance gain.

As early as 1990, experimenters such as Rodney Vaughan [53] have tested the concept of polarization-diversity systems:

For suburban base stations, the dominance of the vertical polarization makes the diversity gain rather small—only a couple of decibels at the 99.5% probability level,

In urban environments, however, the diversity gain is nearly 7 dB at the 99.5% level, offering much promise for system design using polarization diversity.

Polarization diversity is created by the nature of the wireless communication system that the signal energy from one polarization is generally coupled into other polarizations. Indeed, multiple reflections between the transmitter and the receiver lead to depolarization of radio waves, coupling some energy of the transmitted signal into the other polarized wave [22, 24, 53].

Fig. 2.7 Cross-polarized antenna

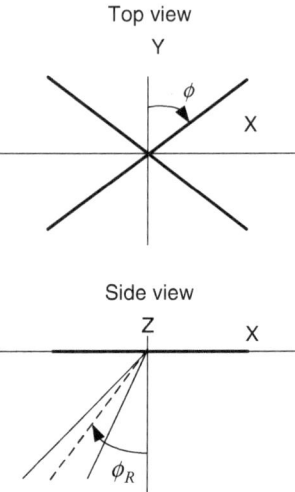

Due to nonideal antennas cross-polar isolation (XPI) and the existence of cross-polar ratio (XPR) in the multipath radio channel, vertically/horizontally polarized transmitted waves have also horizontal/vertical component. When we combine both effects, we obtain the cross-polar discrimination (XPD).

Depolarization caused by nonideal antennas is well known in antenna theory and related to the cross-polar antenna pattern. This can be modelized by a coupling matrix

$$\mathbf{X}_a = \begin{bmatrix} 1 & \sqrt{\chi_a} \\ \sqrt{\chi_a} & 1 \end{bmatrix} \tag{2.65}$$

where χ_a^{-1} is the scalar antenna XPI. For an ideal antenna, we have $\chi_a = 0$.

In the rest of this chapter, we will assume that the co-located antenna have a high XPI and that consequently, we can do the approximation $XPD \approx XPR$. We now consider a downlink communication link with one dual-polarized antenna at both the transmitter and the receiver side with vertical (V_{pol}) and horizontal (H_{pol}) polarization states.

In Fig. 2.7, we show a cross-polarized antenna where a narrow AoA spread is received from an average AoA ϕ_r. The AoA is considered to have an angle spread as given by a probability density function, $p(\phi_r)$.

The system model can be described by the matrix relation $\mathbf{r} = \mathbf{G}\mathbf{x} + \mathbf{n}$, where $\mathbf{x}$ is the 2×1 transmit signal vector and $\mathbf{r}$ is the 2×1 received vector. $\mathbf{G}$ is the 2×2 channel or polarization matrix involving complex Gaussian random variables:

$$\mathbf{G} = \begin{bmatrix} g_{vv} & g_{vh} \\ g_{hv} & g_{hh} \end{bmatrix} \tag{2.66}$$

where g_{hh}, g_{vh}, g_{hv} and g_{vv} represent the complex channel gain coefficients for transmission between horizontal receive and horizontal transmit, vertical and horizontal, horizontal and vertical and vertical and vertical antennas, respectively.

Let's denote $p_{ij} = |g_{ij}|^2$. Different cross-polar ratios can be defined [36]:

Uplink cross-polar ratios

$$XPR_{U^v} = p_{vv}/p_{vh} \qquad XPR_{U^h} = p_{hh}/p_{hv} \qquad (2.67)$$

Downlink cross-polar ratios

$$XPR_{D^v} = p_{vv}/p_{hv} \qquad XPR_{D^h} = p_{hh}/p_{vh} \qquad (2.68)$$

Co-polar ratio (CPR)

$$CPR_{D^v} = p_{vv}/p_{hh} \qquad (2.69)$$

Experimental data collected reveal that the elements of the polarization matrix $\mathbf{G}$ are correlated. We can define the correlation matrix given by $\mathbb{E}[vec\{\mathbf{G}\}vec\{\mathbf{G}^H\}]$. The diagonal elements are, respectively, the average gain $\mathbb{E}[p_{vv}]$, $\mathbb{E}[p_{hv}]$, $\mathbb{E}[p_{vh}]$ and $\mathbb{E}[p_{hh}]$. The cross-polar correlations (XPC) are, respectively, the transmit and receive correlation coefficients

$$\rho^t = <g_{hh}, g_{hv}> = <g_{vh}, g_{vv}> \qquad (2.70)$$

$$\rho^r = <g_{hh}, g_{vh}> = <g_{hv}, g_{vv}> \qquad (2.71)$$

The correlation between g_{hh} and g_{vv} is defined as the co-polar correlation (CPC) while the correlation between g_{vh} and g_{hv} is the anti-polar correlation (APC) [36].

A theoretical model for characterizing correlation between diversity branches in a polarization diversity system was first established by Kozono et al. [22] assuming a narrowband multipath beam arriving from the azimuth. Vaughan [53] further extends Kozono's model to take into account the rotation of two antennas with a fixed angular separation around their phase center. When the angle spread probability distribution is a Laplacian distribution as given in (2.60), the correlation coefficient can be computed from [22, 33] as follows:

$$\rho(\phi_{ant}, \phi_r, XPR) = \left[\frac{-\tan^2(\phi_{ant})\mathbb{E}[\cos^2(\phi_r)] + XPR}{\tan^2(\phi_{ant})\mathbb{E}[\cos^2(\phi_r)] + XPR} \right] \qquad (2.72)$$

where $\mathbb{E}[\cos^2(\phi_r)]$ is the expected value calculated from the probability density function $p(\phi_r)$.

Figure 2.8 shows the correlation in function of the mean AoA. It is obtained using (2.72) by transmitting a vertically polarized signal which suffers depolarizing effects in the environment and received with the cross-polarized subscriber antennas

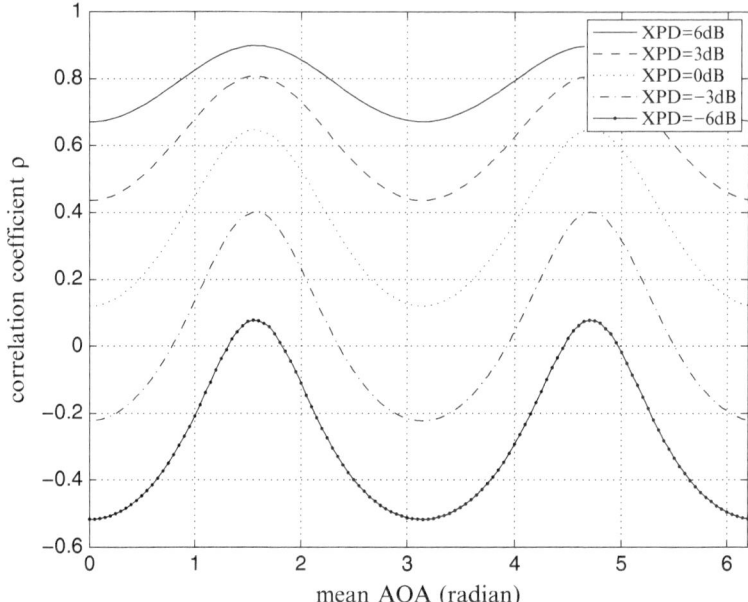

Fig. 2.8 Correlation versus XPR for dual-polarized $\phi = \pm 45°$ antennas, Laplacian PAS $= 35°$

for dual-polarized $\phi = \pm 45°$ antennas and an angle spread of 35° (Laplacian distribution).

We can see that the effect of the antenna orientation produces a significant change in the correlation between the branches. For example, when XPR = 6 dB, the correlation coefficient is quite high (> 0.65) for all mean AoA. While these results give some hints on the link between XPR and correlation coefficient, it is also important to evaluate the distribution of the power ratio XPR for each considered scenario as we will see later.

Correlation between orthogonal components of the electromagnetic field at the mobile has been extended in [6, 46] to a uniform but arbitrarily wide angular spread.

Different experimental results have been conducted and published for both outdoor and indoor scenario in the last decade. We can summarize the main properties of dual-polarized channels as follows:

- The typical values of average CPR vary between 4 and 5.5 dB in urban microcell and between 3 and 11 dB in suburban microcell [30].
- The typical values of average XPR vary in the urban/suburban environment between 3 and 9 dB, with an average of 6 dB. In a rural environment, the average XPR is between 10 and 18 dB, due to the pure LOS connection and the absence of obstacles that couple the signal from the V_{pol} into the H_{pol}. The same results are observed in indoor [22, 23, 28–30, 53].

- Experimental results have also shown that the cross-polar correlation between the V_{pol} and the H_{pol} at the receiver is generally less than 0.2 [19, 28, 30].

2.2.5.1 Dual-Polarized Rayleigh Fading Channels

In Rayleigh fading, it has been shown that spatial correlation can be considered independent of the polarization if the antennas have the same directional spectrum. Under this assumption, the dual-polarized channel matrix denoted as $\mathbf{H}_x$ can be modelized, thanks to the separation of space and polarization effects as follows [36]:

$$\mathbf{H}_x = \mathbf{H} \odot \mathbf{X} \tag{2.73}$$

where $\mathbf{H}$ is the uni-polarized correlated Rayleigh channel matrix and $\mathbf{X}$ models the correlation and power imbalance of the channel depolarization. The relation between $\mathbf{X}$ and $\mathbf{G}$ is

$$\mathbf{G} = h\mathbf{X} \tag{2.74}$$

where h is a scalar complex Gaussian term representing the fading. A general model of $\mathbf{X}$ for VH to VH downlink transmission can be given by

$$\mathrm{vec}\{\mathbf{X}\} = \begin{bmatrix} 1 & \sqrt{\mu\chi}\vartheta & \sqrt{\chi}\sigma & \sqrt{\mu}\delta_1 \\ \sqrt{\mu\chi}\vartheta^* & \mu\chi & \sqrt{\mu}\chi\delta_2 & \mu\sqrt{\chi}\sigma \\ \sqrt{\chi}\sigma^* & \sqrt{\mu}\chi\delta_2^* & \chi & \sqrt{\mu\chi}\vartheta \\ \sqrt{\mu}\delta_1^* & \mu\sqrt{\chi}\sigma^* & \sqrt{\mu\chi}\vartheta^* & \mu \end{bmatrix}^{1/2} \mathrm{vec}\{\mathbf{X}_W\} \tag{2.75}$$

where

- μ and χ are, respectively, the CPR and downlink XPR.
- σ is the receive correlation coefficient (between VV and VH, HV and HH).
- ϑ is the transmit correlation coefficient (between VV and HV, VH and HH).
- δ_1 is the CPC coefficient (between VV and HH).
- δ_2 is the APC coefficient (between HV and VH).
- The elements of the 2×2 matrix $\mathbf{X}_W$ have unit amplitude and angles, ϕ_n $n = 1, \ldots, 4$ uniformly distributed over $[0, 2\pi]$.

2.2.5.2 Dual-Polarized MIMO Fading Channels

When the Tx and Rx array are made of $N_t/2$ and $N_r/2$ dual-polarized arrays, the Rayleigh channel matrix can be written as

$$\mathbf{H}_{x.N_r \times N_t} = \mathbf{H}_{N_r/2 \times N_t/2} \odot \mathbf{X} \tag{2.76}$$

In [34] the authors have investigated the performance of spatial multiplexing and Alamouti scheme in MIMO wireless systems employing dual-polarized antennas. They have shown that while improvements in terms of symbol error rate of up to an order of magnitude are possible in the case of spatial multiplexing, the presence of polarization diversity generally incurs a performance loss compared to spatial diversity for transmit diversity techniques such as Alamouti scheme.

2.3 Orthogonal Frequency Division Multiplexing

Different techniques can cope with the intersymbol interference due to frequency selective channel. Among these techniques, multicarrier modulations avoid the use of complex equalizer at the receiver side. In multicarrier modulation, the data stream is divided into multiple substreams to be transmitted over different orthogonal subchannels centered at different subcarrier frequencies. Consider a linearly modulated system with data rate R and passband bandwidth B_W. The coherence bandwidth for the channel is assumed to be $B_{coh} < B_W$, so the signal experiences frequency-selective fading. The basic idea of multicarrier modulation is to break this wideband system into N linearly modulated subsystems in parallel, each with subchannel bandwidth $B_N = B_W/N$. The number of subchannels N is chosen to make the symbol time on each substream much greater than the delay spread of the channel in order to eliminate intersymbol interference. In the frequency domain this is equivalent to making the substream bandwidth less than the channel coherence bandwidth. For N being sufficiently large, the subchannel bandwidth $B_N << B_{coh}$, which insures relatively flat fading on each subchannel. This can also be seen in the time domain: the symbol time T_N of the modulated signal in each subchannel is equal to $\frac{1}{B_N}(1 + \epsilon)$. ϵ/T_N is the additional bandwidth due to the time limiting of the filter responses. So it implies that $T_N = \frac{1}{B_N}(1 + \epsilon) >> 1/Bc \approx T_m$, where T_m denotes the delay spread of the channel. So each subchannel experiences little ISI. Figure 2.9 illustrates the multicarrier modulation without overlapping subcarriers.

We can improve the spectral efficiency of multicarrier modulation by overlapping the subchannels. The subcarriers must still be orthogonal so that they can be separated out by the demodulator at the receiver. The subcarriers $\cos(2\pi(f_0 + n/T_N)), n = 0, 1, \ldots, N-1$ form a set of approximately orthogonal basis functions on the interval $[0, T_N]$. Figure 2.10 illustrates the multicarrier modulation with overlapping subcarriers.

Multicarrier modulation can be efficiently implemented digitally using the discrete Fourier transform (DFT) and the inverse DFT (IDFT). Using this discrete implementation, called OFDM, the ISI can be completely eliminated through the use of a cyclic prefix [27].

The main idea of OFDM transmission is to turn the channel convolutional effect into a multiplicative one. In order to perform this transformation, it can be noticed that a circular convolution is a multiplication in the frequency domain. Unfortunately, the channel output is not a circular convolution but a linear convolution. However, the linear convolution between the channel input and its impulse response can be turned into a circular convolution by adding a special prefix to the input called a cyclic prefix.

From the sequence $\mathbf{X}[m]$ by applying the IDFT, we obtain

$$x_{i,m} = \frac{1}{\sqrt{N}} \sum_{n=0}^{N-1} X_{n,m} \exp\left(\frac{2j\pi n i}{N}\right) \tag{2.77}$$

In the continuous time domain, we have

$$x_m(t) = \frac{1}{\sqrt{N}} \sum_{n=0}^{N-1} X_{n,m} \exp\left(2j\pi f_n t\right) g(t) \tag{2.78}$$

where $g(t)$ is the pulse shaping. The transmitted signal is

$$x(t) = \sum_m \frac{1}{\sqrt{N}} \sum_{n=0}^{N-1} X_{n,m} \exp\left(2j\pi f_n t\right) g(t - mT) \tag{2.79}$$

where T is the OFDM symbol period. Figure 2.11 gives a description of the OFDM implementation.

The input data are modulated using a QAM or M-PSK modulator resulting in a complex symbol stream $\mathbf{X}[m] = [X_{0,m} X_{1,m} \ldots X_{N-1,m}]$. Each symbol is transmitted over one subcarrier. Thus, the N output symbols from the serial-to-parallel converter are the discrete frequency components of the OFDM modulator output $x(t)$. In order to generate x(t), these frequency components are converted into time samples by performing an IDFT on these N symbols, which is efficiently implemented using the IFFT algorithm. The IFFT yields the OFDM symbol consisting of the sequence

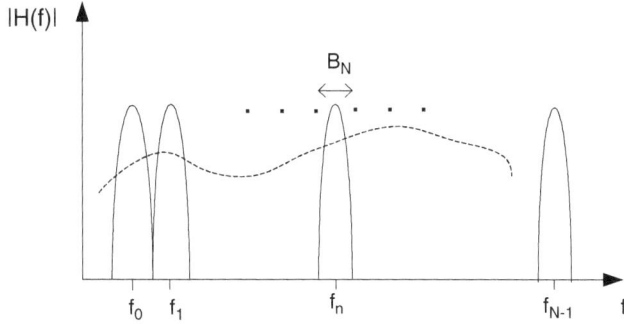

Fig. 2.9 A multicarrier without overlapping subcarriers

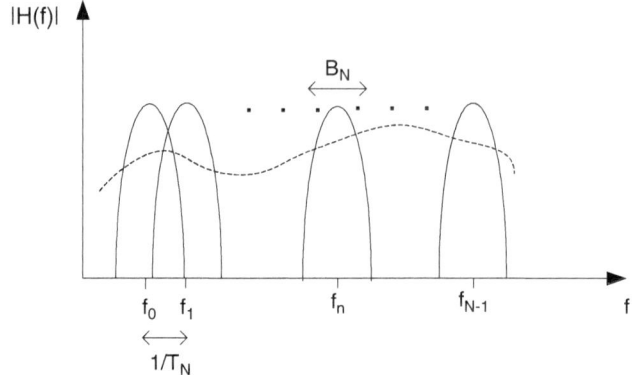

Fig. 2.10 A multicarrier with overlapping subcarriers

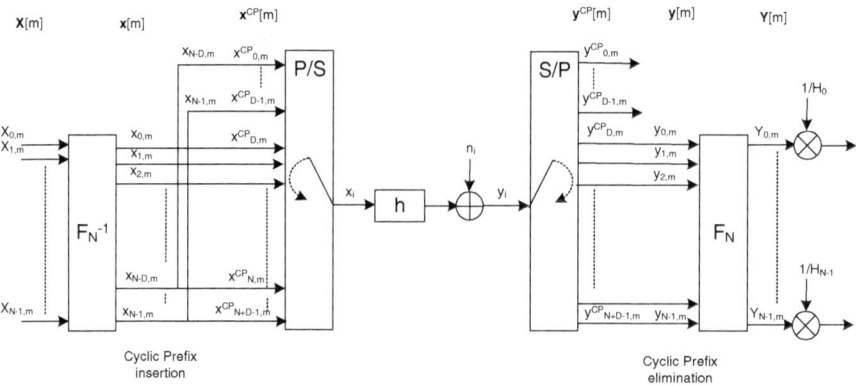

Fig. 2.11 OFDM implementation

$\mathbf{x}[m] = [x_{0,m} x_{1,m} \ldots x_{N-1,m}]$ of length N:

$$\mathbf{x}[m] = \mathbf{F}_N^{-1} \mathbf{X}[m] \tag{2.80}$$

$\mathbf{F}_N^{-1} = \mathbf{F}_N^H$ is the inverse of the Fourier matrix $\mathbf{F}_N$.

At the output of the IFFT, a cyclic prefix (CP) or guard interval of D samples is added at the beginning of each block and the resulting vector is $\mathbf{x}^{CP}[m] = [x_{N-D,m}, \ldots, x_{N-1,m}, x_{1,m}, \ldots, x_{i,m}, \ldots x_{N-1,m}]$. The length D should be longer than the time spread L of the channel ($D > L$). Since the redundancy is $\dfrac{N}{N+D}$, N should be much higher than D in order to limit the negative impact of the prefix cyclic on the efficiency of the scheme. After a parallel to serial conversion (P/S) the sequence is passed through a digital to analog converter (DAC).

The time invariant selective additive white Gaussian noise (AWGN) channel can be described using its equivalent discrete baseband impulse response

$$\mathbf{h}[m] = [h_{0,m}, \ldots, h_{L-1,m}, 0, \ldots, 0] \tag{2.81}$$

At the receiver after analog to digital conversion, the received signal is given by

$$y_{i,m} = \sum_{l=0}^{L-1} h_{l,m} x_{i-l,m} + n_{i,m} \tag{2.82}$$

We assume that the channel response does not change during the OFDM symbol duration. The frequency response of the channel is the Fourier transform of $\mathbf{h}[m]$: $\mathbf{H}[m] = \mathbf{F}_N^{-1} \mathbf{h}[m]$

At the receiver after analog to digital conversion the received signal is given by

$$\mathbf{y}^{CP}[m] = \begin{bmatrix} y_{0,m}^{CP} \\ y_{1,m}^{CP} \\ \vdots \\ \vdots \\ \vdots \\ y_{N+D-1,m}^{CP} \end{bmatrix}_{(N+D)\times 1} = \mathbf{h}^{ISI}[m] \begin{bmatrix} x_{N-D,m} \\ \vdots \\ x_{N-1,m} \\ x_{0,m} \\ \vdots \\ x_{N-1,m} \end{bmatrix}_{(N+D)\times 1}$$

$$+ \mathbf{h}^{IBI}[m] \begin{bmatrix} x_{N-D,m-1} \\ \vdots \\ x_{N-1,m-1} \\ x_{0,m-1} \\ \vdots \\ x_{N-1,m-1} \end{bmatrix}_{(N+D)\times 1} + \mathbf{n}[m] \tag{2.83}$$

with

$$\mathbf{h}^{ISI}[m] = \begin{bmatrix} h_{0,m} & 0 & \cdots & \cdots & \cdots & 0 \\ \vdots & \ddots & \ddots & & & \vdots \\ h_{L,m} & & \ddots & \ddots & & \vdots \\ 0 & \ddots & & \ddots & \ddots & \vdots \\ \vdots & \ddots & \ddots & & \ddots & \vdots \\ 0 & \cdots & 0 & h_{L,m} & \cdots & h_{0,m} \end{bmatrix}_{(N+D)\times(N+D)}$$

$$
\mathbf{h}^{IBI}[m] = \begin{bmatrix} 0 \dots 0 \; h_{L,m} \dots h_{1,m} \\ \vdots \;\ddots \qquad \ddots\;\ddots\;\; \vdots \\ \vdots \qquad \ddots \qquad\quad \ddots\; h_{L,m} \\ \vdots \qquad\qquad \ddots \qquad\quad 0 \\ \vdots \qquad\qquad\qquad \ddots \quad\; \vdots \\ 0 \dots\dots \; \dots \; \dots \; 0 \end{bmatrix}_{(N+D)\times(N+D)}
$$

$\mathbf{h}^{ISI}[m]$ is the ISI due to the selectivity of the channel on the mth OFDM symbol and $\mathbf{h}^{IBI}[m]$ is the inter-block interference between the mth and $(m-1)$th OFDM symbols.

After removing the prefix cyclic corresponding to the D first samples, we obtain

$$
\begin{bmatrix} y_{0,m} \\ \vdots \\ \vdots \\ y_{N-1,m} \end{bmatrix}_{N\times1} = \begin{bmatrix} h_{0,m} \; 0 \; \dots h_{L,m} \dots h_{1,m} \\ \vdots \;\ddots\ddots \qquad\quad \ddots\; \vdots \\ h_{L,m} \qquad\ddots\;\ddots \qquad h_{L,m} \\ 0 \;\;\ddots \qquad\quad \ddots\;\ddots\; \vdots \\ \vdots \;\ddots\ddots \qquad\qquad \ddots\; 0 \\ 0 \;\; \dots \; 0 \; h_{L,m} \dots h_{0,m} \end{bmatrix}_{N\times N} F_N^H \begin{bmatrix} X_{0,m} \\ \vdots \\ \vdots \\ \vdots \\ X_{N-1,m} \end{bmatrix}_{N\times1} + \mathbf{n}[m]
$$

(2.84)

Using the prefix cyclic we have been able to transform the linear convolution into a circular convolution. Since the circular convolution is equivalent to a multiplication in the frequency domain, the vector $\mathbf{X}_m$ has been transmitted over N parallel flat fading channel defined by a complex attenuation $H_{i,m}$.

$$
\begin{bmatrix} Y_{0,m} \\ \vdots \\ \vdots \\ \vdots \\ Y_{N-1,m} \end{bmatrix}_{N\times1} = \begin{bmatrix} H_{0,m} \; 0 \; \dots\dots\dots \;\; 0 \\ 0 \; H_{1,m} \; 0 \; \dots\dots \;\; \dots \\ \vdots \quad 0 \;\;\ddots\; 0 \; \dots \;\; \dots \\ \vdots \quad \dots\;\dots\;\ddots\ddots\; \dots \;\; \dots \\ \vdots \qquad \dots\;\dots\dots\;\ddots \;\; 0 \\ 0 \qquad \dots \; \dots\dots \; 0 \; H_{N-1,m} \end{bmatrix}_{N\times N} \begin{bmatrix} X_{0,m} \\ \vdots \\ \vdots \\ \vdots \\ X_{N-1,m} \end{bmatrix}_{N\times1} + \mathbf{N}[m]
$$

(2.85)

This input–output relation can be written as

$$
\mathbf{Y}[m] = \mathrm{diag}(\mathbf{H}[m])\mathbf{X}[m] + \mathbf{N}[m] \tag{2.86}
$$

where $\mathrm{diag}(\mathbf{H}[m])$ is a diagonal matrix of size $N \times N$ in which the entries outside the main diagonal are all zero and the diagonal entries are composed of the elements of the vector $\mathbf{H}[m]$.

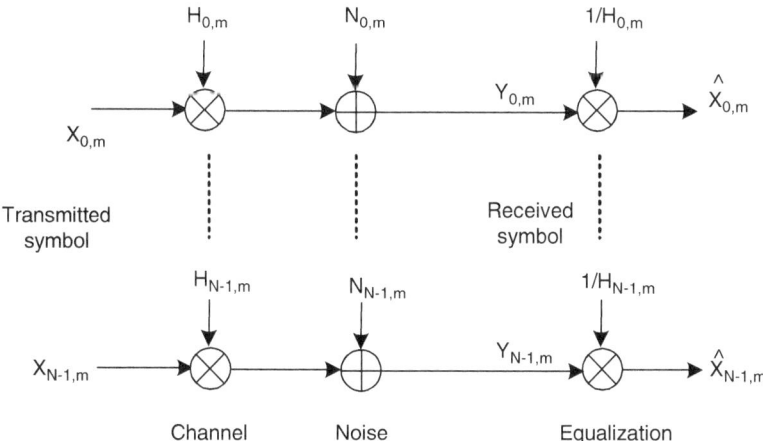

Fig. 2.12 Frequency model of OFDM scheme

Figure 2.12 gives the equivalent model of an OFDM transmission.

If the transmit power on subcarrier n is P_n and the fading on that subcarrier is H_n, then the received signal to noise ratio γ_n is

$$\gamma_n = |H_n|^2 P_n/(N_0 B_N) \tag{2.87}$$

where N_0 is the noise power spectral density.

If $|H_n|^2$ is small then the received SNR on the nth subchannel is quite low, which can lead to a high BER on that subchannel. Since flat fading can seriously degrade performance in each subchannel, it is important to compensate for flat fading in the subchannels.

Since OFDM does not exploit the frequency diversity, it is important to associate to this scheme interleaving and error correcting codes (convolutional codes, block codes, multidimensional constellations, turbo codes) to cope with fading.

2.4 MIMO Systems

2.4.1 Introduction

The next generation wireless communication systems are expected to provide users with mobile multimedia services such as high speed mobile internet access and mobile computing. In order to meet this rapidly growing demand at better quality of services at higher data rates and higher mobility, innovative techniques that improve link reliability and spectral efficiency are needed in mobile propagation environment.

In conventional single transmit and single receive antenna system or SISO system, increasing the data rate can be achieved by increasing either the transmission bandwidth which is expensive and restricted in many cases or the transmission power that requires an expensive amplifier and causes to reduce the battery life of mobile units.

The development of MIMO communication systems that use multiple transmit and multiple receive antennas provides the ability to design spectrally efficient systems without extra power and bandwidth. Assuming that the channels between the transmit and received antennas are independent, Foshini [12] and Telatar [49] have shown that the channel capacity grows linearly with $\min(N_t, N_r)$ where N_t is the number of transmit antennas and N_r is the number of received antennas. This motivates to consider MIMO systems for transmission of high data rates in future wireless communication systems.

MIMO communication systems have two major advantages over SISO systems in wireless channels:

- Multiplexing gain: spatial multiplexing using layered space-time coding techniques (transmission of independent data over different antennas) enables to communicate at higher data rates without increasing the bandwidth or the transmit power.
- Spatial diversity: space-time codes (STC) [35] exploit the spatial diversity (both transmit and receive) available in the multiple spatial channels and improve the performance against fading channels while maintaining a good spectral efficiency. Such techniques include delay diversity, STBC [2, 48] and space-time trellis code (STTC) [47].

A considerable amount of research has addressed the design and implementation of STC and spatial multiplexing systems for wireless flat fading channels. However, many communication channels are frequency selective in nature, for which the STC design problem becomes complicated. On the other hand, the OFDM technique which transforms a frequency selective channel into parallel flat fading subchannels is a potential candidate for high data rate wireless transmission. Hence, the combination of MIMO signal processing with OFDM is regarded as a promising solution for providing diversity and enhancing the data rates of next generation wireless communication systems operating over frequency selective fading channels. STC can be combined with OFDM in the time domain using coded STTC-OFDM [1, 31] or coded STBC-OFDM systems [16, 26] and in the frequency domain using space-frequency block coded (SFBC) OFDM systems [4, 25].

The MIMO systems should be concatenated with outer channel codes in order to achieve near-capacity performance in wireless flat fading channels. Moreover, in multipath MIMO channels, the maximum achievable diversity order is also related to the number of propagation paths which is named as frequency diversity [5]. Since SFBC-OFDM and STBC-OFDM fail to exploit frequency diversity, concatenating them with an outer channel code and interleaver can provide more diversity compared to flat fading channels.

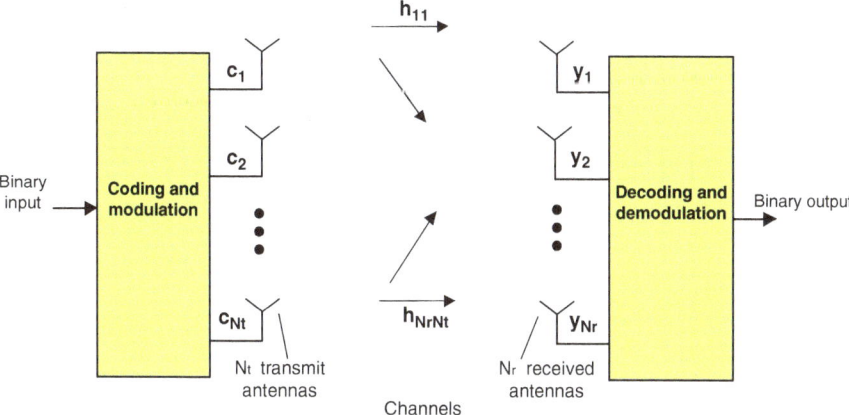

Fig. 2.13 The MIMO wireless transmission system

In this section, we shall present the necessary theories of MIMO systems in adequate depth. We will examine the principles and design criteria of the MIMO system in wireless channels by defining the potential gains. Then, we will present the well-known MIMO codes in detail such as spatial multiplexing code vertical-BLAST (V-BLAST), the orthogonal and non-orthogonal STBC. For these MIMO systems, we will only consider that the channel state information (CSI) is unknown at the transmitter and perfectly known at the receiver.

2.4.2   Capacity of MIMO Systems

We consider a MIMO communication link with N_t transmit antennas and N_r receive antennas, as illustrated in Fig. 2.13.

Assuming that the channel is flat fading, the signal from receive antenna j is a noisy superposition of transmitted signals from N_t transmit antennas as

$$y_j = \sum_{i=1}^{N_t} h_{ji} c_i + n_j, \tag{2.88}$$

where h_{ji} is the complex channel coefficient from transmit antenna i to receive antenna j and n_j is the additive noise which is modeled as independent samples of a zero-mean complex Gaussian random variable. There is a total power constraint, P on the signals from the transmit antennas. The noise power per receive antenna is $N_0 B$ where $N_0/2$ is the noise power spectral density and B is the bandwidth.

We define the channel coefficient matrix $\mathbf{H}$ with the dimension of $N_r \times N_t$ as follows:

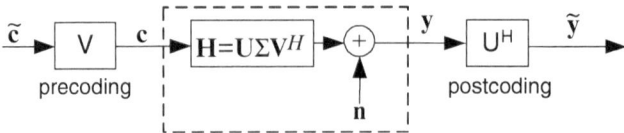

Fig. 2.14 SVD decomposition of the MIMO channel

$$\mathbf{H} = \begin{bmatrix} h_{11} & \dots & h_{1N_t} \\ \vdots & \ddots & \vdots \\ h_{N_r 1} & \dots & h_{N_r N_t} \end{bmatrix}. \tag{2.89}$$

Then, the received signal in (2.88) can be written in a matrix form as

$$\mathbf{y} = \mathbf{H}\mathbf{c} + \mathbf{n}, \tag{2.90}$$

where $\mathbf{y}$ and $\mathbf{n}$ are the received and noise matrices, respectively, with the dimension of $N_r \times 1$.

We decompose the MIMO channel matrix $\mathbf{H}$ into equivalent parallel SISO channels by using singular value decomposition (SVD) as (Fig. 2.14)

$$\mathbf{H} = \mathbf{U}\Sigma\mathbf{V}^H, \tag{2.91}$$

where $\mathbf{U} = [\mathbf{u}_1 \dots \mathbf{u}_{N_r}]$ of size $N_r \times N_r$ and $\mathbf{V} = [\mathbf{v}_1 \dots \mathbf{v}_{N_t}]$ of size $N_t \times N_t$ are unitary matrices, $\mathbf{u}_i$ and $\mathbf{v}_j$ are, respectively, the ith left singular and the jth right singular vector of the channel matrix $\mathbf{H}$. Σ is a rectangular matrix $\Sigma = \text{diag}(\sqrt{\lambda_1}, \sqrt{\lambda_2}, \dots, \sqrt{\lambda_r}, 0, \dots, 0)$ where $\lambda_1 \geq \lambda_2 \geq \dots \lambda_r$ are the nonzero eigenvalues of $\mathbf{H}^H\mathbf{H}$ (assuming $N_t \leq N_r$). The number of eigenvalues r equals to the rank of the channel matrix $\mathbf{H}$ which corresponds to $\min(N_t, N_r)$.

By applying preprocessing to the transmitted symbols ($\mathbf{V}\mathbf{c}$) at the transmitter side and post-processing to the received signal ($\mathbf{U}^H\mathbf{y}$) we obtain the relation

$$\begin{aligned} \mathbf{U}^H\mathbf{y} &= \mathbf{U}^H(\mathbf{U}\Sigma\mathbf{V}^H)\mathbf{V}\mathbf{c} + \mathbf{U}^H\mathbf{n} \\ \tilde{\mathbf{y}} &= \Sigma\tilde{\mathbf{c}} + \tilde{\mathbf{n}} \end{aligned} \tag{2.92}$$

where $\tilde{\mathbf{n}}$ is still Gaussian with the same variance than $\mathbf{n}$ and $||\tilde{\mathbf{c}}||^2 = ||\mathbf{c}||^2$, thus the power is unchanged.

Equation (2.92) represents the system as r equivalent parallel SISO Gaussian channels with signal powers given by the eigenvalues.

$$\tilde{y}_i = \sqrt{\lambda_i}\tilde{c}_i + \tilde{n}_i \qquad 1 \leq i \leq r \tag{2.93}$$

Thus, if the channel is perfectly known at the transmitter and the receiver, the transmitter can optimize its transmission strategy for each fading channel realization. The ergodic capacity under the short power constraint, where the power associated with each channel realization must equal the average power constraint P, is given by

$$C(N_t, N_r) = \max_{\mathbf{Q}:\mathrm{tr}(\mathbf{Q})=P} B_W \log_2 \left(\det \left(\mathbf{I}_{N_r} + \frac{1}{N_0 B_W} \mathbf{HQH}^H \right) \right) \qquad (2.94)$$

where $\mathbf{Q}$ is the covariance of the MIMO channel input $\mathbf{Q} = \mathbb{E}(\mathbf{cc}^H)$.

The capacity can be obtained as the sum of independent parallel channels capacities with the transmit power optimally allocated between these channels. We have

$$C(N_t, N_r) = \sum_{i=1}^{r} B_W \log_2 \left(1 + \frac{P_i}{N_0 B_W} \lambda_i \right) \qquad (2.95)$$

where the power allocation is obtained by waterfilling

$$\frac{P_i}{P} = \begin{cases} \frac{1}{\gamma_0} - \frac{N_0 B_W}{\lambda_i P} & \frac{\lambda_i P}{N_0 B_W} \geq \gamma_0 \\ 0, & \frac{\lambda_i P}{N_0 B_W} < \gamma_0 \end{cases} \qquad (2.96)$$

with the cutoff value γ_0 chosen to satisfy the total power constraint $\sum_{i=1}^{r} P_i = P$.

If the channel is unknown at the transmitter and perfectly known at the receiver, the power should be uniformly distributed over the transmit antennas. The instantaneous capacity can be obtained as

$$C(N_t, N_r) = \sum_{i=1}^{r} B_W \log_2 \left(1 + \frac{P}{N_t N_0 B_W} \lambda_i \right) \qquad (2.97)$$

The instantaneous capacity of a (N_t, N_r) MIMO system can also be given using the well-known "logdet" equation [12, 49] as

$$C(N_t, N_r) = B_W \log_2 \det \left(\mathbf{I}_{N_r} + \frac{P}{N_t N_0 B_W} \mathbf{HH}^H \right) \qquad (2.98)$$

From the capacity in (2.97) and (2.98), we introduce two other capacities: the ergodic capacity and the outage capacity.

The channel capacity of a (N_t, N_r) MIMO system in ergodic channel environment can also be given by [12, 49] as

$$C(N_t, N_r) = \mathbb{E}_{\mathbf{H}} \left\{ B_W \log_2 \det \left(\mathbf{I}_{N_r} + \frac{P}{N_t N_0 B_W} \mathbf{HH}^H \right) \right\} \qquad (2.99)$$

Like the instantaneous capacity, the ergodic capacity is given in bits per second.

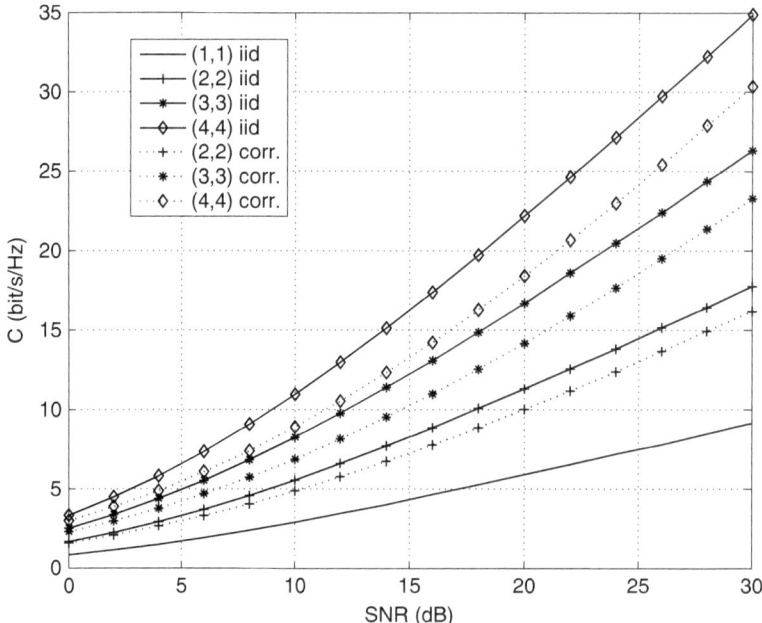

Fig. 2.15 The ergodic capacity for different antenna configurations in correlated and uncorrelated flat fading channels

Figure 2.15 shows the ergodic capacity of several MIMO configurations as a function of SNR over nonselective channels, i.i.d. according to the Rayleigh distribution. We can check that the capacity increases linearly with $\min(N_t, N_r)$ (each increase of 3 dB imply $\min(N_t, N_r)$ more bit/s/Hz). We also give the capacities obtained using correlated MIMO channels. We consider an uplink channel using linearly space antennas. At the transmitter side, the distance between the antennas is 0.5λ; the AoD is $20°$. At the receiver side, the distance between the antennas is 4.0λ, the AoA as $20°$ and the AS as $5°$ which corresponds to an uplink case. The capacity improves by increasing the number of transmit and receive antennas and also SNR. The correlation between the MIMO channels decreases the capacity such as at 20 bps/Hz $(4, 4)$; MIMO system needs 18 dB and 22 dB in the independent and correlated channel environment, respectively.

If the duration of the block is limited compared to the coherence time of the MIMO channels, the capacity is treated as a random variable which depends on the instantaneous channel response and remains constant during the transmission of the block of data. If the channel capacity is below the outage capacity, there is no possibility that the transmitted block of information can be decoded with no errors.

The outage capacity, $C_{out}(q)$, is defined as the information rate that is guaranteed for $(100 - q)\%$ of the channel realizations. We have

$$Pr(C \le C_{out}(q)) = q\% \tag{2.100}$$

2.4.3 Gain of MIMO Systems

MIMO systems can extract various gains considering that the channel is known or unknown at the transmitter side. If the channel is unknown at the transmitter, diversity and multiplexing gain can be achieved by MIMO systems. If the transmitter has the knowledge of channel information using feedback, array gain can also be provided by MIMO systems. We will summarize these gains in the following sections.

2.4.3.1 Array Gain

Considering the single input multiple output (SIMO) system with one transmit and N_r receive antennas, if the channel is known to the receiver, appropriate signal processing techniques can be applied to combine the received signal coherently so that the resultant power of the signal at the receiver is enhanced, leading to an improvement in signal quality. The average increase in received signal power at the receiver, $\mathbb{E}[\|\mathbf{h}\|^2]$, is defined as array gain which is proportional to the number of receive antennas. Array gain can also be exploited in systems with multiple antennas at the transmitter using linear or nonlinear precoding. Extracting the maximum possible array gain in such systems requires channel knowledge at the transmitter, so that the signals may be optimally processed before transmission. The array gain in MIMO systems depends on the number of transmit and receive antennas and is a function of the dominant singular value of the channel matrix.

2.4.3.2 Diversity Gain

As we mentioned before, the signal power in wireless channel fluctuates or fades with time, frequency and space. Diversity is used in wireless systems to combat fading. The basic idea behind diversity is to provide the receiver with several replicas over independently fading links (or diversity branches). As the number of the diversity branches increases, the probability that at any instant of time, one or more branch is not in a fade increases. Thus, diversity helps to stabilize a wireless link.

Diversity is available in SISO links in the form of time, such as channel coding in conjunction with interleaver or frequency diversity at the cost of data rate reduction due to the utilization of more time or more bandwidth. The introduction of multiple antennas at the transmitter and/or receiver provides spatial diversity. Compared to time and frequency diversity, spatial diversity does not incur a penalty in the data rate. The spatial diversity gain in the system is characterized by the number of independently fading diversity branches. The receive diversity can be obtained by using more than one antenna at the receiver and combining the independent signals at each receive antenna. In order to extract transmitter diversity, a suitable design of the transmitted signal is required using STC or precoding scheme. The overall spatial diversity gain d of a MIMO system is defined by

$$d = - \lim_{\text{SNR} \to \infty} \frac{\log P_e(\text{SNR})}{\log \text{SNR}} \qquad (2.101)$$

The spatial diversity gain describes how fast error probability decays with SNR since $P_e \propto SNR^{-d}$. The maximum spatial diversity order is $N_t \times N_r$.

2.4.3.3 Multiplexing Gain or Degrees of Freedom

When the path gains between individual transmit-receive antenna pairs fade independently, the channel matrix is well conditioned with high probability. Consequently, by transmitting independent information streams in parallel through the spatial channels, the data rate can be significantly increased. This effect is known as multiplexing gain and represents the increase in channel capacity. The multiplexing gain or degrees of freedom (DoF) r is defined as

$$r = \lim_{\text{SNR} \to \infty} \frac{R(\text{SNR})}{\log \text{SNR}}. \qquad (2.102)$$

Multiplexing gain is the pre-log of the rate in the high-SNR regime and determines how fast rate increases with SNR. In a point to point MIMO system, the multiplexing gain is limited by $\min(N_t, N_r)$ that is the degree of freedom of this communication channel.

2.4.4 Diversity Multiplexing Tradeoff

Zheng and Tse [57] have proved that there exists a tradeoff between diversity gain and multiplexing gain for a slow fading channel. For a MIMO system with N_t transmit and N_r receive antennas, the optimum tradeoff curve, $d(r)$, has been evaluated in [57]. Assuming that the block size is equal or bigger than $N_t + N_r - 1$, the optimal tradeoff curve $d(r)$ is given by the piece-wise linear function connecting the points $(r, d(r))$; $r = 0, \ldots, \min(N_t, N_r)$ where

$$d(r) = (N_t - r)(N_r - r). \qquad (2.103)$$

This is a fundamental tradeoff which shows that it is not possible to reach at the same time the maximum diversity gain and maximum multiplexing gain. The DMT can be used to evaluate and compare the MIMO schemes. The function $d(r)$ is plotted in Fig. 2.16.

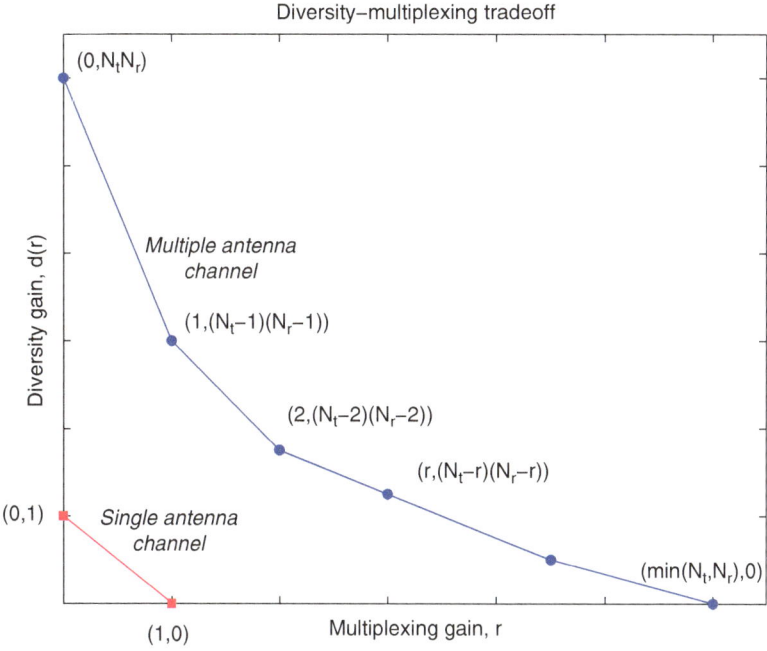

Fig. 2.16 Diversity-multiplexing tradeoff, $d(r)$ for general N_t, N_r

2.4.5 MIMO Codes

For a MIMO wireless transmission system, we consider a communication link with N_t transmit antennas and N_r receive antennas.

At the transmitter, information symbols s belonging to the constellation set Λ_s are parsed into blocks $\mathbf{s} = [s_1 \quad s_2 \quad \dots \quad s_Q]^T$ of size $Q \times 1$ in the symbol vector where Q is the number of information symbols. The MIMO encoder performs the mapping of the vector $\mathbf{s}$ to the matrix code $\mathbf{C}$ of size $N_t \times T$:

$$
\mathbf{C} = \begin{bmatrix} c_{1,1} & \cdots & c_{1,T} \\ \vdots & \ddots & \vdots \\ c_{N_t,1} & \cdots & c_{N_t,T} \end{bmatrix}
\tag{2.104}
$$

where T is the number of time intervals that the information symbols are transmitted in the code matrix. Then, the T columns of $\mathbf{C}$ are sent through the N_t transmit antennas simultaneously. Each symbol in the code matrix belongs to constellation set Λ_c which may be identical or different than Λ_s depending on the code structure. Since Q information symbols are sent from each transmit antenna in T time intervals, the transmission rate of MIMO code is equal to $R_{\text{MIMO}} = Q/T$.

Assuming that the channel is flat fading, the signal from receive antenna j at time t is a noisy superposition of transmitted signals from N_t transmit antennas as

$$y_{j,t} = \sum_{i=1}^{N_t} h_{j,i,t} c_{i,t} + n_{j,t}, \qquad (2.105)$$

where $h_{j,i,t}$ is the complex channel coefficient from transmit antenna j to receive antenna i at time t and $n_{j,t}$ is the additive noise which is modeled as independent samples of a zero-mean complex Gaussian random variable.

Considering the quasi-static flat fading channels ($h_{j,i,t} = h_{ji}$ for $1 \leq t \leq T$), the channel coefficient matrix $\mathbf{H}$ with the dimension of $N_r \times N_t$ is denoted as

$$\mathbf{H} = \begin{bmatrix} h_{1,1} & \cdots & h_{1,N_t} \\ \vdots & \ddots & \vdots \\ h_{N_r,1} & \cdots & h_{N_r,N_t} \end{bmatrix}. \qquad (2.106)$$

Then, the received signal in (2.105) can be written with a matrix form as

$$\mathbf{Y} = \mathbf{HC} + \mathbf{N}, \qquad (2.107)$$

where $\mathbf{Y}$ and $\mathbf{N}$ are the received and noise matrices, respectively, with the dimension of $N_r \times T$. Given $\mathbf{Y}$, the MIMO decoder decodes $\mathbf{s}$ using the unique mapping between $\mathbf{C}$ and $\mathbf{s}$.

There are two main techniques which exploit the potential benefits of MIMO channel: spatial multiplexing (SM), which can be regarded as a special class of STC where streams of independent data are transmitted over different antennas, thus maximizing the average data rate over the MIMO systems and STBC, which offers diversity and coding gains with improved spectral efficiency. In the next section, we shall describe the well-known MIMO schemes such as STBC codes and V-BLAST.

2.4.6 Design Criteria for Space-Time Codes

In this section, we will introduce the criteria to build efficient STC. We define the pairwise block error event for a realization of the channel $\mathbf{H}$, $\Pr(\mathbf{C} \to \mathbf{C}'|\mathbf{H})$ as the event that the receiver decodes the block $\mathbf{C}'$ erroneously when the block $\mathbf{C}$ is actually sent.

Defining the difference matrix

$$\mathbf{D} = \begin{bmatrix} c_{11} - c_{11}' & \cdots & c_{1T} - c_{1T}' \\ c_{21} - c_{21}' & \cdots & c_{2T} - c_{2T}' \\ \vdots & \ddots & \vdots \\ c_{N_t 1} - c_{N_t 1}' & \cdots & c_{N_t 2} - c_{N_t 2}' \end{bmatrix} \qquad (2.108)$$

Since the matrix $\mathbf{E} = \mathbf{DD}^H$ is Hermitian of size $N_t \times N_t$, there exists a unitary matrix $\mathbf{T}$ and a real diagonal matrix $\mathbf{U}$ such as $\mathbf{TET}^H = \mathbf{U}$. The elements of the diagonal of $\mathbf{U}$ are the eigenvalues of $\mathbf{A}$, i.e., $\lambda_i; i = 1, 2, .., N_t$.

We can show that the probability $P(\mathbf{C} \to \mathbf{C}'|\mathbf{H})$ can be upper bounded using the Chernoff bound:

$$Pr(\mathbf{C} \to \mathbf{C}'|\mathbf{H}) \leq \exp\left(-\frac{E_s}{4N_0 N_t} d^2(\mathbf{C}, \mathbf{C}')\right) \qquad (2.109)$$

$$d^2(\mathbf{C}, \mathbf{C}') = \sum_{j=1}^{N_r} \mathbf{h}_j \mathbf{DD}^H \mathbf{h}_j^H$$

$$= \sum_{j=1}^{N_r} \mathbf{h}_j \mathbf{T}^H \mathbf{U} \mathbf{Th}_j^H$$

$$= \sum_{j=1}^{N_r} \sum_{i=1}^{N_t} \lambda_i \|\beta_{ij}\|^2$$

where $\mathbf{h}_j = [h_{j1} \ h_{j2} \ldots h_{jN_t}]$ is the jth line of $\mathbf{H}$. β_{ij} is the ith element of the vector $\beta_j = \mathbf{h}_j \mathbf{T}^H$.

After averaging over the fading channels, we show that the pairwise error probability (PEP) $Pr(\mathbf{C} \to \mathbf{C}')$ can be upper bounded with

$$Pr(\mathbf{C} \to \mathbf{C}') \leq \left(\frac{E_s}{4N_t N_0}\right)^{-r_d N_r} \left(\prod_{k=1}^{r_d} \lambda_k\right)^{-N_r} \qquad (2.110)$$

where r_d is the rank of the matrix $\mathbf{E}$ and λ_k corresponds to the nonzero eigenvalues of the difference matrix $\mathbf{D}$.

The aim is to build an optimal encoder minimizing the PEP $Pr\{\mathbf{C} \to \mathbf{C}'\}$ for all possible pairs [47].

From this expression, we define two criteria for the construction of the code: the rank criterion and the determinant criterion which determine, respectively, diversity and coding gain.

Rank Criterion: in order to achieve maximum diversity gain equal to $N_t N_r$, the difference matrix $\mathbf{D}$ has to be full rank over all pairs of distinct codewords. If the minimum rank is equal to r_d over all possible codewords, the diversity gain of $r_d N_r$ is achieved.

Determinant Criterion: the term $\prod_{k=1}^{r_d} \lambda_k$ is the coding gain. It should be maximized over all pairs of distinct codewords.

The diversity criterion is the most important one of the two since it determines the slope of the performance curve in double-logarithmic scale.

If the difference matrix is of full rank, then the product of eigenvalues is equal to the determinant of matrix $\mathbf{E}$. In this case, the determinant criterion can be written as

$$c_g = \min_{\mathbf{C} \neq \mathbf{C}'} \left| \det[(\mathbf{C} - \mathbf{C}')(\mathbf{C} - \mathbf{C}')^H] \right|^{1/r_d} = \min_{\mathbf{C} \neq \mathbf{C}'} |\det \mathbf{E}|^{1/r_d}. \qquad (2.111)$$

Trace Criterion: For small SNR and/or the high values of $r_d N_r$, in order to minimize the error probability in (2.110), the minimum sum of all eigenvalues $\sum_{r=1}^{r_d} \lambda_r$ of matrix $\mathbf{E}$ among all the pairs of distinct codewords should be also maximized.

Since $\mathbf{E}$ is a $N_t \times N_t$ matrix, the sum of all the eigenvalues is equal to the sum of the all elements on the main diagonal or the trace of $\mathbf{E}$ that can be expressed as

$$\sum_{r=1}^{r_d} \lambda_r = \text{tr}(\mathbf{E}) = \sum_{i=1}^{N_t} E_{i,i}. \qquad (2.112)$$

The trace of $\mathbf{E}$ can be also written as

$$\text{tr}(\mathbf{E}) = \sum_{i=1}^{N_t} \sum_{t=1}^{T} |c_{i,t} - c'_{i,t}|^2, \qquad (2.113)$$

Consequently, this criterion is equivalent to the maximization of the Euclidean distance between the pair of codewords $\mathbf{C}$ and $\mathbf{C}'$. This distance can be bounded by the minimum Euclidean distances of the signal constellations.

2.4.6.1 Orthogonal Space-Time Block Codes

Alamouti proposed a remarkable low complexity open loop transmit diversity scheme for $N_t = 2$ with $Q = 2$ and $T = 2$ [2]. Alamouti's code is the only complex orthogonal STBC that achieves maximum diversity with rate 1 allowing to reach the full capacity.

In this scheme, the input data stream is first mapped into symbols using a constellation mapper, and the symbol stream is then divided into two substreams. At a particular symbol period, the transmitted symbols from the first and second antenna are s_1 and s_2, respectively. In the next symbol period, the symbol transmitted from the first antenna is $-s_2^*$ and the symbol transmitted from the second antenna is s_1^*. Hence, we can write the code matrix as

$$\mathbf{C} = \begin{bmatrix} s_1 & -s_2^* \\ s_2 & s_1^* \end{bmatrix}. \qquad (2.114)$$

Note that it is orthogonal since $\mathbf{C}\mathbf{C}^H = \left(|s_1|^2 + |s_2|^2\right)\mathbf{I}_2$.

Under the assumption that the channel is flat fading and constant over two consecutive symbol periods, the two received signals for $N_r = 1$ are given by

$$[y_{11} \quad y_{12}] = [h_{11} \quad h_{12}] \begin{bmatrix} s_1 & -s_2^* \\ s_2 & s_1^* \end{bmatrix} + [n_{11} \quad n_{12}]. \qquad (2.115)$$

Defining the transmitted and received signal vectors as $\mathbf{S} = [s_1 \quad s_2]^T$ and $\mathbf{y_e} = [y_{11} \quad y_{12}^*]^T$, respectively, we can also write (2.115) in a matrix/vector form as

$$\mathbf{y_e} = \begin{bmatrix} h_{11} & h_{12} \\ h_{12}^* & -h_{11}^* \end{bmatrix} \begin{bmatrix} s_1 \\ s_2 \end{bmatrix} + \begin{bmatrix} n_{11} \\ n_{12}^* \end{bmatrix}$$

$$= \mathbf{H_e s} + \mathbf{n_e} \qquad (2.116)$$

Note that the channel matrix has an orthogonal structure, $\mathbf{H_e}^H \mathbf{H_e} = G_{21} \mathbf{I_2}$, where $G_{21} = |h_{11}|^2 + |h_{21}|^2$ is the diversity gain.

Assuming that all the symbols are equiprobable, and since the noise vector $\mathbf{n_e}$ is assumed to be a multivariate AWGN vector, we can easily write the optimum maximum likelihood (ML) decoder as

$$\hat{\mathbf{s}} = \arg \min_{\mathbf{s} \in \Lambda_s^2} \| \mathbf{y_e} - \mathbf{H_e s} \|^2. \qquad (2.117)$$

Since $\mathbf{H_e}$ has an orthogonal structure, the ML decoding in (2.117) can be simplified considering the modified signal vector which is the multiplication of the Hermitian of the channel matrix by the received vector

$$\tilde{\mathbf{s}} = \mathbf{H_e}^H \mathbf{y_e} = G_{21} \mathbf{s} + \tilde{\mathbf{n}} \qquad (2.118)$$

Since the noise vector $\tilde{\mathbf{n}} = \mathbf{H_e}^H \mathbf{n_e}$ is still zero-mean and covariance $\rho N_0 \mathbf{I_2}$, we can decode separately the two symbols.

In that case, the decoding is performed as

$$\hat{\mathbf{s}} = \arg \min_{\mathbf{s} \in \Lambda_s^2} \| \tilde{\mathbf{s}} - G_{21} \mathbf{s} \|^2. \qquad (2.119)$$

Hence, by using this simple linear combining, the two-dimensional minimization problem in (2.119) is decoupled into two one-dimensional problems as

$$\hat{s}_1 = \arg \min_{s_1 \in \Lambda_s} |\tilde{s}_1 - G_{21} s_1|^2 \quad \text{and} \quad \hat{s}_2 = \arg \min_{s_2 \in \Lambda_s} |\tilde{s}_2 - G_{21} s_2|^2. \qquad (2.120)$$

Compared to a SISO transmission, SNR for each symbol is $(|h_{11}|^2 + |h_{12}|^2) P / N_0 B_W$. Hence a two branch diversity performance (diversity gain of order two) is obtained at the receiver.

For decoding, the $(2, 1)$ STBC requires only two complex multiplications and one complex addition per symbol. Using 2^{M_c} constellation points, where M_c is the number of bits per symbol, this linear combining reduces the number of decoding metrics that has to be computed for ML decoding from 2^{2M_c} to 2^{M_c+1}. Furthermore, $(2, 1)$ STBC can be easily extended to N_r receive antenna case. The diversity order provided by this scheme is $2N_r$.

The (N_t, N_r, T) STBC capacity is given by [12]

$$C(N_t, N_r, T) = \frac{1}{T}\mathbb{E}_{\mathbf{H_e}}\left\{ B \log_2 \det\left(\mathbf{I}_{N_r T} + \frac{P}{N_t N_0 B_W}\mathbf{H_e H_e}^H \right)\right\}. \qquad (2.121)$$

In order to investigate the $(2, 1)$ STBC capacity, we need to compute the mutual information between the transmitted and the received vectors $\mathbf{s}$ and $\mathbf{y_e}$ in the equivalent channel matrix $\mathbf{H_e}$ and compare it with the capacity of multiple antenna system with $N_t = 2$ and $N_r = 1$. Since $\mathbf{H_e H_e}^H = \mathbf{H_e}^H \mathbf{H_e} = G_{21}\mathbf{I}_2$ for $N_r = 1$, the achievable capacity of this orthogonal STBC is

$$C = B \log_2\left(1 + \frac{P}{2N_0 B_W}G_{21} \right) = C(N_t = 2, N_r = 1), \qquad (2.122)$$

which indicates that $\mathbf{C}$ code can achieve the same channel capacity with the multiple antenna system for $N_t = 2$ and $N_r = 1$.

It is proven by the Hurwitz–Radon theorem that except the Alamouti code, there exist only few complex orthogonal STBC with rate less than the Alamouti code [48]. For example, for $N_t = 3$, $N_r = 1, Q = 3$ and $T = 4$ and consequently $R_{MIMO} = 3/4$ we have the following code matrix:

$$\mathbf{C}_{STBC,3} = \begin{bmatrix} s_1 & s_2 & s_3 & 0 \\ -s_2^* & s_1^* & 0 & -s_3 \\ -s_3^* & 0 & s_1^* & s_2 \end{bmatrix} \qquad (2.123)$$

However, it is possible to achieve transmission rate 1 for complex constellations for more than two transmit antennas by sacrificing orthogonality.

2.4.6.2 Nonorthogonal Space-Time Block Codes

Non-orthogonal (NO) STBC have been proposed for three transmit antennas in [52] at different transmission rates and for four transmit antennas with transmission rate 1 in terms of capacity in [38] and diversity in [17] at the expense of performance loss. By sacrificing orthogonality, it is possible to build codes with rate one such as the quasi-orthogonal space-time codes (QO-STBC) [17, 38, 50]. The matrix code of the QO-STBC is given by

$$\mathbf{C}_{QO-STBC,4} = \begin{bmatrix} s_1 & -s_2^* & -s_3^* & s_4 \\ s_2 & s_1^{\,*} & -s_4^* & -s_3 \\ s_3 & -s_4^* & s_1^{\,*} & -s_2 \\ s_4 & s_3^* & s_2^{\,*} & s_1 \end{bmatrix} \tag{2.124}$$

This matrix is obtained from two Alamouti matrices and a Hadamard transform.

Defining the transmitted and received signal vectors as $\mathbf{S} = [s_1 \quad s_2 \quad s_3 \quad s_4]^T$ and $\mathbf{y_e} = [y_{11} \quad y_{12}^* \quad y_{13}^* \quad y_{14}]^T$, respectively, the input–output relation can be written as

$$\mathbf{y_e} = \begin{bmatrix} h_{11} & h_{12} & h_{13} & h_{14} \\ h_{12}^* & -h_{11}^* & h_{14}^* & -h_{13}^* \\ h_{13}^* & h_{14}^* & -h_{11}^* & -h_{12}^* \\ h_{14} & -h_{13} & -h_{12} & h_{11} \end{bmatrix} \begin{bmatrix} s_1 \\ s_2 \\ s_3 \\ s_4 \end{bmatrix} + \begin{bmatrix} n_{11} \\ n_{12}^* \\ n_{13}^* \\ n_{14} \end{bmatrix}$$

$$= \mathbf{H_e s} + \mathbf{n_e} \tag{2.125}$$

Compared to orthogonal codes, multiplying the received vector by $\mathbf{H_e}^H$ does not allow us to obtain the ML performance. Indeed, we have

$$\mathbf{H_e}^H \mathbf{H_e} = \sum_{i=1}^{4} (|h_{1i}|^2)\mathbf{I_4} + \mathbf{J} \tag{2.126}$$

where $\mathbf{J}$ is an interfering matrix composed of some nonzero elements given by

$$\mathbf{J} = \begin{bmatrix} 0 & 0 & 0 & \alpha \\ 0 & 0 & -\alpha & 0 \\ 0 & -\alpha & 0 & 0 \\ \alpha & 0 & 0 & 0 \end{bmatrix} \tag{2.127}$$

where $\alpha = 2\Re(h_{11}^* h_{14} - h_{12}^* h_{13})$. Since this code is not orthogonal, compared to the Alamouti code, linear decoding (ZF or MMSE) is no more optimal. In order to obtain the ML solution, we can apply sphere decoding.

Other nonorthogonal STBC such as the diagonal space-time (DAST) block codes [9] and threaded algebraic space-time (TAST) block codes [10] are using rotated modulations to achieve the optimal tradeoff between diversity, rate and delay.

2.4.6.3 V-BLAST

An architecture which theoretically achieves the capacity in an independent Rayleigh scattering environment was proposed as bell labs space time (BLAST) in [11]. Difficulties in the implementation of the original scheme led to a simplified

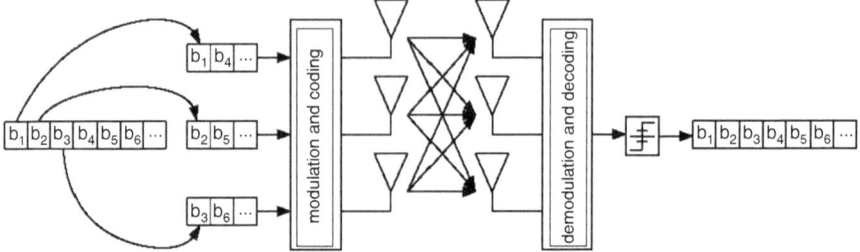

Fig. 2.17 V-BLAST scheme

architecture called V-BLAST [13, 56] where each layer is associated with a certain transmit antenna as shown in Fig. 2.17.

As an example, we will consider the case with $N_t = N_r = 2$, $Q = 2$, $T = 1$ which provides the rate of $R_{MIMO} = 2$; the code matrix is given by

$$\mathbf{C}_{V-BLAST,2} = \begin{bmatrix} s_1 \\ s_2 \end{bmatrix} \qquad (2.128)$$

where s_1, s_2 are obtained from the symbol constellation Λ_s.

Under the assumption that the channel is flat fading and constant over two consecutive symbol periods, the received signal is written in the matrix form as

$$\begin{bmatrix} y_{11} \\ y_{21} \end{bmatrix} = \mathbf{H} \begin{bmatrix} s_1 \\ s_2 \end{bmatrix} + \begin{bmatrix} n_{11} \\ n_{21} \end{bmatrix} \qquad (2.129)$$

In this case we have $\mathbf{C} = \mathbf{s}$ and the relation (2.107) becomes the V-BLAST input–output relation

$$\mathbf{y} = \mathbf{Hs} + \mathbf{n} \qquad (2.130)$$

where $\mathbf{s}$ is the transmit vector of size $N_t \times 1$ and $\mathbf{y}$ is the receive vector of size $N_r \times 1$.

In order to be able to decode, the system should be well conditioned: when $N_r \geq N_t$, the MIMO channel matrix should be full rank with high probability.

The optimum decoding method is ML decoding, where the receiver compares all possible combinations of symbols which could have been transmitted and results in the estimate

$$\hat{\mathbf{s}} = \arg \min_{\mathbf{s} \in \Lambda_s^{N_t}} \|\mathbf{y} - \mathbf{Hs}\|^2 . \qquad (2.131)$$

The complexity of ML decoding becomes high when many antennas or higher-order modulations are used. Sphere decoding which avoids explicit exhaustive search [8] can be used to reduce the complexity of the ML decoding. If the number of receive antennas is at least as large as the number of transmit antennas ($N_r \geq N_t$), there exists linear and nonlinear decoding schemes for V-BLAST.

The zero forcing (ZF) receiver consists in multiplying the received vector by the $N_s \times N_r$ pseudo-inverse matrix of $\mathbf{H}$

$$\tilde{\mathbf{y}} = \mathbf{G}\mathbf{y} = (\mathbf{H})^\dagger \mathbf{y} \qquad (2.132)$$

where $\mathbf{A}^\dagger$ is the Moore–Penrose pseudo-inverse of $\mathbf{A}$.

The minimum mean square error (MMSE) receiver applies a $N_s \times N_r$ matrix to the received vector

$$\tilde{\mathbf{y}} = \mathbf{G}\mathbf{y} = \left[\mathbf{H}^H\mathbf{H} + \frac{N_t N_0 B_W}{P} \mathbf{I}_{N_t} \right]^{-1} \mathbf{H}^H \mathbf{y} \qquad (2.133)$$

The nulling and cancelling (NC) algorithm [14] is a generalization of the decision-feedback algorithm. After having transformed the MIMO channel matrix into an upper triangular matrix, we estimate a first data, then remove its contribution before the estimation of the second one. After the QR decomposition of the channel matrix $\mathbf{H} = \mathbf{Q}\mathbf{R}$ where $\mathbf{Q}$ is an unitary matrix and $\mathbf{R}$ is an upper triangular matrix, we calculate the two following matrices:

$$\begin{cases} \mathbf{G} = \text{diag}^{-1}(\mathbf{R})\mathbf{Q}^H \\ \mathbf{L} = \text{diag}^{-1}(\mathbf{R})\mathbf{R} - \mathbf{I}_{N_r} \end{cases} \qquad (2.134)$$

After multiplying the received vector by $\mathbf{G}$, we get the following relation:

$$\tilde{\mathbf{y}} = \mathbf{G}\mathbf{y} = \text{diag}^{-1}(\mathbf{R})\mathbf{R}\mathbf{s} + \mathbf{G}\mathbf{n} \qquad (2.135)$$

Consequently, it is possible to estimate successively the symbols $s_{N_t}, s_{N_t-1}, \ldots, s_1$:

$$\begin{cases} \hat{\mathbf{s}}_{N_t} & = \text{decision}\,((\tilde{\mathbf{y}})_{N_t}) \\ \hat{\mathbf{s}}_{N_t-1} & = \text{decision}\,((\tilde{\mathbf{y}})_{N_t-1} - \hat{\mathbf{s}}_{N_t}\mathbf{L}_{N_t-1,N_t}) \\ \vdots \\ \hat{\mathbf{s}}_1 & = \text{decision}\,((\tilde{\mathbf{y}})_1 - \hat{\mathbf{s}}_{N_t}\mathbf{L}_{1,N_t} - \ldots - \hat{\mathbf{s}}_2\mathbf{L}_{1,2}) \end{cases} \qquad (2.136)$$

The NC detection has been introduced using a ZF criterion but can be also applied to MMSE criterion. Furthermore, instead of choosing the natural ordering, it is better to decode starting from the most powerful symbol.

References

1. Agrawal D, Tarokh V, Naguib A, Seshadri N (1998) Space-time coded OFDM for high data-rate wireless communication over wideband channels. Proc. of Vehicular Technology Conference (VTC). 3 : 2232–2236
2. Alamouti S M (1998) A Simple Transmit Diversity Technique for Wireless Communications. IEEE Journal on Selec. Areas in Commun. 16 : 1451–1458
3. Bello P A (1963) Characterization of randomly time-varying linear channels. IEEE Transaction on Communications. 11 : 360–393
4. Bolcskei H, Paulraj A J (2000) Space-frequency coded broadband OFDM systems. Proc. of IEEE Wireless Communications and Networking Conference (WCNC). 1 : 1–6
5. Bolcskei H, Paulraj A J (2000) Performance of space-time codes in the presence of spatial fading correlation. Proceedings of IEEE Asilomar Conf. on Signals, Systems, and Computers. Pacific Grove, CA. 1 : 687–693
6. Brown T W C, Saunders S R, Stavrou S, Fiacco M(2007) Characterization of Polarization Diversity at the Mobile. IEEE Transactions on Vehicular Technology. 56 : 2440–2447
7. Clarke R H (1968) A statistical theory of mobile-radio reception. Bell System Technical Journal. 47 : 957–1000
8. Damen M O, Chkeif A, Belfiore J C (2000) Lattice code decoder for space time codes. IEEE Communication Letter. 4 : 161–163
9. Damen M O, Abed-Meraim K, Belfiore J C (2002) Diagonal Algebraic Space Time Block Codes. IEEE Trans. on Information Theory. 48: 628–636
10. El Gamal H, Damen M O (2003) Universal Space Time Coding. IEEE Trans. on Information Theory. 49 : 1097–1119
11. Foschini G J (1996) Layered space-time architecture for wireless communications in a fading environment when using multiple antennas. Bell Laboratories Technical Journal. 1 : 41–59
12. Foschini G J, Gans M J (1998) On the limits of wireless communications in fading environment when using multiple antennas. Wireless Personal Communications. 6 : 311–335
13. Foschini G J, Golden G D, Valenzuela R A, Wolniansky P W (1999) Simplified processing for high spectral efficiency wireless communication employing multi-element arrays. IEEE Journal on Selected Areas on Communications. 17 : 1841–1852
14. Golden G D, Foschini G J, Valenzuela R A, Wolniansky P W (1999) Detection algorithm and initial laboratory results using the V-BLAST space time communication architecture. Electronic Letters. 35 : 14–16
15. Goldsmith A (2005) Wireless Communications. Cambridge University Press.
16. Hong Z, Hughes B L (2002) Robust space-time codes for broadband OFDM systems, Proc. of IEEE Wireless Communications and Networking Conference (WCNC). 1 : 105–108
17. Jafarkhani H (2000) A quasi-orthogonal space-time block code. IEEE Transaction on Communication. 49 : 1–4
18. Jakes W C (1974) Microwave Mobile Communications. Wiley-IEEE Press. 2nd edition.
19. Kainulainen A, Vuokko L, Vainikainen P (2005) Polarization behavior in different urban radio environments at 5.3 GHz. COST 273, Tech. Rep. 05–018
20. Kermoal J P, Schumacher L, Mogensen P E, Pedersen K I (2000) Experimental investigation of correlation properties of MIMO radio channels for indoor picocell scenario. Proc. of Vehicular Technology Conference (VTC). Boston, USA. 1 : 14–21
21. Kermoal J P, Schumacher L, Pedersen K I, Mogenson P E, Frederiksen F (2002) A stochastic MIMO radio channel model with experimental validation. IEEE Journal on Selected Areas on Communications. 20 :1211–1226
22. Kozono S, Tsuruhara T, Sakamoto M (1984) Base station polarization diversity reception for mobile radio. IEEE Trans. Veh. Technol. 33 : 301–306
23. Kyritsi P (2001) Propagation characteristics of horizontally and vertically polarized electric fields in an indoor environment: simple model and results, in Proc. IEEE Vehicular Technology Conference. 3 : 1422–1426

24. Lee W C Y, Yeh Y S (1972) Polarization diversity system for mobile radio. IEEE Transaction on Communications. 20 : 912–923
25. Lee K F, Williams D B (2000) A space-frequency transmitter diversity technique for OFDM systems. Proc. IEEE Global Telecommunications Conference (GLOBECOM). 3 : 1473–1477
26. Lee K F, Williams D B (2000) A space-time coded transmitter diversity technique for frequency selective fading channels. Proceedings of Sensor Array and Multichannel Signal Processing Workshop. 149–152
27. Le Floch, Alard M, Berrou C (1995) Coded orthogonal frequency division multiplex. Proceedings of the IEEE. 83 : 982–996
28. Lempiainen J J A, Laiho-Steffens J K (1998) The Performance of Polarization Diversity Schemes at a Base Station in Small/Micro Cells at 1800 MHz. IEEE Transactions on Vehicular Technology. 47 : 1087–1092
29. Loredo S, Manteca B, Torres R (2002) Polarization diversity in indoor scenarios: an experimental study at 1.8 and 2.5 GHz. Proc. of IEEE International Symposium on Personal, Indoor and Mobile Radio Communications (PIMRC). 2 : 896–900
30. Lotse F, Berg J E, Forssen U, Idahl P (1996) Base station polarization diversity reception in macrocellular systems at 1800 MHz. Proc. IEEE Vehicular Technology Conference. 1643–1646
31. Lu B, Wang X (2000) Space-time code design in OFDM systems. Proc. IEEE Global Telecommunications Conference (GLOBECOM). 2 :1000–1004
32. Molisch A F (2011) Wireless Communications. John Wiley, 2nd edition
33. Motorola (2002) Polarization effects and Path statistics for the spatial channel model, SCM-055, SCM Ad-hoc meeting.
34. Nabar R U, Bölcskei H, Erceg V, Gesbert D, Paulraj A J (2002) Performance of Multiantenna Signaling Techniques in the Presence of Polarization Diversity, in IEEE transactions on signal processing. 50 : 2553–2562
35. Naguib A, Tarokh V, Seshadri N, Calderbank A R (1998) A space-time coding modem for high-data-rate wireless communications. IEEE Journal on Selected Areas on Communications. 16 : 1459–1478
36. Oestges C, Clerckx B, Guillaud M, Debbah M (2008) Dual-polarized wireless communications: from propagation models to system performance evaluation. IEEE Trans. Wireless Commun. 7 : 4019–4031
37. Ozcelik H, Herdin M, Weichselberger W, Wallace J, Bonek E (2003) Deficiencies of Kronecker MIMO Radio Channel Model. Electronics Letters. 39 : 1209–1210
38. Papadias C B, Foschini G J (2001) A Space-time coding approach for systems employing four transmit antenna. Proceedings of IEEE International Conference on Acoustics Speech and Signal Processing (ICASSP). 4 : 2481–2485
39. Pedersen K I, Mogensen P E, Fleury B H (2000) A stochastic model of the temporal and azimuthal dispersion seen at the base station in outdoor propagation environments, IEEE Trans. Veh. Technol. 49 : 437–447
40. Proakis J G (2001) Digital communications. McGraw-Hill,Boston,USA, 4th edition.
41. Rappaport T S (1996) Wireless Communications: Principles and Practice. Upper Saddle River, Prentice Hall.
42. Saleh A A M, Valenzuela R A (1987) A statistical model for indoor multipath propagation. IEEE Journal on Selected Areas in Communications. 5 : 128–137
43. Sayeed A M (2002) Deconstructing multiantenna fading channels, IEEE Trans. Signal Processing 5 : 856–866
44. Schumacher L, Pedersen K I, Mogenson P E (2002) From antenna spacings to theoretical capacities - guidelines for simulating MIMO systems. Proceedings of IEEE PIMRC. 2 : 587–592
45. Steinbauer M, Molisch A F, Bonek E (2001) The doubledirectional radio channel. IEEE Antennas and Propagation Magazine. 43 : 51–63
46. Svantesson T, Jensen M A, Wallace J W (2004) Analysis of electromagnetic field polarizations in multiantenna systems. IEEE Transactions on Wireless Communications. 3 : 641–646

47. Tarokh V, Seshadri N, Calderbank A R (1998) Space-time codes for high data rate wireless communication: Performance criterion and code construction. IEEE Transaction on Information Theory. 44 : 744–765
48. Tarokh V, Jafarkhani H, Calderbank A (1999) Space-time block codes from orthogonal designs. IEEE Transaction on Information Theory. 45 : 1456–1467
49. Telatar E (1995) Capacity of multiple antenna Gaussian channels. AT&T Bell Laboratories, *Technical Report*
50. Tirkkonen O, Boarius A, Hottinen A (2000) Minimal nonorthogonality rate 1 space-time block code for 3+ Tx antennas. in Proc. of IEEE International Symposium on Spread Spectrum Techniques and Applications (ISSSTA). 2 : 429–432
51. Tse D, Viswanath P, Fundamentals of Wireless Communication, Cambridge University Press.
52. Uysal M, Georghiades C N (2002) Non-Orthogonal Space-Time Block Codes for 3TX Antennas. IEE Electronics Letters. 38 : 1689–1691
53. Vaughan R G (1990) Polarization diversity in mobile communications. IEEE Transactions on Vehicular Technology. 39 : 177–186
54. Wallace J, Jensen M, Modeling the Indoor MIMO Wireless Channel (2002) IEEE Trans. on Antennas and Propagation. 50 : 591–599
55. Weichselberger W, Herdin M, Ozcelik H, Bonek E (2006) A Stochastic MIMO Channel Model with Joint Correlation of Both Link Ends. IEEE Trans. Wireless Commun. 5 : 90–100
56. Wolniansky P W, Foschini G J, Golden G D, Valenzuela R A (1998) V-BLAST : An architecture for realizing very high data rates over rich-scattering wireless channel. Proceedings of URSI International Symposium on Signals, Systems, and Electronics. 1 : 295–300
57. Zheng L, Tse D (2003) Diversity and multiplexing: A fundamental tradeoff in multiple antenna channels, IEEE Trans. on Information Theory. 49 : 1073–1096

Chapter 3
Feedback in SISO Single User Wireless Communication

3.1 Introduction

In this chapter, we will consider the different feedback strategies for the case of a single user wireless communication system where both the transmitter and the receiver are equipped with a single antenna. We will first review the information-theoretic concept of the capacity of finite state and then we will restrict our study to the important case of Rayleigh fading channel. We will show that depending on the availability of the channel state information (CSI) at the transmitter (CSIT) and at the receiver (CSIR), the capacity of the channels can significantly increase by exploiting the time variation of the channel. Then, we will consider the important class of frequency selective or wideband fading channel where the power can be optimally shared among the frequencies.

In the second part of this chapter, we will study the adaptive transmission over time and frequency consisting in the adaptation of one or more parameters such as modulation, rate, codes, and power in order to maximize the spectral efficiency of the radio link for a given quality. We will also evaluate the impact of quantization and delay due to the feedback link on the quality of the CSI and consequently on the system's performance.

3.2 Narrowband Single Input Single Output Systems

3.2.1 Introduction

In this section, we will consider channels whose conditional output probability density function depends on a state process and where the CSI is available at the receiver or both at the transmitter and the receiver. These channels have been widely studied over the years and they can serve for modeling a wide range of problems, depending on some assumptions regarding the channel state and on the availability

B. Özbek and D. Le Ruyet, *Feedback Strategies for Wireless Communication*, DOI 10.1007/978-1-4614-7741-9__3, © Springer Science+Business Media New York 2014

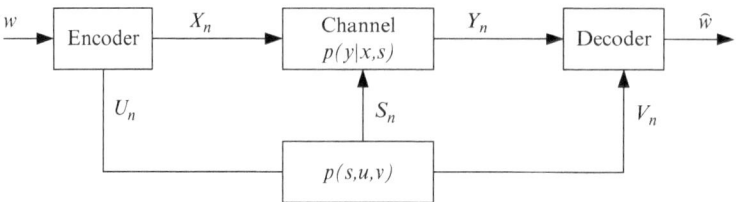

Fig. 3.1 Block diagram of the communication system with finite-state channel

and quality (noiseless or noisy) of the side information at the transmitter and/or the receiver. For CSI available at the transmitter, we can distinguish between channels where the CSIT is causal and channels where it is noncausal.

In the case of noncausal CSIT, the transmitter knows in advance the realization of the entire channel state sequence from the beginning to the end of the block. Non causal CSIT has been mainly considered for storage applications [33]. The capacity of noncausal CSIT channel model was found in 1980 by Gel'fand and Pinsker [51].

In the causal case, the CSIT, at every time instant, depends only on the past and present CSI. Clearly, causal feedback is more suitable to communication systems where channel states are estimated sequentially in the temporal domain. Consequently, in this chapter, we will consider only channels where CSIT is causal.

The causal CSIT channel model was introduced in 1958 by Shannon [62].

3.2.2   Finite-State Channel

We will first study the simplified single user channel with independent and identically distributed (i.i.d.) finite state. Following [10], we will show that when the CSIT is a deterministic function of the CSIR, optimal codes can be constructed directly over the input alphabet, while in general, coding over an expanded alphabet is required. These results will be applied later to the case of fading channels with additive white Gaussian noise (AWGN).

Let's consider the communication system given in Fig. 3.1, with channel input $x_n \in \mathcal{X}$, channel output $y_n \in \mathcal{Y}$ and state $s_n \in \mathcal{S}$ at index n where $\mathcal{X}, \mathcal{Y}$ and $\mathcal{S}$ are, respectively, the set of channel inputs, channel outputs and states.

We assume that the channel is memoryless (x_n, y_n and s_n are independent of past channel inputs). Consequently, the conditional output probability density function $p(y_1^N | x_1^N, s_1^N)$ is given by

$$p(y_1^N | x_1^N, s_1^N) = \prod_{n=1}^{N} p(y_n | x_n, s_n) \qquad (3.1)$$

where x_1^N, y_1^N and s_1^N are the sequences of the N inputs, outputs and states, respectively.

Let's assume that the transmitter and the receiver have an estimate of the channel state at index n denoted by $u_n \in \mathcal{U}$ and $v_n \in \mathcal{V}$, respectively.

The transmitter sends a source message $w \in \mathcal{W}$ to the receiver using N channel uses. This message is uniformly distributed over the set of possible messages $\mathcal{W} = \{1, \ldots, 2^{NR}\}$ where R is the code rate.

A block code of length N and rate R is defined by:

- A sequence of N encoding functions $f_n : \mathcal{W} \times \mathcal{X}^n \to \mathcal{X}$ for $n = 1, \ldots, N$ such that $x_n = f_n(w, u_1^n)$ where w ranges over $\mathcal{W}$ and u_1^n is the CSIT realization from time 1 to time n.
- A decoding function $\phi : \mathcal{Y}^N \times \mathcal{V}^N \to \mathcal{W}$. The decoded message is $\hat{w} = \phi(y_1^N, v_1^N)$

In the case where $\mathcal{S}$ is an i.i.d. sequence and the CSIT is perfect $U_n = S_n$ and no CSIR is available, Shannon in 1958 [62] has shown that the capacity of this channel is equal to the capacity of a discrete memoryless channel (DMT) with the same output alphabet and an extended input alphabet $\mathcal{T}$ of size $|\mathcal{X}|^{|\mathcal{S}|}$ given by

$$C = \max_{q(t)} I(T; Y) \tag{3.2}$$

$T \in \mathcal{T}$ is the equivalent channel input obtained by combining the input X with the state S. It is also called "strategy" with probability density function $q(t)$.

The mutual information $I(T; Y)$ of the memoryless time-invariant channel with input t and output y is given by

$$I(T; Y) = H(T) - H(T|Y)$$

$$= \sum_{t \in \mathcal{T}} \sum_{y \in \mathcal{Y}} p(t, y) \log_2 \left(\frac{p(t, y)}{p(t)p(y)} \right) \tag{3.3}$$

Optimal codes for this channel are constructed over the extended input alphabet $\mathcal{T}$ of size $|\mathcal{X}|^{|\mathcal{S}|}$. Since the capacity is expressed in term of strategies, the complexity of such a solution might be too high especially for large values of $|\mathcal{S}|$.

Caire and Shamai [10] found a special case in which the capacity can be obtained without using the above strategies. In this case, the CSIT is a deterministic function of the CSIR. Let $U_n = g(V_n)$ where $g(.)$ is a deterministic function from $\mathcal{V}$ to $\mathcal{U}$. Then the capacity is given by

$$C = \sum_{u \in \mathcal{U}} p(u) \max_{p(x|u)} I(X; Y|V, U = u) \tag{3.4}$$

This capacity can be achieved by a multiplexed multiple codebook scheme [25]. For each value $u \in \mathcal{U}$, a codebook composed of $2^{p(u)NR_j}$ codewords achieving

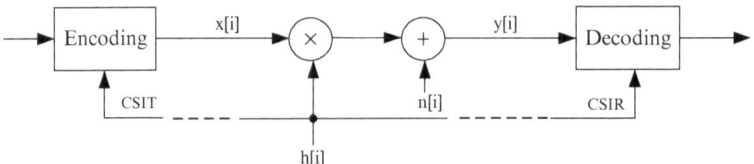

Fig. 3.2 Block diagram of the narrowband fading communication system

the rate $R_j \approx I(X;Y|V,U = u)$ is generated. For the message w, a set of $|\mathcal{U}|$ codewords is selected. At index n, assuming $U_n = u$, the corresponding encoder/decoder pair is connected through the channel and the transmitter sends the first not-transmitted symbol of the uth codeword. Since $g(.)$ is deterministic, the receiver can perform the demultiplexing and the $|\mathcal{U}|$ codewords can be independently decoded. The average rate is the sum of the rates R_j weighted by $p(u)$.

The case where the CSI is perfectly known at both the receiver and transmitter, originally considered by [73], can be obtained from (3.4). Assuming an ergodic state $\{S_n\} = \{U_n\} = \{V_n\}$, we have

$$C = \sum_{s \in \mathcal{S}} p(s) \max_{p(x|s)} I(X;Y|S = s) \tag{3.5}$$

It should be noted that while CSIT does not increase the capacity of memoryless channels, we have seen that the capacity of time-varying channels can be increased by adapting to channel conditions. However, the optimal power control strategy assumes that the channel state estimated at the receiver is instantaneously fed back to the transmitter.

Other works regarding capacity of channels with CSI have been published. In [52], Medard and Goldsmith have considered the capacity of time-varying inter symbol interference (ISI) channel for both perfect and unperfect CSI.

Another important case is when the CSI perfectly estimated at the receiver is available with a certain latency at the transmitter. Indeed, in practice, CSI is fed back through another auxiliary channel (essentially noiseless, as it operates at very low rates) from the receiver to the transmitter. In [70], Viswanathan has calculated the capacity of Markov channels with receiver CSI and delayed feedback.

3.2.3 Fading Channel

In this section, we will address the narrowband single antenna wireless links (Fig. 3.2). We will consider a discrete-time complex baseband model of this narrowband system mathematically modeled as

$$y[i] = h[i]x[i] + n[i] \tag{3.6}$$

where $x[i]$ is the complex transmitted symbol, $y[i]$ is a complex received symbol and $n[i]$ is the circularly symmetric i.i.d. Gaussian noise. $h[i]$ denotes the time-varying stationary and ergodic fading process. The channel is normalized such that $\mathbb{E}_h[|h[i]|^2] = 1$. We will consider that a long-term constraint power is imposed to the transmitted signal $x[i]$ as follows:

$$\mathbb{E}_{h,i}[|x[i]|^2] \leq P_{av} \tag{3.7}$$

where P_{av} denotes the average transmit power.

Other constraints are also possible such as the short-term power constraint if the power is constrained for each transmitted symbol.

The instantaneous received signal to noise ratio is given by

$$\gamma[i] = \frac{|h[i]|^2 P_{av}}{N_0 B_W}, \tag{3.8}$$

where N_0 is the noise power spectral.

From (2.13), the probability density function of the signal to noise ratio when the channel h follows a Rayleigh distribution is an exponential distribution given by

$$p_\gamma(\gamma) = \frac{N_0 B_W}{P_{av}} \exp\left(-\frac{N_0 B_W \gamma}{P_{av}}\right) \tag{3.9}$$

and the associated cumulative distribution function

$$F_\gamma(\gamma) = 1 - \exp\left(-\frac{N_0 B_W \gamma}{P_{av}}\right) \tag{3.10}$$

To allow the receiver to perform coherent detection, channel estimation techniques are usually performed using known transmitted symbols called pilots. This technique is also called pilot symbol assisted modulation (PSAM) [14]. Thanks to this channel estimation, an estimate of the CSI is available at the receiver. Assuming that the pilots have an amplitude a_p, based on the observation $y[i]$, the fading estimate of $h[i]$ is given by

$$\overline{h}[i] = h[i] + \frac{n[i]}{a_p} \tag{3.11}$$

Like previously, we will consider the same three different scenarios regarding the knowledge of CSIR.

Since the aim of the transmitter is to adapt the signal to the channel conditions, modeling how the channel varies during the transmission of a codeword is important. According to the variation of $h[i]$, we have seen in Chap. 2 that the channel can be categorized as slow/fast fading and frequency/non-frequency-selective channels.

Another popular model that we will consider is the block-fading channel model where the channel state remains constant for several transmit symbols and then changes independently. For this model, the lth channel block satisfies $h[lK_b] = h[lK_b + 1] = \cdots = h[lK_b + K_b - 1]$ where K_b is the length of the fading block. A codeword of length $N = K_d K_b$ spans K_d blocks. Those blocks can be thought of as separated in time like in time division multiplexing access (TDMA) or separated in frequency like in orthogonal frequency division multiple access (OFDMA).

In this book, we will mainly consider the block-fading channel model with $K_d = 1$ where the channel state remains constant for the duration of one codeword of length N and then changes independently.

We will also often consider Jakes' model with Doppler spread f_D in order to modelize the time variation of the fading channel. As described in Chap. 2, in this model, an isotropic environment is assumed and the channel autocorrelation function is $\mathbb{E}[h^*[i]h[i']] = J_0(2\pi f_D |i - i'| T_s)$ where $J_0(.)$ is the zeroth order Bessel of first kind.

3.2.3.1 Knowledge Only of the Channel Distribution Information

We first consider the case where only the channel gain probability density function $p(\gamma)$ is known. For i.i.d. fading the capacity is given by

$$C = \max_{p(x)} I(X;Y) \tag{3.12}$$

However, finding the probability density function $p(x)$ maximizing the mutual information is quite complicated depending on the fading distribution. For i.i.d. Rayleigh fading channels the optimal input distribution for this channel has a discrete i.i.d. power and irrelevant phase. For low-average signal to noise ($\gamma < 8\,\mathrm{dB}$), the distribution has two signaling levels $x = 0$ and $x = \sqrt{\alpha}$ with probabilities $1 - p_\alpha$ and p_α, respectively, where $\alpha p_\alpha = P_{av}$ [1]. This optimal distribution and its corresponding capacity must be found numerically. Generally speaking, finding the capacity-achieving input distribution and corresponding capacity of fading channels under only channel distribution information remains an open problem for almost all channel distributions.

3.2.3.2 Perfect CSIR and No CSIT

Assuming perfect knowledge of the channel at the receiver only, the ergodic capacity can be obtained from the results of [25,51]. We have

$$C = \mathbb{E}_h \left[B_W \log_2 \left(1 + \gamma \right) \right]$$

$$= \int_0^\infty B_W \log_2 \left(1 + \gamma\right) p_\gamma(\gamma) d\gamma \tag{3.13}$$

where $p_\gamma(\gamma)$ is the probability density function of the signal to noise ratio γ.

This ergodic capacity can be achieved by using a standard long Gaussian codebook. Compared to the AWGN channel, the length of the codewords depends also on the dynamics of the fading process (the time duration of the transmission of a codeword must be much higher than the coherence time of the channel).

When the ergodic assumption is not satisfied like in communication systems requiring small delay constraints, the capacity with outage is more relevant [57]. This capacity defines the maximum rate that can be transmitted over a channel with some outage probability corresponding to the probability that the transmission cannot be decoded with negligible error probability. The capacity-versus-outage performance is determined by the probability that the channel cannot support a given rate.

We consider the case of the flat fading with CSI available at the receiver only. In that case, the signal to noise ratio γ is given by

$$\gamma = \frac{P_{av}|h|^2}{N_0 B_W} \tag{3.14}$$

The channel capacity, viewed as a random variable since it depends on γ, is given by

$$C(\gamma) = B_W \log_2 \left(1 + \gamma\right) \tag{3.15}$$

The outage probability for a given rate is

$$P_{out} \triangleq \Pr(C(\gamma) \leq R)$$

$$= \Pr\left(B_W \log_2 \left(1 + \gamma\right) \leq R\right) \tag{3.16}$$

$$= \Pr\left(|h|^2 \leq \frac{N_0 B_W}{P_{av}} \left(2^{\frac{R}{B_W}} - 1\right)\right) \tag{3.17}$$

For Rayleigh fading, the outage probability is

$$P_{out} = 1 - \exp\left(-\frac{N_0 B_W}{P_{av}} \left(2^{\frac{R}{B_W}} - 1\right)\right) \tag{3.18}$$

3.2.3.3 Perfect CSIT and Perfect CSIR

Assuming that the channel coefficients $h[i]$ are known by both the transmitter and the receiver at each time i, the capacity of time-varying channels is a special case of the capacity given in (3.5) since this continuous case can be treated as the limiting case of the discrete ones under the average power constraint (3.7). The capacity is given by

$$C = \max \mathbb{E}\left[B_W \log_2 \left(1 + \gamma\right)\right]$$
$$= \max \int_0^\infty B_W \log_2 \left(1 + \gamma\right) p(\gamma) d\gamma \qquad (3.19)$$

where the maximum is performed over the power distribution $P_w(\gamma)$ that should satisfy the following constraint:

$$\mathbb{E}\left[P_w(\gamma)\right] \le P_{av} \qquad (3.20)$$

or equivalently

$$\int_0^\infty P_w(\gamma) p(\gamma) d\gamma \le P_{av} \qquad (3.21)$$

To find the optimal power adaptation which maximizes (3.19) [25] we form the Lagrangian

$$J(P_w(\gamma)) = \int_0^\infty B_W \log_2 \left(1 + \frac{P_w(\gamma)\gamma}{P_{av}}\right) p(\gamma) d\gamma - \lambda \int_0^\infty P_w(\gamma) p(\gamma) d\gamma \quad (3.22)$$

Then we differentiate the Lagrangian and fix the derivative equal to zero as follows:

$$\frac{\partial J(P_w(\gamma))}{\partial P_w(\gamma)} = \left[\left(\frac{B_W/\ln(2)}{1 + P_w(\gamma)\gamma/P_{av}}\right)\frac{\gamma}{P_{av}} - \lambda\right] p(\gamma) = 0 \qquad (3.23)$$

Solving the above relation with the constraint that $P_w(\gamma)$ should be positive yields the optimal power adaptation which maximizes (3.19) as

$$\frac{P_w(\gamma)}{P_{av}} = \begin{cases} \frac{1}{\gamma_0} - \frac{1}{\gamma} & \gamma \ge \gamma_0 \\ 0, & \gamma < \gamma_0 \end{cases} \qquad (3.24)$$

for some constant γ_0. If $\gamma[i]$ is lower than this threshold, no data is transmitted at time i. This optimal power control algorithm gives rise to a "water-pouring" in time, analogous to the water-pouring used to achieve capacity on frequency selective fading channels.

Substituting (3.24) into (3.19) the resulting capacity is

$$C = \int_{\gamma_0}^{\infty} B_W \log_2 \left(\frac{\gamma}{\gamma_0} \right) p(\gamma) d\gamma \tag{3.25}$$

The value of the constant γ_0 can be computed from the power constraint given in (3.21) where the inequality becomes an equality since the capacity is an increasing function of the power. We have

$$\int_0^{\infty} \frac{P_w(\gamma)}{P_{av}} p(\gamma) d\gamma = 1 \tag{3.26}$$

By substituting (3.24) into (3.26) γ_0 must satisfy the following equation:

$$\int_{\gamma_0}^{\infty} \left(\frac{1}{\gamma_0} - \frac{1}{\gamma} \right) p(\gamma) d\gamma = 1 \tag{3.27}$$

Goldsmith and Varaiya [25] have shown that varying both power and rate leads to a negligibly higher Shannon capacity over varying just the rate alone. They have shown that a multiplexed multiple codebook scheme, where different codebooks with rate $\log_2(1 + \gamma)$ are used when the fading realization is γ and the associated power is $P_w(\gamma)$, can achieve this capacity. It has been shown in [10] that this capacity can also be achieved using a fixed-rate coding scheme called single codebook with dynamic power allocation. In that scheme, a long Gaussian block code is used and the code symbol is dynamically scaled by an appropriate gain $P_w(\gamma)/P_{av}$.

In both cases, the block codes should be long enough to reveal the ergodic property of the fading channel.

Suboptimal power control strategies such as channel inversion and truncated channel inversion have also been proposed by Goldsmith and Varaiya [25].

It is important to point out that the systems designed from the capacity criterion and from the uncoded probability of error criterion are quite different. For example, when minimizing the probability of error, the power is allocated proportionally to the inverse of the channel in order to mitigate the fading effect. On the other hand, when maximizing the capacity, the power is allocated proportionally to the channel magnitude.

As shown in Fig. 3.3, the availability of CSIT gives only little gain in terms of ergodic capacity. CSIT provides a performance improvement over constant power transmission only at low SNR.

On the other hand, when considering capacity versus outage, the performance improvement is significant. Let's consider the case of a block flat fading channel with $K_d = 1$ (single block). The optimal power allocation for the outage minimization is given by [11]

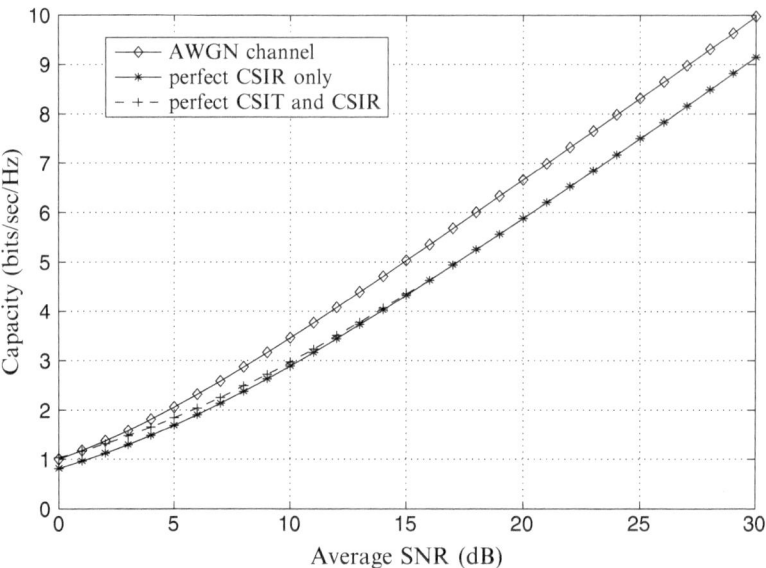

Fig. 3.3 Ergodic capacity for Rayleigh fading channel

$$P_w(\gamma) = \begin{cases} \left(2^{\frac{R}{B_W}} - 1\right)/\gamma & \gamma \geq \gamma^* \\ 0, & \gamma < \gamma^* \end{cases} \qquad (3.28)$$

where for a given long-term power constraint P_{av} and rate R the signal to noise ratio γ^* is the solution of

$$\left(2^{\frac{R}{B_W}} - 1\right)\mathrm{Ei}\left(1, \frac{2^{\frac{R}{B_W}} - 1}{\gamma^*}\right) = P_{av} \qquad (3.29)$$

where $\mathrm{Ei}(n, x)$ is the exponential integral of order n:

$$\mathrm{Ei}(n, x) \triangleq \int_1^\infty (e^{-xt}/t^n)dt \qquad \text{for } \mathrm{Re}(x) > 0 \qquad (3.30)$$

The resulting outage probability for Rayleigh fading is given by

$$P_{out} = 1 - \exp\left(-\left(2^{\frac{R}{B_W}} - 1\right)/\gamma^*\right) \qquad (3.31)$$

3.2.3.4 Perfect CSIR and Unperfect CSIT

In this section, we consider a slowly fading channel with perfect CSIR and quantized CSIT. We assume that the CSIT is quantized in L fading regions. The encoding function is implemented in the receiver using a deterministic index mapping as follows:

$$L(\gamma) = n, \gamma \in [\gamma_n^b, \gamma_{n+1}^b], n = 0, \ldots, L - 1, \text{ where } \gamma_0^b = 0 \text{ and } \gamma_L^b = +\infty \tag{3.32}$$

where γ_n^b denotes the boundary points of the quantization regions. The index n is sent to the transmitter via a noiseless, zero-delay feedback link. We will consider two cases depending on the power constraint, namely short-term and long-term power constraint.

Short-Term Power Constraint

Due to hardware or complexity limitations, there are cases where the power can not exceed a maximum value. Under the short-term power constraint, it is clear that the power has no impact on the limited feedback design.

Given the received index n, the transmitter will choose an operating rate R_n associated to a SNR γ_n. Since we know that $\gamma_n \in [\gamma_n^b, \gamma_{n+1}^b]$, to maximize the average rate R_T, it is necessary that $R_n \in [B_W \log_2 \gamma_n^b, B_W \log_2 \gamma_{n+1}^b]$. Assuming that the actual rate $R = B_W \log_2 \gamma$, the system will be in outage if $R_n > R$ (or equivalently if $\gamma_n > \gamma$). Consequently, the adaptive transmission policy is defined by a set of L boundary points $\gamma_0^b, \ldots \gamma_{L-1}^b$ and the corresponding set of rates $R_0, \ldots R_{L-1}$ (or equivalently the set of SNR $\gamma_0, \ldots \gamma_{L-1}$).

The average rate is determined as

$$R_T(\gamma_0, \ldots, \gamma_{L-1}, \gamma_0^b, \ldots, \gamma_{L-1}^b) = \sum_{n=0}^{L-1} \left(F_\gamma(\gamma_{n+1}^b) - F_\gamma(\gamma_n) \right) B_W \log_2(1 + \gamma_n) \tag{3.33}$$

where $F_\gamma(.)$ is the cdf of γ.

The outage probability is given by

$$P_{out}(\gamma_0 \ldots, \gamma_{L-1}, \gamma_0^b, \ldots, \gamma_{L-1}^b) = \sum_{n=0}^{L-1} \left(F_\gamma(\gamma_n) - F_\gamma(\gamma_n^b) \right) \tag{3.34}$$

It has been shown in [38, 46] that with no outage probability constraint the maximum average rate is obtained by considering the SNR to be equal to its worst case within each region except the first region, i.e., $\gamma_n = \gamma_n^b$ for $n = 1, \ldots, L - 1$ as shown in Fig. 3.4. Consequently, in this case, optimizing the two above sets is equivalent to optimizing the set composed of the $L - 1$ boundary points $\gamma_1^b, \ldots \gamma_{L-1}^b$ and the SNR γ_0.

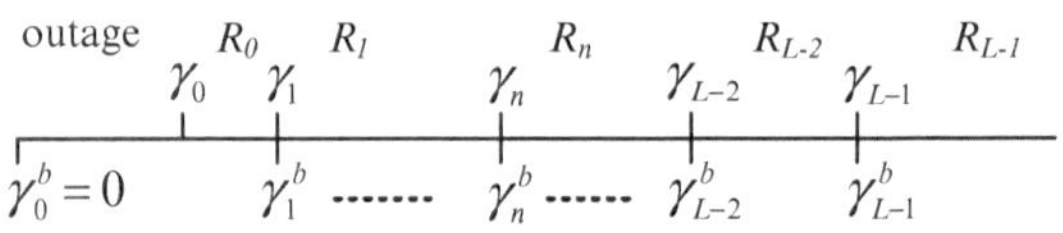

Fig. 3.4 Worst case
quantization regions

The determination of the set implies solving the following optimization problem:

$$\max_{\gamma_0, \gamma_1^b, \ldots, \gamma_{L-1}^b} R_T(\gamma_0, \gamma_1^b, \ldots, \gamma_{L-1}^b) \tag{3.35}$$

where

$$R_T(\gamma_0, \gamma_1^b, \ldots, \gamma_{L-1}^b) = \Big(F_\gamma(\gamma_1^b) - F_\gamma(\gamma_0) \Big) B_W \log_2(1 + \gamma_0)$$

$$+ \sum_{n=1}^{L-1} \Big(F_\gamma(\gamma_{n+1}^b) - F_\gamma(\gamma_n^b) \Big) B_W \log_2(1 + \gamma_n^b) \tag{3.36}$$

The objective function is non-convex and one way of finding the optimum policy is brute-force maximization over a quantized space of all possible terms. A practical solution is to perform an iterative adjustment of the partitions using the local optimization technique called water spilling [46] that varies boundaries one at a time. The description of the algorithm is given here below:

- *Step 1 Initialization*: $k = 0$, initialize the boundary points $\gamma_n^{b(k)} = n/L$ for $n = 1, \ldots, L$ and $\gamma_0^{(k)} = 1/L$.
- *Step 2 Water-spilling*:
    For $n = L - 1, L - 2, \ldots, 1$ compute

$$\gamma_n^b = \arg \max_{\gamma_n^b : \gamma_{n-1}^b \leq \gamma_n^b \leq \gamma_{n+1}^b} \Bigg[\Big(F_\gamma(\gamma_n^b) - F_\gamma(\gamma_{n-1}^b) \Big) B_W \log_2(1 + \gamma_{n-1}^b)$$

$$+ \Big(F_\gamma(\gamma_{n+1}^b) - F_\gamma(\gamma_n^b) \Big) B_W \log_2(1 + \gamma_n^b) \Bigg] \tag{3.37}$$

Then compute

$$\gamma_0 = \arg \max_{\gamma_0 : 0 \leq \gamma_0 \leq \gamma_1^b} \Big(F_\gamma(\gamma_1^b) - F_\gamma(\gamma_0) \Big) B_W \log_2(1 + \gamma_0) \tag{3.38}$$

Let $\mathbf{a}^{(k)} = (\gamma_0, \gamma_1^b, \ldots, \gamma_{L-1}^b)$
- *Step 3*: If $R_T(\mathbf{a}^{(k)}) - R_T(\mathbf{a}^{(k-1)}) < \epsilon$, stop the algorithm, $\mathbf{a}^{(k)}$ is the solution. Else $k = k + 1$ and go back to step 2.

Long-Term Power Constraint

Under the long-term power constraint, after receiving the index n, the transmitter will choose an operating rate R_n that will depend not only on the partition but also of the allocated power P_n. In this case, maximizing the ergodic rate is equivalent to optimizing the set composed of the $L - 1$ boundary points $\gamma_1^b, \dots \gamma_{L-1}^b$ and the SNR γ_0 and the set of the L allocated powers $P_0 \dots P_{L-1}$. The determination of these sets implies solving the following optimization problem

$$\max_{\gamma_0, \gamma_1^b, \dots, \gamma_{L-1}^b, P_0, \dots P_{L-1}} R_L(\gamma_0, \gamma_1^b, \dots, \gamma_{L-1}^b, P_0, \dots P_{L-1}) \tag{3.39}$$

where

$$R_L(\gamma_0, \gamma_1^b, \dots, \gamma_{L-1}^b P_0, \dots P_{L-1}) = \left(F_\gamma(\gamma_1^b) - F_\gamma(\gamma_0)\right) B_W \log_2(1 + \frac{P_0 \gamma_0}{P_{av}})$$
$$+ \sum_{n=1}^{L-1} \left(F_\gamma(\gamma_{n+1}^b) - F_\gamma(\gamma_n^b)\right) B_W \log_2(1 + \frac{P_n \gamma_n^b}{P_{av}}) \tag{3.40}$$

s. t.

$$P_0\left(F_\gamma(\gamma_1^b) - F_\gamma(\gamma_0)\right) + \sum_{n=1}^{L-1} P_n\left(F_\gamma(\gamma_{n+1}^b) - F_\gamma(\gamma_n^b)\right) \leq P_{av} \tag{3.41}$$

Given the partition, by introducing the Lagrange multiplier λ, the optimum power allocation is obtained by the water-filling algorithm

$$P_0 = \left[\frac{F_\gamma(\gamma_1^b) - F_\gamma(\gamma_0)}{F_\gamma(\gamma_1^b)} \frac{1}{\lambda} - \frac{1}{\gamma_0}\right]^+ \tag{3.42}$$

$$P_n = \left[\frac{1}{\lambda} - \frac{1}{\gamma_n^b}\right]^+ \qquad n = 1, \dots, L - 1 \tag{3.43}$$

where $[x]^+ \triangleq \max(x, 0)$ and λ is chosen such that the power constraint is satisfied. The short-term power constraint algorithm described above can still be used by adding a waterfilling step after each update of the set of partitions.

In Figs. 3.5 and 3.6 we give the average rate versus SNR for $L = 0, 2, 4, 8$ and 16 over Rayleigh fading channels and under short- and long-term power constraints, respectively. The difference between the no CSIT and full CSIT curves is about 7–8 dB. For a 2 b/s/Hz rate, a 3 dB gain is obtained using a feedback scheme with

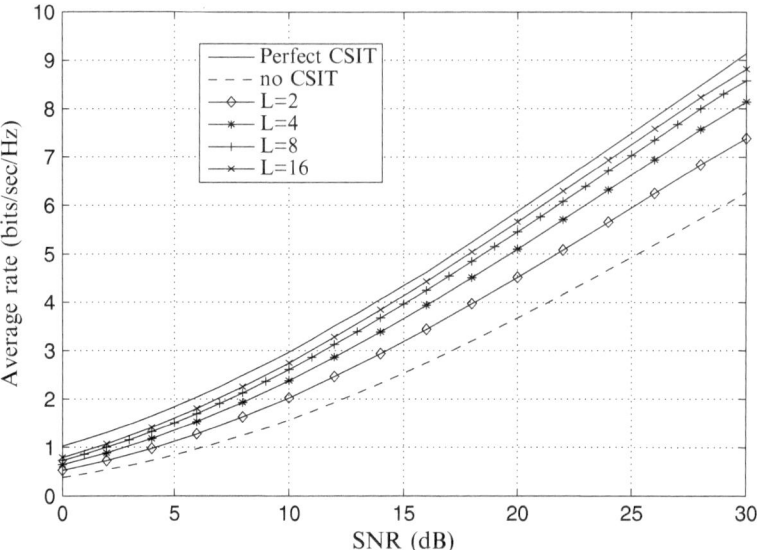

Fig. 3.5 Average rate with different numbers of quantization regions under a short-term power constraint over a Rayleigh fading channel

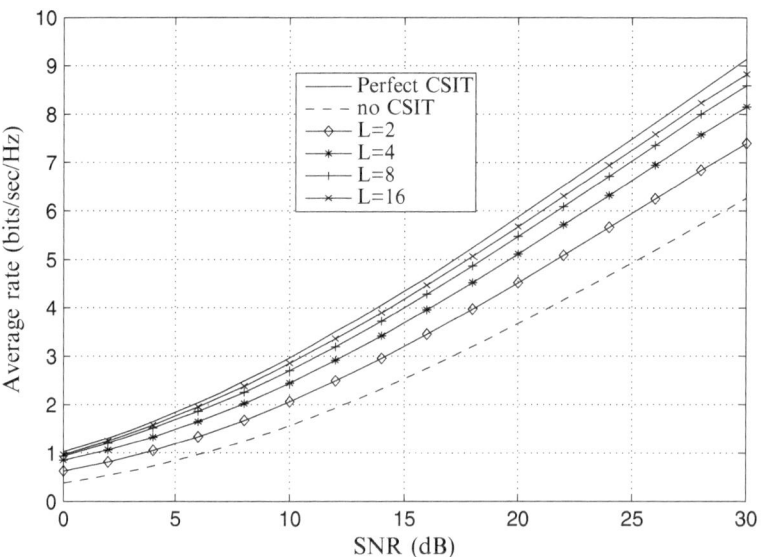

Fig. 3.6 Average rate with different numbers of quantization regions under a long-term power constraint over a Rayleigh fading channel

two quantization regions. The figures indicate also that thanks to the water-filling algorithm, the long-term power allocation increases significantly the average rate at low SNR (at SNR $= 0$ dB and for $L = 16$, 0.95 b/s/Hz compared to 0.8 b/s/Hz for short-term power constraint). However, at medium to high SNR, there is no performance gain between short- and long-term power constraint.

Liu et al. [48] found the average rate of slowly fading channels considering that the CSIR is perfect and that the quantized channel SNR information is transmitted at the transmitter via a noiseless channel. Lin et al. [46] and later Kim and Skoglund [38] showed that the worst case is the best value which can be considered in each quantization region, maximizing the total achievable rate. The power allocation strategy uses the quantized channel realization subject to either a short-term power constraint or a long-term power constraint. More recently, considering noisy feedback channels, Ekbatani et al. [19] studied the effect of feedback channel noise on the average rate of the slowly fading channel.

3.3 Wideband Systems

In this section we will address the capacity of frequency selective fading channels assuming that the channel is known at both the transmitter and receiver.

Due to the difficulty to deal with the ISI coming from the selectivity of the channel, we have shown in Chap. 2 that a classical approach consists in dividing the total bandwidth B_W into a large number of small band using multicarrier communication such as OFDM. We have shown that after removing the cyclic prefix the input–output vector relation can be written as

$$\mathbf{Y}[m] = \text{diag}(\mathbf{H}[m])\mathbf{X}[m] + \mathbf{N}[m] \tag{3.44}$$

where m is the OFDM index. The frequency selective fading channel thus consists of a set of N AWGN subchannels of bandwidth B_N in parallel. To make the notations simpler we will in this section omit the OFDM index. Using a scalar notation we can write for the nth subchannel $0 \le n \le N - 1$ the input–output relation as

$$Y_n = H_n X_n + N_n \tag{3.45}$$

The power P_n allocated to the nth subchannel is subject to the power constraint $\sum_n P_n \le P$ where the maximum total transmit power is P.

The capacity of this parallel set of channels is the sum of rates associated with each channel with power optimally allocated over all channels [28]

$$C = \max_{P_1, P_2 \dots P_N} \sum_{n=0}^{N-1} B_N \log_2 \left(1 + \frac{|H_n|^2 P_n}{N_0 B_N}\right) \tag{3.46}$$

subject to

$$\sum_{n=1}^{N} P_n \leq P \tag{3.47}$$

This equation is similar to the capacity and optimal power allocation for a flat-fading channel, with power and rate changing over frequency in a deterministic way rather than over time in a probabilistic way. Solving the optimization problem we can show that it leads to the water-filling power allocation like for the time-varying channel:

$$\frac{P_n}{P} = \begin{cases} \frac{1}{\gamma_0} - \frac{1}{\gamma_n} & \gamma_n \geq \gamma_0 \\ 0, & \gamma_n < \gamma_0 \end{cases} \tag{3.48}$$

where $\gamma_n = |H_n|^2 P / N_0 B_N$. γ_n is the SNR associated to the nth subchannel assuming that the total power P is allocated to this subchannel. The value of the constant γ_0 can be computed from the power constraint by substituting (3.48) into (3.47).

γ_0 must satisfy the following equation:

$$\sum_{n:\gamma_n \geq \gamma_0} \left(\frac{1}{\gamma_0} - \frac{1}{\gamma_n} \right) = 1 \tag{3.49}$$

where the inequality becomes an equality since the capacity is an increasing function of the power. The resulting capacity is given by

$$C = \sum_{n:\gamma_n \geq \gamma_0} B_N \log_2 \left(\frac{\gamma_n}{\gamma_0} \right) \tag{3.50}$$

3.4 Adaptive Modulation and Coding

3.4.1 Introduction

Adaptive transmission is a method to adapt the spectral efficiency of a radio link for a given required quality constraint such as error probability, delay or packet loss.

In order to perform adaptive transmission, the transmitter must know some information about the CSI such as the channel amplitude $|h[i]|$ or the signal to noise ratio $\gamma[i]$. From a training sequence or pilot symbols, the receiver can estimate the channel response between the transmitter and the receiver and can design feedback to be sent as overhead on the feedback link as shown in Fig. 3.7. To make the notations simpler we will, in this section and when the context is clear, omit the time index i.

The idea of varying the transmission power level has been proposed in the 1960s by Hayes [32]: *an efficient approach to mitigating these detrimental effects is to*

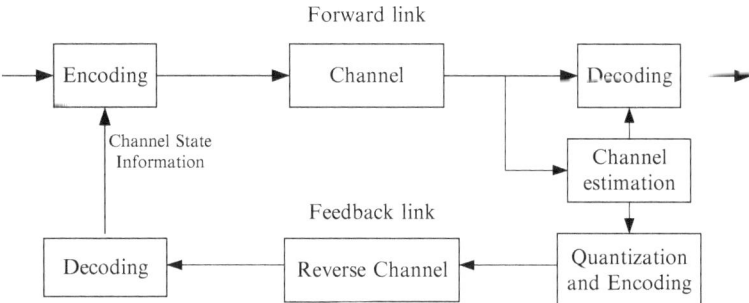

Fig. 3.7 Generic adaptive transmitter with feedback link

adaptively adjust the modulation and/or the channel coding format as well as a range of other system parameters based on the near-instantaneous channel quality information perceived by the receiver, which is fed back to the transmitter with the aid of a feedback channel.

Among the earlier works in the field, Bello and Cowan [7] have studied the performance of an on/off transmission, which can be considered as a two-rate transmission system depending on channel conditions, and Cavers [13] has proposed to continuously adjust the data rate depending on the channel variations of a fading channel.

In this section we will describe the different possible adaptive techniques depending on the parameters that are adjusted. We will assume that the channel is a flat Rayleigh fading channel and that the receiver has a perfect knowledge of the CSI. We will first consider that the feedback link is error-free and has no latency. Then we will consider the impact of delayed information on adaptive techniques.

3.4.2 Adaptive Discrete Rate Technique

In adaptive rate technique, the data rate $R(\gamma[i])$ is adjusted by using different modulation schemes depending on the signal to noise ratio $\gamma[i]$ measured by the receiver.

Like in Sect. 3.2.3.4, the CSIR is quantized in L fading regions corresponding to L different modulation schemes. The encoding function is implemented in the receiver using a deterministic index mapping as follows:

$$L(\gamma) = n, \gamma \in [\gamma_n, \gamma_{n+1}], n = 0, \ldots, L-1, \text{ where } \gamma_0 = 0 \text{ and } \gamma_L = +\infty \quad (3.51)$$

In adaptive discrete rate technique, the receiver must only feedback the index n to the transmitter via the feedback link. The feedback load is consequently $\lceil \log_2 L \rceil$ per channel use. The transmitter will use the constellation of size M_n when $L(\gamma) = n$ is received. The associated spectral efficiency is $\log_2 M_n$ bit/s/Hz for $n > 0$. The modulation scheme is selected in order to maintain the bit error rate (BER) below a given maximum value.

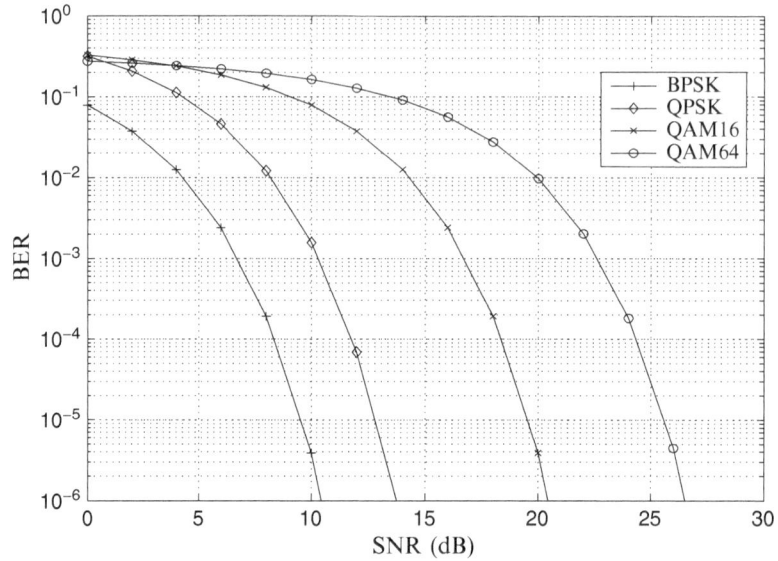

Fig. 3.8 BER versus SNR for BPSK, QPSK, QAM16 and QAM64 modulations

Consequently the boundaries of the region are determined by the performance of the different modulations. The BER of rectangular M-QAM constellation for a given SNR γ can be approximated by

$$BER(M,\gamma) \approx \frac{4(\sqrt{M}-1)}{\sqrt{M}\log_2 M} Q\left(\sqrt{\frac{3\gamma}{(M-1)}}\right) \tag{3.52}$$

Another common approximation of the BER to within 1 dB for M-QAM with $M \geq 4$ and $0 \leq \gamma \leq 30$ dB is given by [4, 28]

$$BER(M,\gamma) \approx 0.2\exp\left(-\frac{3\gamma}{2(M-1)}\right) \tag{3.53}$$

Figure 3.8 provides BER versus SNR for modulations BPSK, QPSK, 16-QAM and 64-QAM using the exact expression given in (3.52).

The set of boundaries of the region γ_n $1 \leq n \leq N-1$ can be obtained by reading the SNR points corresponding to a target BER. For example, if the target BER is 10^{-3}, the thresholds are 6.7 dB for BPSK, 10.3 dB for QPSK, 16.7 dB for QAM16, and 22.6 dB for QAM64.

γ_1 is the minimum required SNR in order to guarantee that the BER will be lower than the maximum allowed value considering the most robust modulation scheme. When the SNR is lower than γ_1, no constellation is selected and the transmitted power is 0. Using the adaptive rate technique, the transmitted power is constant when $\gamma \geq \gamma_1$ and such as $\int_{\gamma_1}^{\infty} p(\gamma)d\gamma = P_{av}$:

$$P(\gamma) = \begin{cases} \frac{P_{av}}{1-F_\gamma(\gamma_1)} & \gamma \geq \gamma_1 \\ 0, & \gamma < \gamma_1 \end{cases} \tag{3.54}$$

Assuming Nyquist data pulses at the bandwidth $1/T_s$ (roll-off factor $\beta = 0$), the spectral efficiency becomes

$$\frac{R}{B_W} = \sum_{n=1}^{L-1} \log_2(M_n)\Big(F_\gamma(\gamma_{n+1}) - F_\gamma(\gamma_n)\Big) \tag{3.55}$$

The average BER associated to region n is given by

$$\overline{BER}(M_n) = \int_{\gamma_n}^{\gamma_{n+1}} BER(M_n,\gamma)p(\gamma)d\gamma \tag{3.56}$$

and the overall system BER is the ratio of the average number of bits in error over the average number of transmitted bits, i.e., the spectral efficiency:

$$\overline{BER} = \frac{\sum_{n=1}^{L-1} \log_2(M_n)\overline{BER}(M_n)}{\sum_{n=1}^{L-1} \log_2(M_n)\Big(F_\gamma(\gamma_{n+1}) - F_\gamma(\gamma_n)\Big)} \tag{3.57}$$

For comparison, given a fixed SNR γ and BER P_b, the spectral efficiency curves of continuous-rate M-QAM can be obtained by inverting equation (3.53) as follows:

$$\frac{R}{B_W} = \log_2(M) = \log_2\left(1 - \frac{3\gamma}{2\ln(5BER)}\right) \tag{3.58}$$

Figure 3.9 provides the spectral efficiency versus SNR curves using (3.55) for the three-regions case (no modulation, BPSK, and QPSK), four-regions case (no modulation, BPSK, QPSK, and QAM16) and five-regions case (no modulation, BPSK, QPSK, QAM16, and QAM64). The schemes have been designed for BER = 10^{-3}. For comparison, the continuous-rate M-QAM spectral efficiency is also given.

3.4.3   Adaptive Power Technique

Instead of keeping the power constant and changing the data rate according to the channel power gain, it is possible to keep the data rate constant and adapt the power in order to fulfill the BER constraint or equivalently to maintain a fixed received SNR ξ. The power adaptation for the truncated channel inversion is given by

$$\frac{P(\gamma)}{P_{av}} = \begin{cases} \frac{\xi}{\gamma} & \gamma \geq \gamma_1 \\ 0, & \gamma < \gamma_1 \end{cases} \tag{3.59}$$

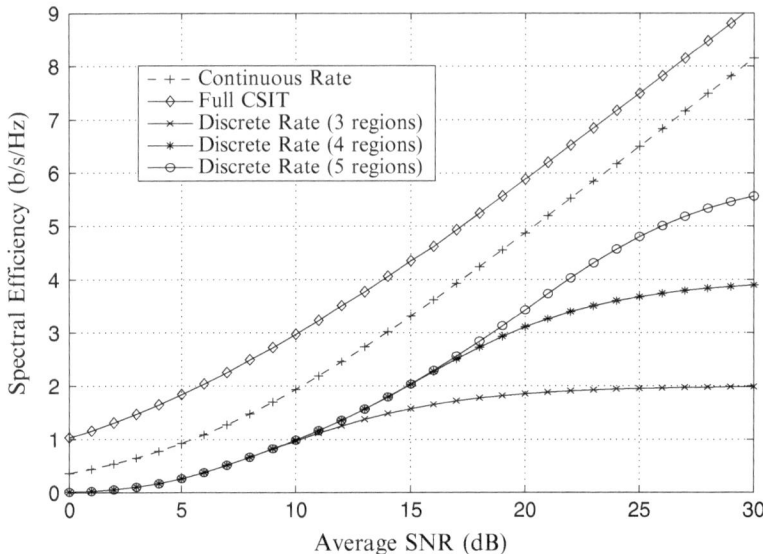

Fig. 3.9 Spectral efficiency of adaptive rate schemes for 3, 4 and 5 regions in Rayleigh fading

where γ_1 can be chosen to satisfy a given BER constraint or an outage probability P_{out}.

The average power constraint is given by

$$\int_{\gamma_0}^{\infty} \frac{P_{\gamma}(\gamma)}{P_{av}} p(\gamma) d\gamma = \int_{\gamma_0}^{\infty} \frac{\xi}{\gamma} p(\gamma) d\gamma = 1 \qquad (3.60)$$

Consequently, we have

$$\xi = \frac{1}{\int_{\gamma_0}^{\infty} \frac{1}{\gamma} p(\gamma) d\gamma} \qquad (3.61)$$

3.4.4 Adaptive Discrete Rate and Power

Let's consider now the more general case where both discrete rate and power can be adapted according to the channel power gain in order to fulfill a BER constraint. Compared to the adaptive discrete rate technique, here the receiver must have the knowledge of the channel magnitude of the direct link or the received SNR. This CSI is estimated by the receiver and is then sent back to the transmitter using the feedback link.

Considering the same L fading regions corresponding to L different modulation schemes given in (3.51) and an instantaneous BER constraint, the power adaptation is given by

$$\frac{P(\gamma)}{P_{av}} = \begin{cases} \frac{\xi_n}{\gamma} & \gamma_n \leq \gamma \leq \gamma_{n+1} \quad n = 1 \ldots L - 1 \\ 0 & \gamma < \gamma_1 \end{cases} \tag{3.62}$$

where ξ_n is received SNR required to satisfy the BER constraint for a constellation of size M_n. Using the approximation (3.53) for M-QAM modulation, we have

$$\xi_n = \frac{M_n - 1}{-1.5} \ln(5BER) \tag{3.63}$$

The boundaries of the regions can be obtained by solving the following optimization problem:

$$(\gamma_1, \ldots, \gamma_{L-1}) = \arg\max \sum_{n=1}^{L-1} \log_2(M_n)\left(F_\gamma(\gamma_{n+1}) - F_\gamma(\gamma_n)\right) \tag{3.64}$$

subject to

$$\sum_{n=1}^{L-1} \int_{\gamma_n}^{\gamma_{n+1}} \frac{\xi_n}{\gamma} p(\gamma)d\gamma = 1 \tag{3.65}$$

The Lagrange equation can be used to solve this problem

$$J(\gamma_1, \ldots, \gamma_{L-1}) = \sum_{n=1}^{L-1} \log_2(M_n)\left(F_\gamma(\gamma_{n+1}) - F_\gamma(\gamma_n)\right) + \lambda\left[\sum_{n=1}^{L-1} \int_{\gamma_n}^{\gamma_{n+1}} \frac{\xi_n}{\gamma} p(\gamma)d\gamma - 1\right] \tag{3.66}$$

where λ is the Lagrange multiplier. The region boundaries $\gamma_1, \ldots, \gamma_{L-1}$ are obtained by solving the equation system

$$\frac{\partial J(\gamma_1, \ldots, \gamma_{L-1})}{\partial \gamma_n} = 0 \quad 1 \leq n \leq L - 1 \tag{3.67}$$

The solution is given by

$$\gamma_1 = \frac{\xi_1}{\log_2(M_1)}\rho \tag{3.68}$$

and

$$\gamma_n = \frac{\xi_n - \xi_{n-1}}{\log_2(M_n) - \log_2(M_{n-1})}\rho \quad 2 \leq n \leq L - 1 \tag{3.69}$$

where ρ is found from the average power constraint (3.65). There is no closed-form solution for ρ and a numerical search technique must be used to obtain it.

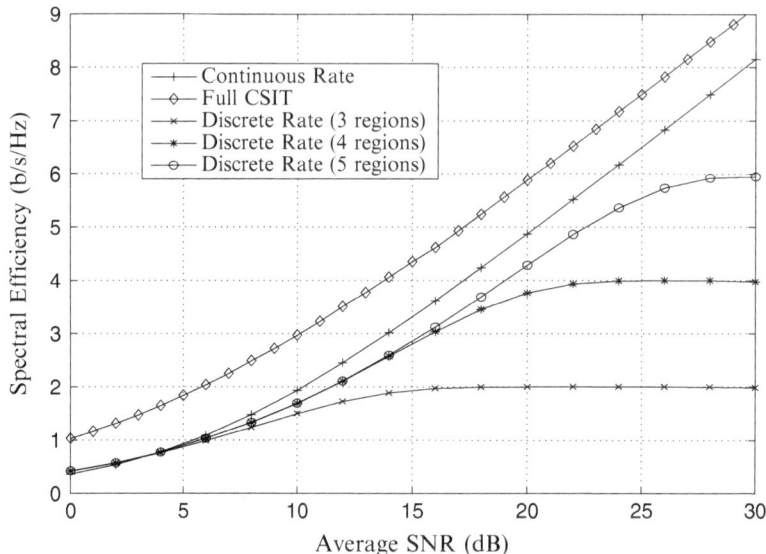

Fig. 3.10 Spectral efficiency of adaptive rate and power schemes for 3, 4 and 5 regions in Rayleigh fading

Figure 3.10 provides the spectral efficiency versus SNR curves of adaptive discrete rate and continuous power schemes for the 3, 4 and 5 regions in Rayleigh fading and a fixed $BER = 10^{-3}$. We see that the spectral efficiency for the 5 regions case results is within 2 dB of the one of the continuous-rate M-QAM. In order to simplify the algorithm and reduce the amount of exchanged information on the feedback link, it is possible to use a constant transmit power for each region. In that case, we fix the power to satisfy the BER constraint for the lowest SNR in each region. As a consequence, the effective BER will be lower than the target BER.

The works of [67, 71] investigate adapting constellation size and [39] considers adapting power, rate, and instantaneous BER. In [3, 26], the authors provide a systematic study on the increase in spectral efficiency obtained by optimally varying combinations of the transmission rate, power, and instantaneous BER. More recently, other works on rate adaptation have been published [12, 58]. An overview of feedback constraints and imperfections for adaptive transmission can be found in [20, 66].

3.4.5 Adaptive Modulation and Coding

To further improve the system performance, codes can be combined with the adaptive M-QAM. The resulting method is usually called adaptive modulation and coding (AMC) or adaptive trellis coded modulation (ATCM). ATCM has been studied in [2, 27, 29] assuming perfect CSIT. These works consider joint optimization of the coding rate and modulation based on a target average error rate or average throughput requirement.

For ATCM, the number of subsets is kept constant for all the possible coding rates and modulations. The k information bits at the input of the convolutional encoder of rate k/n, produce n bits for the selection of one of the 2^n cosets from a partition of the signal constellation. The trellis codes designed for AWGN channel can be used in ATCM. The additional uncoded bits select one of the signal points in the selected coset, which is then transmitted. The size of the M-QAM signal constellation from which the signal point is selected is determined by the instantaneous SNR and the desired BER like in the uncoded case described previously. The effective coding gain G_c of the trellis codes is a function of the minimum squared distance between the signal point sequences, which is determined by the convolutional code properties and the subset partitioning. Compared to the uncoded adaptive rate scheme, the BER approximation given in (3.53) for coded M-QAM modulation becomes

$$BER(M, \gamma) \approx 0.2 \exp\left(-\frac{3\gamma G_c}{2(M-1)}\right) \tag{3.70}$$

The adaptive discrete rate and power technique introduced previously can still be used by taking into account the coding gain G_c in the computation of ξ_n in order to derive the new boundary regions.

Goeckel [24] has considered ATCM in fast fading channels. He has shown that the effective channel conditioned on outdated fading estimates becomes Rayleigh or Rician, and thus a performance degradation is observed in ATCM schemes due to the use of uncoded bits that only achieve a diversity of one in this scheme. By designing with this channel variability in mind, it is possible to build adaptive coding schemes that display significant gains over conventional schemes. More precisely, he proposed to maintain the intersubset and intrasubset differences.

Interleaving was introduced to maintain the spectral efficiency of AMC [24]. In this scheme, the coded bits are interleaved before mapping to the M-QAM symbols.

In order to extend the AMC schemes to systems with high mobility, Ormeci et al. [56] have proposed to employ bit-interleaved coded modulation (BICM) [9, 74] instead of trellis coded modulation (TCM). In BICM, all the information bits are encoded by a convolutional encoder and bit-by-bit interleaver. Consequently, it provides an independent fading component for each channel bit out of the convolutional encoder. Compared to TCM, there is no uncoded bits for the adaptation of the constellation size. Due to the time-diversity caused by interleaving, this scheme performs significantly better than the original ATCM scheme considering outdated CSI.

3.4.6   Channel Prediction

The above adaptive schemes such as adaptive power or AMC rely on ideal assumptions assuming zero delay in the feedback path. In practice, the difficulties to obtain an exact relation between the estimated instantaneous SNR, the BER and the impact of the feedback delay will have to be taken into account in this adaptive transmission scheme. In this section, we will consider the impact of the latency on the performance and channel prediction.

The latency due to channel estimation at the receiver and the transmission of the CSI using the feedback link is a key factor when designing an adaptive transmission scheme. For fast fading channel the latency can result in outdated channel estimation at the transmitter if the channel has changed during this period. In order to compensate the latency, the CSI needs to be predicted at the transmitter. This predicted CSI will then be used to perform the adaptive transmission.

Several fading prediction methods have been proposed to improve the adaptive transmission performance. These methods can be separated into three main categories [18]:

- Autoregressive (AR) model-based techniques.
- Sum of sinusoids model-based algorithm: since the channel response can be described by a sum of weighted complex sinusoids, spectral estimation methods such as MUSIC and ESPRIT [64] are used to determine the parameters associated with each sinusoid and to accomplish the prediction of the fading process.
- Basis expansion algorithm: the basis functions of the subspace of time-concentrated and band-limited sequences are determined using the autocorrelation function of the fading channel. The extrapolated basis functions are then used to obtain the predicted fading coefficients.

In this section, we will focus only on channel prediction using the AR model-based technique based on PSAM [17, 55]. The data stream is parsed into blocks of length L_B, and one known symbol is inserted per block to estimate the channel coefficient.

We assume that all the pilot symbols have the same amplitude a_p and transmitted power P_p. From any pilot symbol, the estimate of the complex channel is obtained by dividing the noisy received signal by the corresponding pilot symbol. Consequently, the fading estimate of $h[iL_B]$ based on one received observation from (3.6) is given by

$$\bar{h}[iL_B] = h[iL_B] + \frac{n[iL_B]}{a_p}$$

$$= h[iL_B] + b[iL_B] \tag{3.71}$$

where the noise $b[iL_B]$ has zero mean and variance $N_0 B_W / P_p$. To improve the quality of the estimation, the data bits can also be used in decision-directed mode.

Assuming that the feedback delay is a multiple of the duration $L_B T_S$ and that we wish to predict the channel d blocks ahead using a linear predictor of order Pth, the channel predicted response is given by

$$\hat{h}[(i + d)L_B] = \sum_{p=0}^{P-1} w_p^* \bar{h}[(i - p)L_B]$$

$$= \mathbf{w}^H \bar{\mathbf{h}}[iL_B] \tag{3.72}$$

where $\mathbf{w} = [w_0, \ldots, w_P]^T$ is the predictor filter coefficient vector and $\bar{\mathbf{h}}[iL_B] = [\bar{h}[iL_B], \ldots, \bar{h}[iL_B - p]]^T$ is the vector of the memoryless estimates of the complex fading amplitude from the last p pilot symbols.

The prediction error is given by

$$\epsilon[(i + d)L_B] = \hat{h}[(i + d)L_B] - h[(i + d)L_B] \tag{3.73}$$

It is shown that the prediction filter coefficient vector that minimizes the mean square prediction error (MSE) $\sigma_\epsilon^2 = \mathbb{E}(|\epsilon[(i + d)L_B]|^2)$ for the complex fading amplitude on a Rayleigh channel can be expressed as

$$\mathbf{w} = \mathbf{R}^{-1}\mathbf{r} \tag{3.74}$$

where $\mathbf{R}$ is the autocorrelation matrix of size $P \times P$ of the fading at the pilot-symbol instants

$$\mathbf{R} = \mathbb{E}[\bar{\mathbf{h}}[iL_B]\bar{\mathbf{h}}^H[iL_B]] \tag{3.75}$$

and $\mathbf{r}$ is the correlation vector given by

$$\mathbf{r} = \mathbb{E}[h^*[(i + d)L_B]\bar{\mathbf{h}}[iL_B]] \tag{3.76}$$

The direct computation of $\mathbf{w}$ in (3.74) requires to compute the inversion of $\mathbf{R}$ and has a complexity of $O(P^3)$. By exploring the structure of the matrix, Levinson-Durbin recursion can reduce complexity to $O(P^2)$. An important factor to evaluate the performance of the system is the correlation between the predicted and the true SNR given by

$$\rho = \frac{\text{Cov}(\hat{\gamma}, \gamma)}{\sqrt{\text{Var}(\hat{\gamma})\text{Var}(\gamma)}} \tag{3.77}$$

where $\hat{\gamma} = |\hat{h}|^2 P_{av}/N_0 B_W$ is the predicted overall SNR.

Thanks to the orthogonality principle between $\hat{h}[(i + d)L_B]$ and $\epsilon[(i + d)L_B]$, it can be shown [55] that the MSE $\sigma_\epsilon^2 = 1 - \mathbf{r}^H \mathbf{R}^{-1}\mathbf{r}$ and that the predicted fading channel has zero mean and variance $\mathbf{r}^H \mathbf{R}^{-1}\mathbf{r}$. From this result we can further simplify ρ as

$$\rho = \mathbf{r}^H \mathbf{R}^{-1} \mathbf{r} \tag{3.78}$$

When we use the Jakes model [28] the entries of the autocorrelation matrix $\mathbf{R}$ and $\mathbf{r}$ are given by

$$[\mathbf{R}]_{p,q} = J_0(2\pi f_D |p-q| L_B T_s) + \frac{N_0 B_W}{P_p} \delta(p-q) \tag{3.79}$$

$$[\mathbf{r}]_p = J_0(2\pi f_D (d+p) L_B T_s) \tag{3.80}$$

where $J_0(x)$ is the zeroth-order Bessel function of the first kind and f_D is the Doppler spread.

The computation of $\mathbf{w}$ requires the knowledge of the correlation matrix $\mathbf{R}$ function of the channel coefficients. Since this function is a priori unknown and can be time-variant, it has to be estimated from the channel observations. Efficient adaptive techniques such as least mean squares (LMS), recursive least squares (RLS) or QR-decomposition based RLS (QR-RLS) algorithms can be used to track the changes of $\mathbf{w}$ and considerably reduce the computational complexity [17, 64].

In the above derivation, we have predicted both the phase and the amplitude of the complex flat fading channel coefficient. However for the SISO case, it is generally enough to predict the fading power since the SISO adaptive transmission techniques require only the knowledge of the power [30]. An overview of long-range prediction of fading signals with application to adaptive transmission is given in [17].

3.4.6.1 BER Analysis

In this section we consider the extension of the adaptive discrete rate introduced above to the case of time delay. The boundary region are computed like previously assuming perfect CSI. However, the choice of the M-QAM constellation will be done based on the value of $\hat{\gamma}$ instead of γ. We can see that the average spectral efficiency will not be affected by this time delay but the average BER will be modified. The average BER associated to each region and the overall system BER computed in (3.56) and (3.57) can be significantly modified in case of time delay [4]. By taking into account this delay, the average BER associated to region n is given by

$$\overline{BER}(M_n) = \int_{\gamma_n}^{\gamma_{n+1}} \int_0^{+\infty} BER(M_n, \gamma | \hat{\gamma}) p(\hat{\gamma}, \gamma) d\gamma d\hat{\gamma} \tag{3.81}$$

$BER(M_n, \gamma | \hat{\gamma})$ is the BER experienced when applying the modulation of size M_n and signal to noise γ.

$p_{\hat{\gamma},\gamma}(\hat{\gamma}, \gamma)$ is the joint distribution of the actual and predicted SNR. If we assume that the two SNRs are Rayleigh distributed, $p_{\hat{\gamma},\gamma}(\hat{\gamma}, \gamma)$ is a bivariate Rayleigh distribution given by

$$p_{\hat{\gamma},\gamma}(\hat{\gamma},\gamma) = \frac{4\sqrt{\rho\gamma\hat{\gamma}}}{1-\rho}\left(\frac{1}{\overline{\gamma}}\right)^2 I_0\left(\frac{2\rho\sqrt{\hat{\gamma}\gamma}}{(1-\rho)\overline{\gamma}}\right)\exp\left(-\frac{\rho\gamma+\hat{\gamma}}{(1-\rho)\overline{\gamma}}\right) \qquad (3.82)$$

where $I_0(.)$ is the $0th$ order modified Bessel function of the first kind and ρ is the correlation factor between γ and $\hat{\gamma}$.

Using the generalized Marcum Q-functions the average BER associated to each region n can be derived as

$$\overline{BER}(M_n) = \frac{0.2}{\overline{\gamma}}\frac{\Gamma(1,b_n\gamma_n) - \Gamma(1,b_n\gamma_{n+1})}{b_n} \qquad (3.83)$$

where $\Gamma(m,x)$ is the complementary incomplete gamma function and

$$b_n = \begin{cases} \frac{1}{\overline{\gamma}} + \frac{3\rho}{3(1-\rho)\overline{\gamma}+2(2^n-1)} & n = 2,\ldots,L-1 \\ \frac{1}{\overline{\gamma}} + \frac{\rho}{(1-\rho)\overline{\gamma}+1} & n = 1 \end{cases} \qquad (3.84)$$

using the BER approximation (3.53) for M-QAM modulation and $BER(2,\gamma) \approx 0.1\exp(-\gamma)$ for BPSK.

It is clear that the correlation coefficient depends on L_B, P_p, f_D, T_S, P and d. Following [55] we consider a communication system using a Rayleigh fading channel with carrier frequency 2 GHz, symbol rate of 400.10^3 symbol per second, and a terminal speed of 30 m/s resulting in $f_D = 200$ Hz. The normalized Doppler spread is consequently $f_D T_S = 5.10^{-4}$.

In Fig. 3.11 we show the advantage of using a predictor. We plot the dependence of the correlation coefficient in function of the normalized delay given $f_D T_S L_B d$ for different orders of predictor P. We fix $P_p/N_0 B_W = 20$ dB and $L_B = 10$. While L_B should be chosen lower such that $L_B < 1/2 f_D T_S$ to satisfy the sampling theorem, in reality, due to the noise, pilot symbols need to be transmitted with L_B several orders of magnitude smaller than $1/2 f_D T_S$. When d is large, the pilot symbols must be spaced further apart to have a large memory span in order to obtain autocorrelation samples with large values. In fact, for each predictor filter length P, there is an optimal sampling rate to maximize the correlation coefficient. As expected, we can observe in Fig. 3.11 that the prediction accuracy decreases when the normalized delay increases. Furthermore, as the predictor filter length P increases, the correlation coefficient and prediction accuracy increase significantly.

In Fig. 3.12 we show the dependence of the correlation coefficient with the signal to noise ratio $P_p/N_0 B_W$ for a predictor filter length $P = 10$ and $P = 50$. Like previously, the increase of P improves significantly prediction accuracy.

Finally, Fig. 3.13 illustrates the average BER as a function of the normalized time delay. We fix $P_p/N_0 B_W = 20$ dB, and $L_B = 10$. Without delay, the average BER is $1.5.10^{-4}$ and a target BER $= 10^{-3}$ since the BER reaches the target BER only near the boundary regions. Depending on the predictor filter length, the degradation

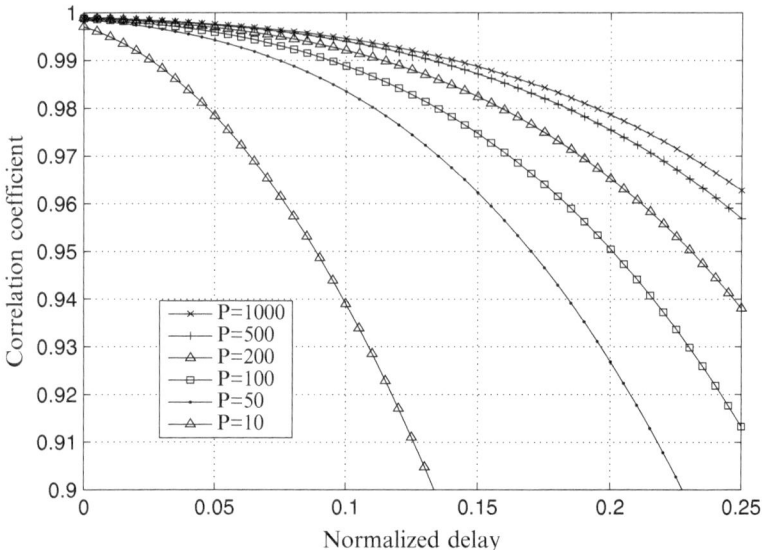

Fig. 3.11 Correlation coefficient ρ versus normalized delay

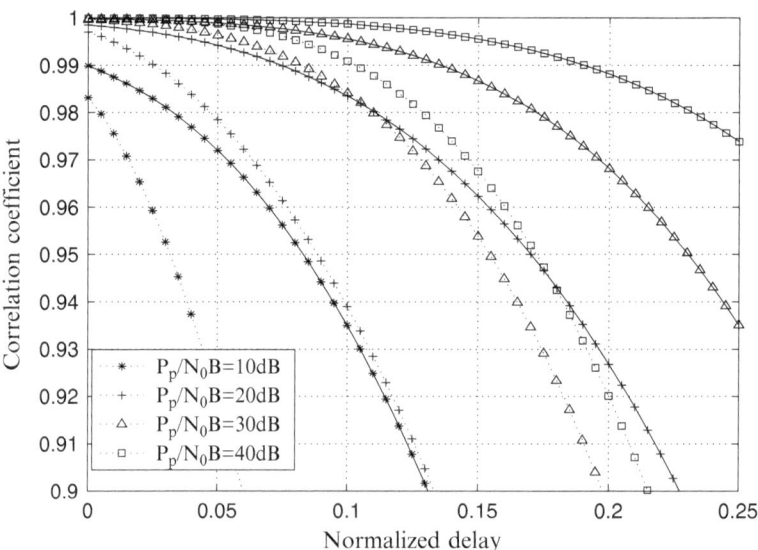

Fig. 3.12 Correlation coefficient ρ versus normalized delay for $P = 10$ (*dashed lines*) and $P = 50$ (*solid curves*)

of BER due to the time delay can be important. For a maximum BER of 10^{-3} the maximum normalized time delay is 0.125 for $P = 50$ and 0.22 for $P = 1,000$.

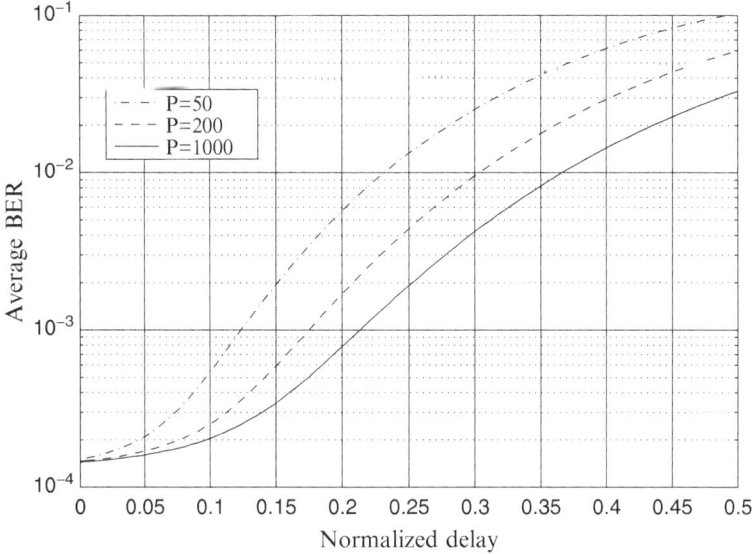

Fig. 3.13 Average BER versus normalized delay $f_D T_S L_B d$

As a rule of thumb, the feedback time delay should be kept to within $0.001/f_D$ in order to keep the BER near its desired target. The effects of estimation error and feedback delay on adaptive modulation were first analyzed in [26]. The extension to Nakagami and gamma distribution has been considered in [4] and in [55], respectively.

3.4.7 Wideband Channel

We have shown in the previous chapter that OFDM is an efficient scheme to deal with the intersymbol interference coming from the frequency selective channels. In conventional OFDM, the same modulation scheme is employed for all subchannels. However due to the frequency selectivity, the BER can be impacted by a small set of faded subchannels. Assuming that CSI is available at the transmitter, we have seen that to maximize the data rate, power and data rate should be adjusted across subcarriers according to (3.48).

In order to approach this capacity, power and the data rate in each subchannel can be adapted. The combination of adaptive modulation with OFDM was initially proposed by Kalet [36] and further developed in [16]. The extension of AMC for the bit-interleaved coded OFDM (BICM-OFDM) transmission has been studied in [63].

These algorithms usually assume perfect knowledge of the CSIT. While this assumption is reasonable in wireline channels since the channel is time invariant, for

wireless channels, the channel is randomly varying over time and it is not possible
to guarantee perfect CSI at the transmitter.

A direct extension of the adaptive discrete rate technique to the OFDM scheme
consists in selecting the modulation scheme for each subcarrier n or group of
subcarriers also called cluster according to the received SNR γ_n. The adaptive
power and discrete rate is generally not feasible due to the huge amount of feedback
required for the power adaptation. In this section, we will assume that the transmitter
allocates equal power across the full set or a subset of all subcarriers.

Considering an OFDM scheme with N subchannels grouped into Q clusters and
a quantization of the SNR using L regions, the total number of signaling bits per
OFDM symbol is $\lceil Q \rceil \lceil \log(L) \rceil$. Since this amount of feedback is high and since
the channel is correlated in time and frequency, data compression algorithms can
be designed to reduce the feedback overhead by exploiting these correlations. Two
main classes of data compression algorithm can be applied [21]:

- The discrete-amplitude compression where the modulation and coding indexes
  are fed back using lossless source coding such as Lempel–Ziv–Welch (LZW),
  Run-Length Coding, Huffman codes, or arithmetic coding
- The continuous-amplitude compression where the SNR values of each subcarrier
  are fed back using lossy source coding such as vector quantization, linear
  prediction coding, or transform coding

3.4.7.1 Discrete-Amplitude Compression

Let's consider a system with a bandwidth of $20\,\text{MHz}$ and $N = 1,200$ subcarriers
($15\,\text{KHz}$ per subcarrier). Following the LTE standard, we assume that the 1,200
subcarriers are grouped into $Q = 100$ clusters of 12 subcarriers, each cluster
corresponding to a bandwidth of $180\,\text{kHz}$. We will also assume, like previously,
that the subcarriers in a cluster can be modulated using BPSK, QPSK, QAM16,
QAM64 and no modulation ($L = 5$). Since the amount of data to feedback is still
too high, we will consider different compression algorithms.

A first solution consists in exploiting the frequency correlation by using run-
length encoding (RLE) as the compression algorithm [54]. RLE does not require
any statistics of the source and its implementation complexity is low.

Another solution is to use Huffman or arithmetic codes. However, in order to
obtain high compression rates the information should be correctly conditioned since
without transformation, the distribution of the discrete rate is close to uniform
and consequently Huffman or arithmetic codes will perform poorly. In [35] the
authors have proposed to add a frequency differential encoder in order to exploit
the frequency correlation before using Huffman or arithmetic encoder. Other lossy
source coding methods such as LZW algorithm have been proposed in the literature
[21, 42].

Considering the system described above and the Vehicular A channel model
introduced in Chap. 2, in Fig. 3.14 we can see that with the frequency differential

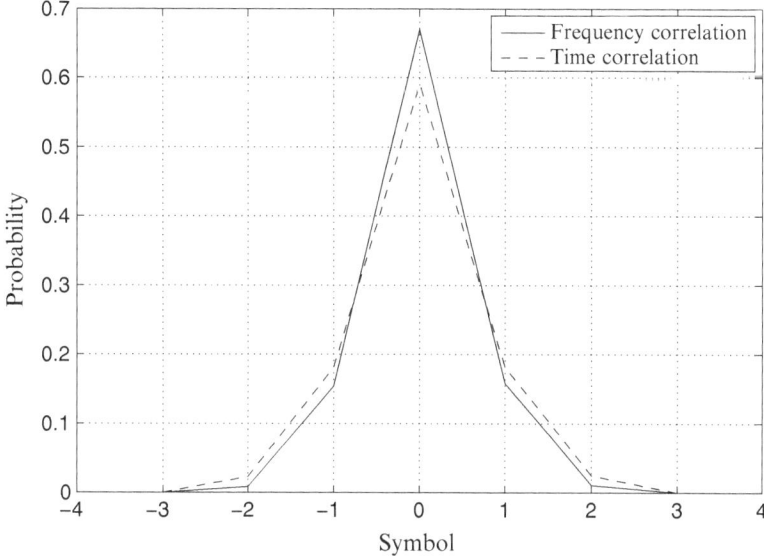

Fig. 3.14 Time and frequency correlation using differential encoder for vehicular A channel

Table 3.1 Comparison of different compression algorithms

Algorithm	Nbr of bits
No compression	300
Run-length encoding	267
Huffman code	171
Arithmetic code	159

encoder, symbol 0 (the current modulation scheme index is the same as the one of the previous cluster), symbol 1 (the current modulation scheme index is increased by 1 compared to the one of the previous cluster) and symbol −1 are the most probable (94% of the realizations).

In Table 3.1, we compare the average number of bits required for the different discrete-amplitude solutions exploiting the frequency correlation. The entropy of the source, computed assuming that the source is a first order Markov source, is 1.59 Sh/cluster. In this context, compared to RLE, Huffman and arithmetic codes almost achieve the entropy rate, but these solutions require a priori statistics that can be hard to estimate.

In Fig. 3.14 we also show the time correlation due to the coherence time of the Vehicular A channel model assuming a time transmission interval (TTI) of 1 ms and a speed of 60 km/h. Consequently, depending on the periodicity of the feedback and Doppler spread, we can also exploit the time correlation to reduce the feedback load. By exploiting the time correlation we can only update the assignments of clusters that have changed with respect to the previous update. Signaling a cluster index change from one period to the next one requires $\lceil \log_2 N \rceil + 1$ bits assuming that the

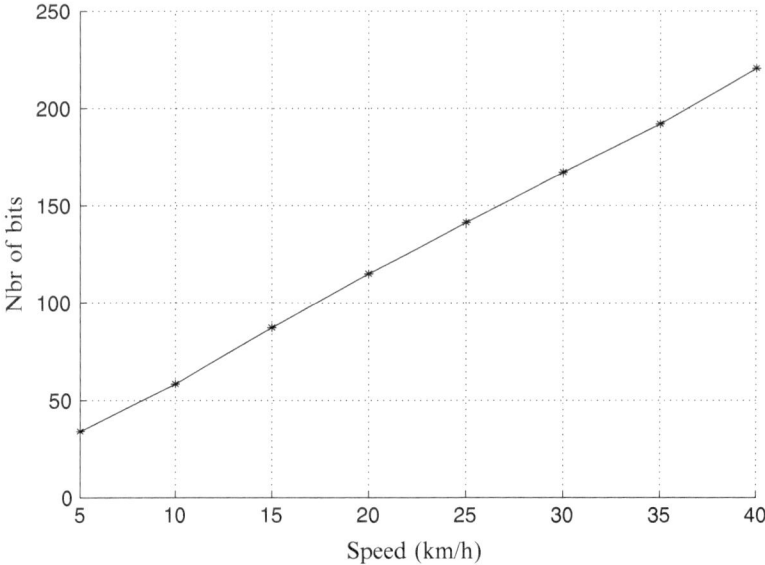

Fig. 3.15 Average number of bits required versus speed of the vehicle

current modulation scheme index is only increased or decreased by 1 compared to the previous one. If only a few assignments are needed, the overall cost for signaling can be reduced significantly compared to the full feedback.

In Fig. 3.15, we plot the average number of bits required for the update versus the speed of the vehicle and assuming again that the periodicity of the feedback is 1 ms. Here, since $N = 100$, seven bits are required for one cluster update. While the reduction in feedback load is significant when the speed of the vehicle is low, there is no gain compared to the use of a simple Huffman code over the frequency when the speed is higher than 30 km/h.

Another method to reduce the amount of feedback for rate optimization called on-off feedback [41, 60, 65] consists in allowing the transmitter to allocate equal power P for each subchannel when the corresponding received SNR γ_n is above a given SNR threshold γ_0. The power constraint is given by

$$\begin{cases} P_n = P_{av}/N_u & \text{if} \quad \gamma_n \geq \gamma_0 \\ P_n = 0 & \text{if} \quad \gamma_n < \gamma_0 \end{cases} \tag{3.85}$$

where N_u is the number of subchannels with nonzero power allocations satisfying $\gamma_n \geq \gamma_0$. The SNR threshold γ_0 is adjusted to maximize the data rate.

In [60], Rong et al. have compared the uncoded BER of three adaptive approaches where the transmitter has only one bit of CSIT per cluster: on-off feedback, adaptive power allocation (APA) and adaptive modulation selection (AMS). While the on-off feedback has the best performance when the feedback channel is perfect, the authors have shown that when the feedback is imperfect the performance of OFDM systems with one bit per cluster feedback can be even worse than the performance of the OFDM system without any feedback if the feedback channel is noisy or delayed. The performance of the APA and AMS approaches can be significantly improved by performing channel prediction and carrying out robust modulation selection.

3.4.7.2 Continuous-Amplitude Compression

In contrast to the lossless schemes described above, the continuous-amplitude compression algorithm relies on lossy compression of the real-valued SNR values.

Vector quantization (VQ) using Lloyd algorithm has been proposed in [49] to generate the codebook of the power loading at the receiver. While the performance is good, the complexity is quite high in practice due to the high number of clusters.

Transform coding (TC), where a de-correlating transform is applied to the channel vector prior to quantization has been proposed by different authors [22,54]. In most of these works, the discrete cosine transform (DCT) has been used to perform the de-correlation due to its low complexity. After the DCT, an optimized scalar quantization is performed. The obtained coefficients are finally compressed for example by a RLE.

In Chap. 5, we will consider multiuser feedback strategies for wideband channels where the above techniques will be extended considering not only quantization and data compression but also reduced feedback strategies.

3.4.7.3 Effective SINR Mapping

In OFDM system, it is often necessary to construct a link to system (L2S) mapping table in order to link SINR to block error rate (BLER) for a given modulation and coding scheme (MCS) to reduce the complexity and duration of system level simulations.

Another very important and different use of the so-called SINR mapping tables arises in the generation of the channel quality information (CQI) by the users. The CQI is calculated and fed back by the users via the feedback link and is then used by the BS for frequency domain resource scheduling and link adaptation using adaptive power control and AMC. Since feedback at subcarrier level consumes considerable uplink bandwidth, it is common to feedback CQI only for a group of N_c subcarriers. Therefore, the users will perform SINR mapping to determine their appropriate MCS for the full group of subcarriers. Since a codeword is usually encoded over multiple subcarriers this mapping table is not easy to derive. If the number of involved subcarriers is N_c, this determination requires a characterization of the BLER as a function of the different subcarrier gains. In order to avoid to generate and use a N_c-dimensional look-up table, approximations have been proposed such as the effective SINR mapping (ESM) that maps the instantaneous values of SINRs obtained from the different subcarriers to a single effective SINR. The ESM first maps the sequence of SINRs of a codeword using a given compression function to an effective SNR SNR_{eff} value which is then used to read the equivalent BLER from the AWGN performance curves of a particular MCS. The general expression of SNR_{eff} is given as follows:

$$SNR_{eff} = -\delta_1 I^{-1}\left(\frac{1}{N_c}\sum_{i=1}^{N_c} I\left(\frac{SNR_i}{\delta_2}\right)\right) \tag{3.86}$$

where $I(SNR_i)$ is a mapping function which transforms the SINR of each subchannel to some information measure that is then linearly averaged over the N_c subcarriers. This averaged value is then transformed into the SNR domain using the inverse mapping function. The parameters δ_1 and δ_2 are calibration factors used to compensate for the different modulations and code rates.

Different L2S mapping methods have been proposed in the literature such as exponential effective SINR mapping (EESM) [75, 76] and mutual information based effective SINR mapping (MI-ESM) [69, 77].

EESM is a simple mapping method that was introduced in system level evaluations and since then has been extensively used for link quality modeling. In EESM, the mapping function $I(SNR_i)$ is calculated using Chernoff Union bound of error probabilities $I(SNR_i) = 1 - \exp(-SNR_i)$

The compression function that maps the instantaneous values of SINRs to SNR_{eff} is given by

$$SNR_{eff} = -\beta \log\left(\frac{1}{N_c}\sum_{i=1}^{N_c} exp\left(\frac{SNR_i}{\beta}\right)\right) \tag{3.87}$$

where the single parameter $\beta = \delta_1 = \delta_2$ needs to be empirically fine tuned for each MCS.

In MI-ESM, the approximation of mapping function and the reverse mapping functions are the mutual information function and its inverse. According to [77] we have

$$SNR_{eff} = I^{-1}\left(\frac{1}{N_c}\sum_{i=1}^{N_c} I\left(SNR_i\right)\right) \tag{3.88}$$

where $I(SNR_i)$ is the constrained capacity function of the applied modulation alphabet.

A comparison of EESM and MI-ESM has been performed in [8]. The authors have shown that MI-ESM technique is more accurate than EESM. In chapter 8, we will show that the CQI is computed using effective SINR mapping.

3.5 Automatic Repeat Request

While modern error correcting codes (ECC) allow to obtain performance close to the Shannon capacity, they do not guarantee error-free data transmission. In most of the communication systems, in order to build a reliable data transmission (an error-free data transmission when allowing some delay), an automatic repeat request (ARQ) scheme using acknowledgement (ACK)/negative acknowledgement (NACK) signaling exchange in the upper layers must be implemented [40, 45]. A drawback of ARQ scheme is the delay added for data retransmission. Consequently, ARQ schemes are more suitable for data or still-image transmissions rather than real-time voice or video transmissions. ARQ can be seen as a special class of feedback schemes using one bit feedback per frame (the ACK/NACK bit).

In the classical ARQ schemes, frame errors are detected using a high-rate error detection code (N, K) also called cyclic redundancy check (CRC) code, where K is the number of information bits and N is the number of bits of the codeword.

After reception of a word, the receiver computes its syndrome. If the received word is a codeword (meaning that the syndrome is zero), the receiver sends an ACK of successful transmission to the receiver. If the receiver detects that the word is not a codeword, it sends a NACK and a retransmission of the frame is done.

The three classical ARQ schemes are, namely, the stop-and-wait, go-back-N, and selective repeat protocols.

The stop-and-wait ARQ protocol is the simplest class of ARQ schemes. A stop-and-wait ARQ transmitter sends one frame at a time. After sending a frame, the transmitter does not send any further frames until it receives an ACK from the receiver. If the ACK is not received by the transmitter before a given timeout, the transmitter sends the same frame again. The stop-and-wait ARQ protocol is illustrated in Fig. 3.16a. The stop-and-wait ARQ is the most inefficient ARQ scheme due to the high delay between the frames.

In the go-back-N ARQ protocol, the transmitter is allowed to transmit consecutively a maximum number of frames N_f when available with the only condition to have no more than N_f unacknowledged frames in its pipeline. The transmitter stores the transmitted frames, pending on the reception of an ACK for each of them. The receiver process keeps track of the sequence number of the next frame it expects to receive and sends that number with every ACK it sends. Whenever the transmitter receives a NACK associated to the codeword i, it stops transmitting new codewords and then proceeds to the retransmission of the codeword i and of the $N_f - 1$ next codewords. At the receiving end, the receiver discards the erroneously received word i and all the subsequent $N_f - 1$ previously received words no matter whether they were correct or not. Transmissions are repeated until codeword i is positively acknowledged and the process is repeated again. The go-back-N ARQ protocol is illustrated in Fig. 3.16b for $N_f = 4$. Since the go-back-N ARQ protocol is designed to operate without a reordering buffer at the receiver this protocol is inefficient when the error rate is high. The results for both reliable and unreliable feedback are expressed in terms of probability matrices in [5].

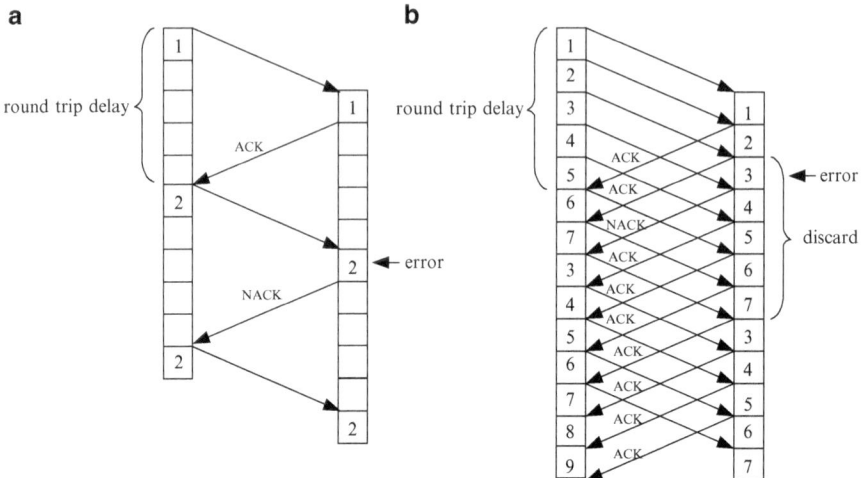

Fig. 3.16 (a) Stop-and-wait protocol and (b) go-back-N ARQ protocol with $N_f = 4$

Fig. 3.17 Selective and repeat protocol

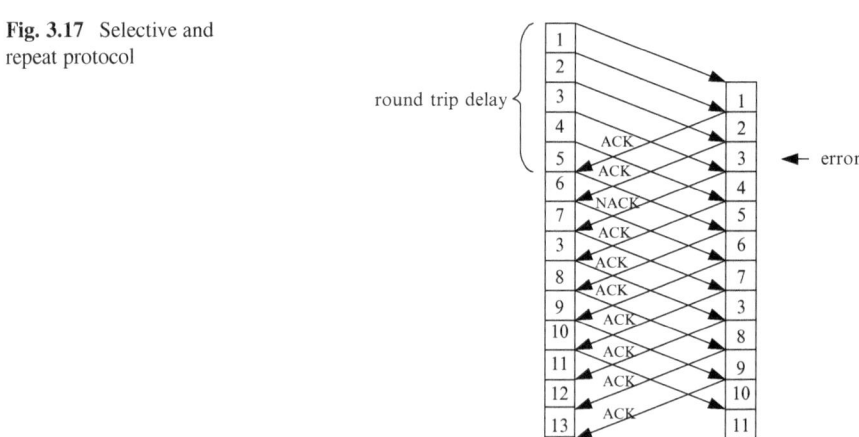

In the selective repeat protocol, only lost frames are retransmitted. Subsequent frames, which are correctly received, are stored in the receiver buffer so that the frames are sent to the higher layers in the correct order. Since only the lost frames are retransmitted, the selective repeat protocol is more efficient [6,31]. The selective repeat protocol is illustrated in Fig. 3.17.

While a new frame is transmitted only when an ACK for the current frame is received in the stop-and-wait protocol, for both go-back-N and selective repeat protocols the frames are continuously transmitted without waiting for ACK messages. Both go-back-N and selective repeat protocols are consequently called continuous ARQ protocols.

These classical ARQ methods can be modified to obtain the so-called truncated ARQ methods where the maximum delay can be adjusted by limiting the maximum number of allowed retransmissions.

To improve the reliability and the system throughput performance of the communication systems, these classical ARQ schemes are applied in conjunction with ECC, yielding to the so-called hybrid automatic repeat request (HARQ) scheme [44,72].

HARQ schemes can be classified into four categories: synchronous nonadaptive, synchronous adaptive, asynchronous nonadaptive and asynchronous adaptive. Synchronous HARQ means that the retransmissions happen at fixed time intervals while in asynchronous HARQ, the retransmission can be scheduled at any time after the NACK is received. Adaptive HARQ protocol allows to change the frequency resource allocation and MCS at each retransmission.

Synchronous nonadaptive HARQ needs the lowest overhead in terms of control signaling, but it implies that the retransmissions must occur at the same frequency resources and consequently does not allow scheduling retransmissions at time-frequency resources with good channel conditions. At the opposite, asynchronous adaptive HARQ provides full flexibility of retransmission but with the constraint that the full control information must be sent at each retransmission. It also prevents collisions with persistently allocated resources.

In a type I HARQ scheme, the ECC allows to reduce the frequency of retransmission by correcting the most frequent error patterns. When the receiver cannot decode the frame, the receiver requests a retransmission and discards the corrupted frame.

If the receiver can buffer the previously received frame, the optimal solution is to use maximum ratio combining (MRC) to combine these multiple frames. While a frame transmitted twice cannot be individually decoded without error, the combination of the two received frames allows us to correctly decode the original data. The type I HARQ with MRC is referred as "Chase combining" (CC) scheme [15].

In the type II HARQ scheme also called "incremental redundancy" (IR) [23, 37, 50, 68], when a NACK is received, instead of sending the same coded frames, the transmitter sends additional coded frames. Rather than discarding the original frame, the different received frames are exploited to provide a stronger error correction capability and increase the probability of correct frame reception.

The IR protocol performs the following steps:

1. Encode information bits using a linear block code (N, K).
2. Puncture the codewords into M subcodewords of size N_k where $N = \sum_{k=1}^{M} N_k$. Store the resulting systematic and parity streams at the transmitter for potential transmission.
3. Initialize $m = 1$.
4. Transmit the mth subcodeword.
5. At the reception, attempt to decode the code using all the symbols received so far. Perform a hard decision on the vector of likelihood of the information

sequence. If the information sequence is successively decoded, send an ACK to the transmitter. Otherwise, $m = m + 1$ and send a NACK to the transmitter and proceed to step 4. If the information sequence is not successively decoded when $m = M$, then declare an outage.

The code rate after the mth transmission is $R_m = \frac{K}{\sum_{k=1}^{m} N_k}$.
The average rate is

$$R_{av} = \sum_{k=1}^{M} R_k Pr(\bar{S}_1, \bar{S}_2, \ldots, S_k) \tag{3.89}$$

and the outage probability

$$P_{out} = Pr(\bar{S}_1, \bar{S}_2, \ldots, \bar{S}_M) \tag{3.90}$$

where S_k is the event that the data is correctly decoded at the kth slot. $\bar{S}_k$ is the complement of S_k.

While the complexity of IR scheme is higher than the other ARQ schemes, it significantly improves the overall throughput. IR schemes using turbo codes and rate compatible punctured turbo (RCPT) codes have been proposed, respectively, in [53] and [61]. The extension to low density parity check (LDPC) codes has been proposed in [43].

Like classical ARQ schemes, HARQ schemes can be used in stop-and-wait or selective repeat modes. Chase combining and incremental redundancy are both supported in the HSDPA standard.

In [34], the authors propose and analyze HARQ type I scheme with power ramping. Power ramping allows the transmission power to be adapted to the number of retransmission.

In [47, 59] instead of considering AMC at the physical layer and ARQ at the data link layer separately, the authors have combined AMC and truncated ARQ by designing the SNR regions for AMC modes taking into consideration the effect of retransmission under the prescribed delay and error performance constraints. Based on extensive simulation, it has been demonstrated that bandwidth efficiency can be improved as much as 5.8 dB compared with the conservative AMC design approach with truncated HARQ.

3.6   Conclusion

In this chapter, we have considered the different feedback strategies for the case of a single user wireless communication system where both the transmitter and the receiver are equipped with a single antenna for narrowband and wideband cases.

We have seen that depending on the availability of the CSIT and CSIR, the capacity of the fading channels can significantly increase by exploiting the time and frequency variation of the channel using power control. Under perfect CSIR and quantized CSIT using L levels of quantization of the SNR, we have evaluated the average rate for both short- and long-term power constraints. We have shown that the long-term power constraint allows a significant increase of the average rate at low SNR.

Then, we have studied the adaptive transmission over time and frequency where one or more parameters such as rate and power are adapted in order to maximize the spectral efficiency of the radio link for a given quality. For the adaptive discrete rate technique and adaptive discrete rate and power we have evaluated spectral efficiency versus SNR. For example, we have shown that the spectral efficiency for the five regions case is within only 2 dB of the one of the continuous-rate M-QAM. The practical implementation of adaptive discrete rate is the so-called AMC where a joint optimization of the coding rate and modulation is performed based on a target average error rate or average throughput requirement.

Due to the delay of the feedback link, it is often necessary to perform channel prediction at the transmitter to compensate this delay. We have shown that if the average spectral efficiency is not affected by the time delay, there is a significant degradation of the average bit error depending on the length of the prediction filter.

For wideband channel, due to correlation of the channel in frequency, the amount of feedback can be reduced by performing data compression based on discrete-amplitude compression such as the LZW algorithm or on continuous-amplitude compression algorithms. In this chapter, we have compared the average number of bits required for different discrete-amplitude solutions. We have also shown that for limited time channel variation, a frequency differential encoder can significantly outperform the non-differential encoders.

Finally, we have reviewed the different ARQ schemes including the important class of HARQ that allows to build a reliable data transmission by allowing ACK/NACK signaling exchange and data retransmission.

Some of the techniques and algorithms introduced in this chapter will be extended to the case of multiuser systems in Chap. 5 and to the case of multiple antenna systems in Chaps. 4 and 6.

References

1. Abou-Faycal I C, Trott M D, Shamai S (2001) The capacity of discrete-time memoryless Rayleigh fading channels. IEEE Trans. Inform. Theory. 1290–1301
2. Alamouti S M, Kallel S (1994) Adaptive Trellis-Coded Multiple-Phase-Shift Keying for Rayleigh Fading Channels. IEEE Trans. Commun. 42 : 2305–2314
3. Alouini M S, Tang X, Goldsmith A J (1999) An adaptive modulation scheme for simultaneous voice and data transmission over fading channels. IEEE Jour. Select. Areas in Commun. 17 : 837–850

4. Alouini M S, Goldsmith A J (2000) Adaptive modulation over Nakagami fading channels. Kluwer Journal on Wireless Personal Communications. 119–143
5. Ausavapattanakun K, Nosratinia A (2007) Analysis of Go-Back-N ARQ in block fading channels. IEEE Trans. Wireless Comm. 6 : 2793–2797
6. Ausavapattanakun K, Nosratinia A (2007) Analysis of selective-repeat ARQ via matrix signal-flow graphs. IEEE Trans. Commun. 55 : 198–204
7. Bello P A, Cowan W M (1962) Theoretical study of on/off transmission over Gaussian multiplicative circuits. Proc. IRE Nat. Commun. Symp. Utica, New York.
8. Brueninghaus K, Astely D, Salzer T, Visuri S, Alexiou A, Karger S, Seraji G A (2005) Link Performance Models for System Level Simulations of Broadband Radio Access Systems. 16th Annual IEEE International Symposium on Personal, Indoor and Mobile Radio Communications. Germany. 2306–2311
9. Caire G, Taricco G, Biglieri E (1998) Bit-interleaved coded modulation. IEEE Trans. Inform. Theory. 44 : 927–946
10. Caire G, Shamai S (1998) On the capacity of some channels with channel state information. Proc. IEEE Int. Symp. Information Theory, Cambridge, MA. 42–42
11. Caire G, Taricco G, Biglieri E (1999) Optimum power control over fading channels. IEEE Trans. Info. Theory. 5 :1468–1489
12. Caire G, Kumar K R (2007) Information theoretic foundations of adaptive coded modulation. Proc. IEEE. 85 : 2274–2298
13. Cavers J (1972) Variable-rate transmission for Rayleigh fading channels. IEEE Trans. Commun. 20 : 15–22
14. Cavers J (1991) An analysis of pilot symbol assisted modulation for rayleigh fading channels. IEEE Transactions on Vehicular Technology. 40 : 686–693
15. Chase D (1973) A combined coding and modulation approach for communications over dispersive channels. IEEE Trans. Commun. 21 : 159–174
16. Chow P, Cioffi J, Bingham J (1995) A practical discrete multitone transceiver loading algorithm for data transmission over spectrally shaped channels. IEEE Trans. Commun. 43 : 772–775
17. Duel-Hallen A, Hu S, Hallen H (2000) Long-range prediction of fading signals : Enabling adaptive transmission for mobile radio channels. IEEE Signal Process. Mag. 17 : 62–75
18. Duel-Hallen A (2007) Fading channel prediction for mobile radio adaptive transmission systems. Proc. IEEE. 95 : 2299–2313
19. Ekbatani S, Etemadi F, Jafarkhani H (2009) Throughput maximization over slowly fading channels using quantized and erroneous feedback, IEEE Trans. Commun. 57 : 2528–2533
20. Ekpenyong A E, Huang Y (2007) Feedback constraints for adaptive transmission. IEEE Sig. Proc. Mag. 34 : 69–78
21. Eriksson T, Ottosson T (2007) Compression of Feedback for adaptive modulation and scheduling. Proceeding of the IEEE. 85 : 2314–2321
22. Eriksson T, Ottosson T (2007) Compression of feedback in adaptive OFDM-based systems using scheduling. IEEE Commun. Letters. 11 : 859–861
23. Frenger P, Parkvall S, Dahlman E (2001) Performance comparison of HARQ with Chase combining and incremental redundancy for HSDPA. in Proc. IEEE Veh. Technol. Conf. 1829–1833
24. Goeckel D L (1999) Adaptive coding for time-varying channels using outdated fading estimates. IEEE Trans. Commun. 47 : 844–855
25. Goldsmith A, Varaiya P (1993) Capacity of fading channels with channel side information. Proc. Int. Conf. Communications, Geneva, Switzerland. 600–604
26. Goldsmith A, Chua S G (1997) Variable-Rate Variable-Power MQAM for Fading Channels. IEEE Trans. Commun. 45 : 1218–1230
27. Goldsmith A, Chua S G (1998) Adaptive coded modulation for fading channels. IEEE Trans. Commun. 46 : 595–602
28. Goldsmith A (2005) Wireless Communications. New York, Cambridge University Press
29. Falahati S, Svensson A, Ekman T, Sternad M (2003) Adaptive modulation systems for predicted wireless channels. Proc. IEEE VTC. 3 : 1532–1536

30. Falahati S, Svensson A, Sternad M, Mei H (2004) Adaptive trellis-coded modulation over predicted flat fading channels. IEEE Trans. Commun. 42 : 307–316
31. Haleem M, Chandramouli R (2006) Adaptive downlink scheduling and rate selection: Cross layer design. IEEE Jour. Select. Areas in Commun. 23 : 1572–1581
32. Hayes J F (1968) Adaptive feedback communications. IEEE Trans. Commun. Technol. 16 : 29–34
33. Heegard C, Gamal A E (1983) On the capacity of computer memory with defects. IEEE Trans. Inform. Theory. 29 : 731–739
34. Hwang S, Kim B H, Kim Y (2001) A hybrid ARQ scheme with power ramping. Proceedings of IEEE Vehicular Technology Conference. 3 : 1579–1583
35. Jimenez V, Eriksson T, Armada A, Garcia M, Ottosson T, Svensson A (2007) Methods for compression of feedback in adaptive multicarrier 4G schemes. Wireless Personal Commun. 47 : 101–112
36. Kalet I (1989) The multitone channel. IEEE Trans. Commun. 37 : 119–124
37. Kallel S (1992) Analysis of memory and incremental redundancy ARQ schemes over a nonstationary channel. IEEE Trans. Commun. 1474–1480
38. Kim T T, Skoglund M (2007) On the expected rate of slowly fading channels with quantized side information. IEEE Trans. Commun. 55 : 820–829
39. Köse C, Goeckel D L (2000) On power adaptation in adaptive signaling systems. IEEE Trans. Commun. 48 : 1769–1773
40. Kurose J F, Ross K W (2013) Computer Networking: A Top-Down Approach. 6th Edition, Addison-Wesley
41. Leke A, Cioffi J M (1998) Multicarrier systems with imperfect channel knowledge. Proc. of IEEE International Symposium on Personal, Indoor and Mobile Radio Communications (PIMRC). 549–553
42. Lestable T, Bartelli M (2002) LZW adaptive bit loading. Proc. IEEE Int. Symp. on Advances Wireless Commun. (ISAWC), Victoria, Canada.
43. Li J, Narayanan K (2002) Rate-compatible low density parity check codes for capacity-approaching ARQ scheme in packet data communications. in Proc. Int. Conf. Communications, Internet, Information Technology (CIIT), U.S. Virgin Islands. 201–206
44. Lin S, Yu P S (1982) A hybrid ARQ scheme with parity retransmission for error control of satellite channels. IEEE Trans. Commun. 30 : 1701–1719
45. Lin S, Costello Jr D J, Miller M J (1984) Automatic-repeat-request error-control schemes. IEEE Commun. Magazine. 22 : 5–17
46. Lin L, Yates R, Spasojevic P (2003) Adaptive transmission with discrete code rates and power levels. IEEE Trans. Inf. Theory. 51 : 2115–2125
47. Liu Q, Zhou S, Giannakis G B (2004) Cross-layer combining of adaptive modulation and coding with truncated ARQ over wireless links. IEEE Trans. Commun. 3 : 1746–1755
48. Liu X, Yang H, Guo W, Yang D (2006) Capacity of fading channels with quantized channel side information. IEICE Trans. Commun. 89 : 590–593
49. Love D J, Heath Jr R W (2005) OFDM power loading using limited feedback. IEEE Trans. on Veh. Technol. 54 : 1773–1780
50. Mandelbaum D M (1974) Adaptive-feedback coding scheme using incremental redundancy. IEEE Trans. Inf. Theory 20 : 388–389
51. McEliece R J, Stark W E (1984) Channels with block interference. IEEE Trans. Inform. Theory. 30 : 44–53
52. Medard M, Goldsmith A (1997) Capacity of time-varying channels with channel side information. IEEE Int. Symp. on Information Theory (ISIT). 372–372
53. Narayanan K R, Stuber G L (1997) A novel ARQ technique using the turbo coding principle. IEEE Commun. Letters. 1 : 49–51
54. Nguyen H, Lestable T (2004) Compression of bit loading power vectors for adaptive multi-carrier systems. Proc. IEEE Int. Midwest Symp. Circuits Syst. Hiroshima, Japan. 243–246

55. Øien G E, Holm H, Hole K J (2002) Channel prediction for adaptive coded modulation on Rayleigh fading channels. Proc. European Signal Processing Conference (EUSIPCO). Toulouse, France.

56. Ormeci P, Liu X, Goeckel D L, Wesel R D (2001) Adaptive bit-interleaved coded modulation. IEEE Trans. Commun. 49 : 1572–1581

57. Ozarow L, Shamai S, Wyner A D (1994) Information-theoretic considerations for cellular mobile radio. IEEE Transactions on Vehicular Technology. 43 : 359–378

58. Paris J F, del Carmen Aguayo-Torres M, Entrambasaguas J T (2001) Optimum discrete-power adaptive QAM scheme for Rayleigh fading channels. IEEE Commun. Lett. 5 : 281–283

59. Park S H, Kim J W, Kang C G (2007) Design of adaptive modulation and coding scheme for truncated HARQ. Proc of 2nd International Symposium on Wireless Pervasive Computing (ISWPC).

60. Rong Y, Vorobyov S A, Gershman A B (2006) Adaptive OFDM Techniques With One-Bit-Per-Subcarrier Channel-State Feedback. IEEE Trans. Commun. 54 : 1993–2003

61. Rowitch D N, Milstein L B (2000) On the performance of hybrid FEC/ARQ systems using rate compatible punctured turbo (RCPT) codes. IEEE Trans. Commun. 48 : 55–67

62. Shannon C E (1958) Channels with side information at the transmitter. IBM Journal Research and Dev. 2 : 289–293

63. Song K B, Ekbal A, Chung S T, Cioffi J M (2006) Adaptive Modulation and Coding (AMC) for Bit-Interleaved Coded OFDM (BIC-OFDM). IEEE Transactions on Wireless Communications. 5 : 1685–1694

64. Stoica P, Moses R (2005) Spectral Analysis of Signals. Upper Saddle River, NJ: Prentice Hall

65. Sun Y, Honig M L (2003) Minimum feedback rates for multi-carrier transmission with correlated frequency-selective fading. Proc. IEEE Global Telecommunications Conf. San Francisco, CA. 3 : 1628–1632

66. Svensson A (2007) An introduction to adaptive QAM modulation schemes for known and predicted channels. Proc. IEEE. 85 : 2322–2336

67. Torrance J M, Hanzo L (1996) Upper bound performance of adaptive modulation in a slow Rayleigh fading channel. Electron. Lett. 32 : 718–719

68. Tuninetti D (2007) Transmitter channel state information and repetition protocols in block fading channels. Proc. Information Theory Workshop. 505–510

69. Tuomaala E, Wang H (2005) Effective SINR Approach of Link to System Mapping in OFDM/Multi-carrier Mobile Network. Proc. 2nd International Conference on Mobile Technology, Applications and Systems.

70. Viswanathan H (1999) Capacity of Markov channels with receiver CSI and delayed feedback. IEEE Trans. Inf. Theory. 45 : 761–771

71. Webb W T, Steele R (1995) Variable rate QAM for mobile radio. IEEE Trans. Commun. 43 : 2223–2230

72. Wicker S (1995) Error Control Systems for Digital Communication and Storage. Englewood Cliffs: Prentice Hall

73. Wolfowitz J (1964) Coding Theorems of Information Theory. 2nd Ed. New York: Springer-Verlag

74. Zehavi E (1992) 8-PSK trellis codes for a Rayleigh channel. IEEE Trans. Commun. 40 : 873–884

75. 3GPP TSG-RAN-1 (2004) OFDM Exponential Effective SIR Mapping Validation, EESM Simulation Results (R1-040089)

76. 3GPP TSG-RAN-1 (2004) System-Level Performance Evaluation for OFDM WCDMA in UTRAN (R1-040090)

77. 3GPP2, WG3 (2003) Effective-SNR Mapping for Modeling Frame Error Rates in Multiple-state Channels (C30-20030429-010)

Chapter 4
Feedback in MIMO Wireless Communication

4.1 Introduction

In Chap. 3, we have shown that by exploiting CSI at the transmitter, we can improve the bit error rate or increase the data rate of SISO systems. In this chapter, we will extend this work to the case of MIMO systems where the transmitter and the receiver are equipped with more than one antenna. Multiple antenna systems are playing an increasing role in wireless communications. While the diversity gain can be extracted without the need of CSIT feedback using space-time codes, when CSI is available at the transmitter, transmit beamforming and linear precoding can considerably increase the potential gain. In time division duplex (TDD) systems, assuming that the duplexing time delay is lower than the channel coherence time, it is theoretically possible to estimate the uplink channel without transmission of the CSI by exploiting the reciprocity of the channel. In practice, this solution requires an accurate transceiver calibration since the amplifiers and filters are not the same in uplink and downlink transmission. Furthermore, this solution implies strong constraints on uplink signaling when assuming a downlink transmission. Due to these difficulties, there are only few MIMO communication systems that have considered the exploitation of the channel reciprocity.

In frequency division duplex (FDD) system, due to the phase difference between the uplink and downlink channel except for long-term channel variation it is generally not possible to directly estimate the uplink channel. Then, the terminal must estimate the CSI and then transmit it to the BS using the feedback channel. In this chapter, we will consider different feedback strategies for the case of a single user wireless communication system where both the transmitter and the receiver are equipped with multiple antennas.

In Sect. 4.2, we will consider beamforming schemes where only one stream is sent from the transmitter equipped with multiple antennas to the receiver equipped with a single antenna or multiple antennas using receiver combining. In that case, the channel direction information (CDI) is the vector that should be fed back to the transmitter. We will study the different solutions to feed back this information

B. Özbek and D. Le Ruyet, *Feedback Strategies for Wireless Communication*, DOI 10.1007/978-1-4614-7741-9_4, © Springer Science+Business Media New York 2014

including quantized and analog feedback. Then we will consider linear precoding schemes for MIMO systems where multiple streams are transmitted in Sect. 4.3. In that case, the CDI is a unitary matrix and the codebook should be designed according to the selected criterion. In Sect. 4.4, we will see that space-time codes can also benefit from CSIT when they are associated with a precoder. We will study the different possible antenna selection strategies in MIMO systems in Sect. 4.5. We will show that antenna selection can also be seen as a special case of linear precoding. In the last part of this chapter, we will study the different techniques to exploit the spatial, time and frequency correlation of the channel. In Sect. 4.6 we will consider the presence of spatial correlation where instead of building a codebook covering all the space, we can restrict the search space. In Sect. 4.7, we will focus on time correlation where the same idea can be exploited. Similarly to coding for correlated source, the class of differential feedback strategies are the most promising. Finally in Sect. 4.8, we will consider the case of selective MIMO channels where we will show that the frequency selectivity can be exploited to further reduce the quantity of feedback information or improve the wideband channel estimation quality.

4.2 Beamforming

Early works on transmit beamforming with feedback started in 1994 with the paper of Gerlach and Paulraj [22] and later [30, 34, 66].

4.2.1 System Model

Let's first consider a flat fading point-to-point single user communication system employing transmit beamforming with N_t transmit antennas and a single receive antenna. This MISO communication system is presented in Fig. 4.1.

The baseband input–output discrete relationship for a given complex transmitted symbol s is represented by

$$y = \mathbf{h}^H \mathbf{w} s + n \tag{4.1}$$

where $\mathbf{h} = [h_1 \ h_2 \ \dots \ h_{N_t}]^T$ is the channel vector and $\mathbf{w} = [w_1 \ w_2 \ \dots \ w_{N_t}]^T$ is the precoding vector with $||\mathbf{w}||^2 = 1$. The noise n is a circularly symmetric independent identically distributed (i.i.d.) Gaussian process. We assume that the entries of $\mathbf{h}$ as i.i.d. zero-mean complex Gaussian random variables with variance one.

The instantaneous signal to noise ratio (SNR) is given by

$$\gamma = \frac{P}{N_0 B_W} |\mathbf{h}^H \mathbf{w}|^2 \tag{4.2}$$

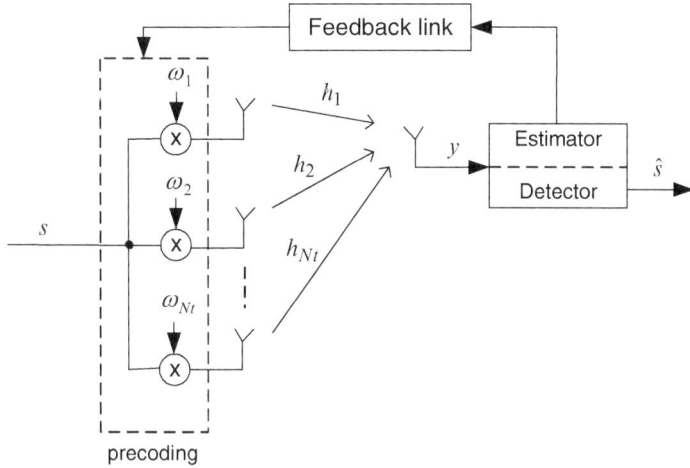

Fig. 4.1 Transmit beamforming for MISO scheme

We define the normalized channel vector:

$$\mathbf{g} = \frac{\mathbf{h}}{||\mathbf{h}||} \tag{4.3}$$

Assuming that the entries of $\mathbf{h}$ are i.i.d. distributed, $\mathbf{g}$ is isotropically distributed on the surface of an N_t dimensional complex hypersphere of unit radius centered at the origin $\mathbb{O}^{N_t}$.

$$\mathbb{O}^{N_t} = \{\mathbf{x} \in \mathbb{C}^{N_t} : ||\mathbf{x}|| = 1\} \tag{4.4}$$

$\mathbf{w}$ should be chosen to maximize the SNR γ, in order to maximize the capacity or to minimize the average probability of error. When full CSIT is available, and given no other design constraints, the optimal precoding vector is

$$\mathbf{w}_{full} = \mathbf{g} \tag{4.5}$$

The ergodic capacities of the point-to-point single user MISO communication link, respectively, with full CSIT and no CSIT are given by

$$C_{fullCSIT} = \mathbb{E}_{\mathbf{h}} \left[B_w \log_2 \left(1 + \frac{P}{N_0 B_w} ||\mathbf{h}||^2 \right) \right]$$

$$C_{noCSIT} = \mathbb{E}_{\mathbf{h}} \left[B_w \log_2 \left(1 + \frac{P}{N_t N_0 B_w} ||\mathbf{h}||^2 \right) \right]$$

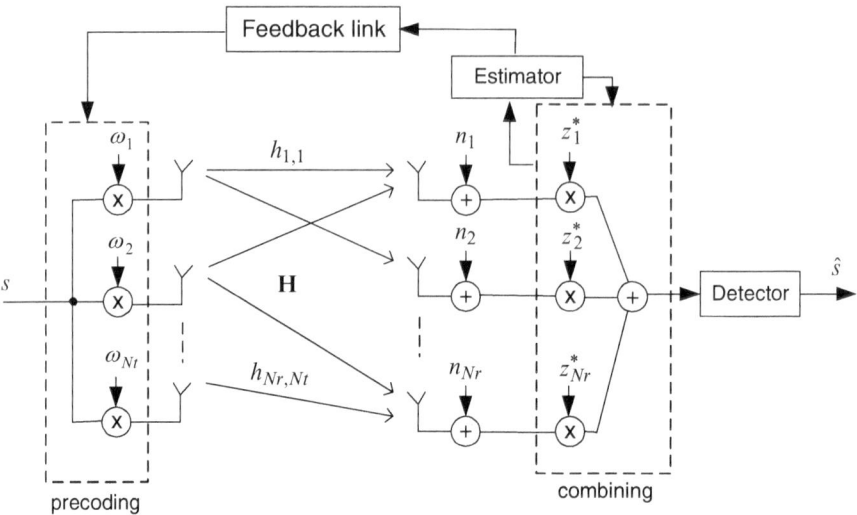

precoding   combining

Fig. 4.2 Transmit beamforming and receive combining for MIMO scheme

Let's now consider the point-to-point single user communication system employing transmit beamforming and receive combining with N_t transmit antennas and N_r receive antennas. This communication system is presented in Fig. 4.2.

The baseband input–output discrete relationship for a given complex transmitted symbol s is now presented by

$$x = \mathbf{z}^H \mathbf{H} \mathbf{w} s + \mathbf{z}^H \mathbf{n} \tag{4.6}$$

where $\mathbf{H}$ is the channel matrix of size $N_r \times N_t$, $\mathbf{w} = [w_1 \ w_2 \ \ldots \ w_{N_t}]^T$ is the precoding vector, and $\mathbf{z} = [z_1 \ z_2 \ \ldots \ z_{N_r}]^T$ is the combining vector. The entries of the noise vector $\mathbf{n}$ are circularly symmetric i.i.d. Gaussian. For noncorrelated MIMO channel $\mathbf{H}$, we model the entries of $\mathbf{H}$ as i. i. d. zero-mean complex Gaussian random variables with variance one.

The instantaneous SNR is given by

$$\gamma = \frac{P}{N_0 B_W} |\mathbf{z}^H \mathbf{H} \mathbf{w}|^2 \tag{4.7}$$

Compared to the MISO case, both $\mathbf{z}$ and $\mathbf{w}$ should be chosen. Given $\mathbf{w}$, $\mathbf{z}$ is chosen to maximize the SNR γ or equivalently $|\mathbf{z}^H \mathbf{H} \mathbf{w}|^2$. Such a receiver is the maximum ratio combining (MRC) receiver and $\mathbf{z}$ is given by

$$\mathbf{z} = \frac{\mathbf{H} \mathbf{w}}{||\mathbf{H} \mathbf{w}||} \tag{4.8}$$

$\mathbf{w}$ is chosen to maximize the SNR given by

$$
\begin{aligned}
\mathbf{w}_{full} &= \arg\max_{w} |\mathbf{z}^H \mathbf{H} \mathbf{w}|^2 \\
&= \arg\max_{w} |\mathbf{w}^H \mathbf{H}^H \mathbf{H} \mathbf{w}|^2 \\
&= \arg\max_{w} ||\mathbf{H}\mathbf{w}||^2 \\
&= \mathbf{v}_1
\end{aligned}
\tag{4.9}
$$

where $\mathbf{v}_1$ is the dominant right singular vector corresponding to the eigenvalue $\sqrt{\lambda_1}$ of the channel matrix $\mathbf{H} = \mathbf{U}\boldsymbol{\Sigma}\mathbf{V}^H$ with $\mathbf{V} = [\mathbf{v}_1 \ldots \mathbf{v}_{N_t}]$ and $\boldsymbol{\Sigma} = \mathrm{diag}(\sqrt{\lambda_1}, \sqrt{\lambda_2}, \ldots, \sqrt{\lambda_r}, 0, \ldots, 0)$ as shown in Chap. 2.

The ergodic capacity of the point-to-point single user MISO communication link employing transmit beamforming and receive combining with full CSIT is given by

$$
\begin{aligned}
C_{fullCSIT} &= \mathbb{E}_{\mathbf{H}} \left[B_w \log_2 \left(1 + \frac{P}{N_0 B_w} ||\mathbf{H}\mathbf{v}_1||^2 \right) \right] \\
&= \mathbb{E}_{\mathbf{H}} \left[B_w \log_2 \left(1 + \frac{P}{N_0 B_w} \lambda_1 \right) \right]
\end{aligned}
\tag{4.10}
$$

In the next section, we will consider the case where there exists a low-rate feedback link and the precoding vector should be chosen among a given set of beamforming vectors.

4.2.2 Codebook Design

4.2.2.1 Criterion and Distortion Function

Due to the limited bandwidth of the feedback channel and in order to address the lack of perfect CSIT, a classical solution is to quantize the CDI (i.e., the normalized channel vector $\mathbf{g}$ for MISO case or the dominant right singular vector $\mathbf{v}_1$ for MIMO with receive combining) before transmission over the finite rate feedback link. The CDI codebook is known by both the user side and the base station (BS). The user quantizes its CDI to the closest codeword. At the BS, CDI should be constructed by minimizing the maximum inner product between codewords [67].

In the quantized feedback scheme, the vector $\mathbf{w}_{opt}$ is taken from a set or codebook of $N = 2^B$ unit vectors where B is the number of feedback bits. Let's define the set of the precoding vectors $\mathscr{W} = \{\mathbf{w}_1, \mathbf{w}_2, \ldots, \mathbf{w}_N\}$. The challenge now is to design the set $\mathscr{W}$ in order to maximize a performance criterion such as the bit error rate or the system capacity. The selected beamforming vector $\mathbf{w}_{opt}$ is chosen according to a given performance criterion.

For point-to-point MISO single user the achievable data rate is approximated as:

$$R_{quantCSIT} = \mathbb{E}_{\mathbf{h}} \left[B_w \log_2 \left(1 + \frac{P}{N_0 B_w} ||\mathbf{h}||^2 |\mathbf{g}^H \mathbf{w}_{opt}|^2 \right) \right] \qquad (4.11)$$

To measure the average distortion due to quantization, we define the distortion function. For the MISO case, we have

$$G(\mathscr{W}) = \mathbb{E}_{\mathbf{h}} \left[1 - |\mathbf{g}^H \mathbf{w}_{opt}|^2 \right]$$

For the point-to-point MIMO case with receive combining, we can derive the achievable data rate. Let the eigen-decomposition of $\mathbf{H}^H \mathbf{H}$ be given by $\sum_{i=1}^{N_t} \lambda_i \mathbf{v}_i \mathbf{v}_i^H$ where $\lambda_1 \geq \lambda_2 \geq \ldots \lambda_{N_t}$. We have

$$R_{quantCSIT} = E_{\mathbf{H}} \left[B_w \log_2 \left(1 + \frac{P}{N_0 B_w} ||\mathbf{H} \mathbf{w}_{opt}||^2 \right) \right]$$

$$\geq \mathbb{E}_{\mathbf{H}} \left[B_w \log_2 \left(1 + \frac{P}{N_0 B_w} |\mathbf{v}_1^H \mathbf{w}_{opt}|^2 \right) \right] \qquad (4.12)$$

The result in Eq. (4.12) follows from zeroing $\lambda_2, \ldots, \lambda_{N_t}$. The average distortion function is then given by

$$G(\mathscr{W}) = \mathbb{E}_{\mathbf{H}} \left[\lambda_1 - ||\mathbf{H} \mathbf{w}_{opt}||^2 \right]$$

$$= \mathbb{E}_{\mathbf{H}} \left[\lambda_1 - \sum_{i=1}^{N_t} \lambda_i |\mathbf{v}_1^H \mathbf{w}_{opt}|^2 \right]$$

$$(4.13)$$

An upper bound of $G(\mathscr{W})$ is given by

$$G_{up}(\mathscr{W}) = \mathbb{E}_{\mathbf{H}} \left[\lambda_1 - \lambda_1 |\mathbf{v}_1^H \mathbf{w}_{opt}|^2 \right] \qquad (4.14)$$

Since eigenvalues and eigenvectors of complex Wishart matrices are independent, for i.i.d. channel, the upper bound can be decomposed into two factors [54]:

$$G_{up}(\mathscr{W}) = \mathbb{E}_{\mathbf{H}} [\lambda_1] \, \mathbb{E}_{\mathbf{H}} \left[1 - |\mathbf{v}_1^H \mathbf{w}_{opt}|^2 \right] \qquad (4.15)$$

For correlated channel matrices, an upper bound has been proposed in [72] and given by

$$G_{up}(\mathscr{W}) = \left(\mathbb{E}_{\mathbf{H}}[\lambda_1] + \sqrt{2 \mathrm{Var}_{\mathbf{H}}(\lambda_1)} \right) \sqrt{\mathbb{E}_{\mathbf{H}}[1 - |\mathbf{v}_1^H \mathbf{w}_{opt}|^2]} \qquad (4.16)$$

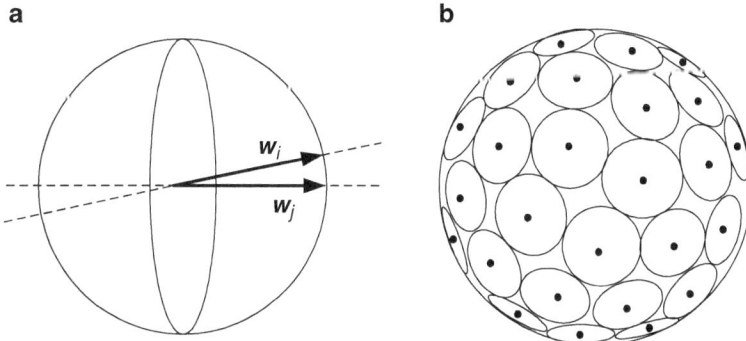

Fig. 4.3 (**a**) Precoding vectors. (**b**) Example of line packing in three real dimensions with $N = 50$

The last two expressions are made of two components: the right factor representing the effect of the codebook design and the left factor representing the channel impact. Consequently, the design of the codebook must minimize the term $1 - |\mathbf{v}_1^H \mathbf{w}_{opt}|^2$.

4.2.2.2 Grassmannian Line Packing

The set $\mathscr{W}$ can be seen as a collection of lines in the Euclidean space $\mathbb{C}^{N_t}$. $\mathbf{w}_i$ can be seen as the coordinates of a point situated at the surface of an hypersphere with a unit radius centered at the origin. Mathematically, $\mathbf{w}_i$ is a point in the Grassmann manifold or Grassmannian space $G(N_t, 1)$ where $G(n, p)$ is the set of all p dimension subspaces of complex Euclidian n-dimensional space $\mathbb{C}^n$.

For i.i.d. channels, in order to minimize the outage probability, to maximize the data rate or to maximize the SNR, the construction of the set $\mathscr{W}$ should maximize the angular separation between the two closest lines [54,65]. This problem is called the Grassmannian line packing problem. A graphical illustration of the line packing problem in three real dimensions is given in Fig. 4.3b.

In 1930, the Dutch botanist Tammes [87] posed the problem of packing when studying the distribution of pores on the surface of pollen spherical grains. The Grassmannian line packing problem has been studied in detail for the last 80 years in applied mathematics and information theory. Researchers have studied the best packings of points in Grassmannian manifolds equipped with different metrics. An overview of packing in Grassmannian spaces is available in [16, 17].

For i.i.d. channels, we have

$$
\begin{aligned}
\mathscr{W}_{opt} &= \min_{\mathscr{W} \in \mathbb{C}^{N_t \times N}} \max_{1 \le i < j \le N} |\mathbf{w}_i^H \mathbf{w}_j| \\
&= \max_{\mathscr{W} \in \mathbb{C}^{N_t \times N}} \min_{1 \le i < j \le N} d(\mathbf{w}_i, \mathbf{w}_j)
\end{aligned}
\tag{4.17}
$$

where $d(\mathbf{w}_i, \mathbf{w}_j)$ is the chordal distance between the unitary vectors $\mathbf{w}_i$ and $\mathbf{w}_i$:

$$d^2(\mathbf{w}_i, \mathbf{w}_j) = 1 - |\mathbf{w}_i^H \mathbf{w}_j|^2$$
$$= \sin^2(\angle(\mathbf{w}_i, \mathbf{w}_j)) \tag{4.18}$$

where $\angle(\mathbf{w}_i, \mathbf{w}_j)$ is the angle between the lines generated by the column spaces of $\mathbf{w}_i$ and $\mathbf{w}_j$.

In the literature of Grassmannian line packing, the minimum chordal distance or the sine of the smallest angle between any pair of lines is defined as

$$\delta(\mathscr{W}) = \min_{1 \le i < j \le N} d(\mathbf{w}_i, \mathbf{w}_j)$$
$$= \min_{1 \le i < j \le N} \sqrt{1 - |\mathbf{w}_i^H \mathbf{w}_j|^2} \tag{4.19}$$

In Eq. (4.4), we have defined the unit sphere $\mathbb{O}^{N_t}$ lying in $\mathbb{C}^{N_t}$ and centered at the origin. $\mathbb{O}^{N_t}$ is also known as the unidimensional Stiefel manifold $U(N_t, 1)$.

For any $0 < \epsilon < 1$, we can define a spherical cap on $\mathbb{O}^{N_t}$ with center $\mathbf{o}$ and square radius ϵ as the open set as follows:

$$\mathscr{B}_\epsilon(\mathbf{o}) = \{\mathbf{f} \in \mathbb{O}^{N_t} : d^2(\mathbf{f}, \mathbf{o}) \le \epsilon\} \tag{4.20}$$

Note that for each pair of vectors $\mathbf{w}_i$ and $\mathbf{w}_j$, with $i \ne j$, and $\epsilon \le \delta^2(\mathscr{W})/4$, we have $\mathscr{B}_\epsilon(\mathbf{w}_i) \cap \mathscr{B}_\epsilon(\mathbf{w}_j) = \varnothing$

$A(\mathscr{B}_\epsilon(\mathbf{w}_j))$ the surface area of the spherical cap is given by [65]

$$A(\mathscr{B}_\epsilon(\mathbf{w}_j)) = \frac{2\pi^{N_t} \epsilon^{N_t-1}}{(N_t - 1)!} \tag{4.21}$$

The density $\Delta(\mathscr{W})$ of the line packing associated to $\mathscr{W}$ is defined as the percentage of $\mathbb{O}^{N_t}$ covered by the spherical caps $\mathscr{B}_{\delta^2(\mathscr{W})/4}(\mathbf{w}_i)$ $\forall i$ of the line packing. Mathematically, the density $\Delta(\mathscr{W})$ can be given by

$$\Delta(\mathscr{W}) = \frac{A\left(\bigcup_{i=1}^N \mathscr{B}_{\delta^2(\mathscr{W})/4}(\mathbf{w}_i)\right)}{A_1} \tag{4.22}$$

where A_1 is the surface area of the hypersphere with unit radius:

$$A_1 = \frac{2\pi^{N_t}}{(N_t - 1)!} \tag{4.23}$$

Since the surfaces are not overlapping, we have [65, 101]

$$A\left(\bigcup_{i=1}^{N} \mathscr{B}_{\delta^2(\mathscr{W})/4}(\mathbf{w}_i)\right) = \sum_{i=1}^{N} A(\mathscr{B}_{\delta^2(\mathscr{W})/4}(\mathbf{w}_i))$$

$$= N\left(\frac{2\pi^{N_t}(\frac{\delta(\mathscr{W})}{2})^{2(N_t-1)}}{(N_t-1)!}\right) \qquad (4.24)$$

Consequently we have

$$\Delta(\mathscr{W}) = N\left(\frac{\delta(\mathscr{W})}{2}\right)^{2(N_t-1)} \qquad (4.25)$$

Different bounds on the minimum chordal distance have been proposed in the literature [5, 86].

Theorem 1 (Hamming upper bound).

$$\delta(\mathscr{W}) \leq 2\left(\frac{1}{N}\right)^{1/(2(N_t-1))} \qquad (4.26)$$

Proof. This follows by using Eq. (4.25) and since the packing density $\Delta(\mathscr{W}) \leq 1$.

$\square$

Theorem 2 (Welch bound [86, 91]). *For any codebook $\mathscr{W}$ and $N \geq N_t$,*

$$\delta(\mathscr{W}) \leq \sqrt{\frac{(N_t-1)N}{N_t(N-1)}} \qquad (4.27)$$

Proof. Let's define the maximum cross-correlation amplitude of the codebook $\mathscr{W}$ given by

$$I_{max}(\mathscr{W}) = \max_{1 \leq i < j \leq N} |\mathbf{w}_i^H \mathbf{w}_j| \qquad (4.28)$$

We define the $N \times N$ Gram matrix $\mathbf{G}$ whose entries are $G_{ij} = |\mathbf{w}_i^H \mathbf{w}_j|$. The nonzero eigenvalues $\lambda_1, \ldots, \lambda_{N_t}$ of the Gram matrix $\mathbf{G}$ satisfy $\text{tr}[\mathbf{G}] = \sum_{k=1}^{N_t} \lambda_k = N$. By the Cauchy–Schwarz inequality we have

$$||\mathbf{G}||^2 = \sum_{i=1}^{N_t}\sum_{j=1}^{N_t} |G_{ij}|^2 \sum_{k=1}^{N_t} \lambda_k^2 \geq \frac{|\text{tr}[\mathbf{G}]|^2}{N_t} = \frac{N^2}{N_t} \qquad (4.29)$$

Since

$$\sum_{i=1}^{N_t}\sum_{j=1}^{N_t} |G_{ij}|^2 = N_t + \sum_{i \neq j} |G_{ij}|^2 \qquad (4.30)$$

we have

$$\sum_{i \neq j} |G_{ij}|^2 \geq \frac{N(N - N_t)}{N_t} \tag{4.31}$$

$I_{\max}^2(\mathcal{W})$ is the largest term among the $N(N - 1)$ non-negative terms of G_{ij} for $i = 1, \ldots, N_t \; j = 1, \ldots, N_t, i \neq j$. We can consequently derive the lower bound on $I_{\max}^2(\mathcal{W})$ as follows:

$$I_{\max}^2(\mathcal{W}) \geq \frac{1}{N(N - 1)} \sum_{i \neq j} |G_{ij}|^2 \geq \frac{N - N_t}{N_t(N - 1)} \tag{4.32}$$

Since $\delta^2(\mathcal{W}) = 1 - I_{\max}^2(\mathcal{W})$, the result then follows.   □

Although the Welch bound original derivation was analytical, Strohmer and Heath [86] have noted that the Welch bound has a geometric character. This bound is also known as the Rankin upper bound for subspace packings with respect to the chordal distance [73]. Codebooks achieving the equality are called maximum welch bound equality (MWBE) codebooks [93].

Theorem 3 (Composite upper bound).

$$\delta(\mathcal{W}) \leq \min \left(\sqrt{\frac{(N_t - 1)N}{N_t(N - 1)}}, 2\left(\frac{1}{N}\right)^{1/(2(N_t - 1))} \right) \tag{4.33}$$

This upper bound is the combination of the two above upper bounds.

Recently, Pitaval et al. in [71] have proposed refinements of the Hamming upper bound for Grassmannian codes.

In Fig. 4.4, we plot the different upper bounds and the maximum achieved values of $\delta(\mathcal{W})$ for $N_t = 2$ and $N = 4, 8, 16$ and 32 obtained using codebook constructions based on vector quantization.

For the MISO case, the beamforming vector is chosen as follows to maximize the instantaneous SNR or equivalently to minimize the chordal distance metric:

$$\mathbf{w}_{opt} = \arg \max_{\mathbf{w}_i \in \mathcal{W}} |\mathbf{h}^H \mathbf{w}_i|^2$$

$$= \arg \min_{\mathbf{w}_i \in \mathcal{W}} (1 - |\mathbf{g}^H \mathbf{w}_i|^2) \tag{4.34}$$

while for the MIMO case with receive combining, the beamforming vector is chosen as follows:

$$\mathbf{w}_{opt} = \arg \max_{\mathbf{w}_i \in \mathcal{W}} ||\mathbf{H}\mathbf{w}_i||^2 \tag{4.35}$$

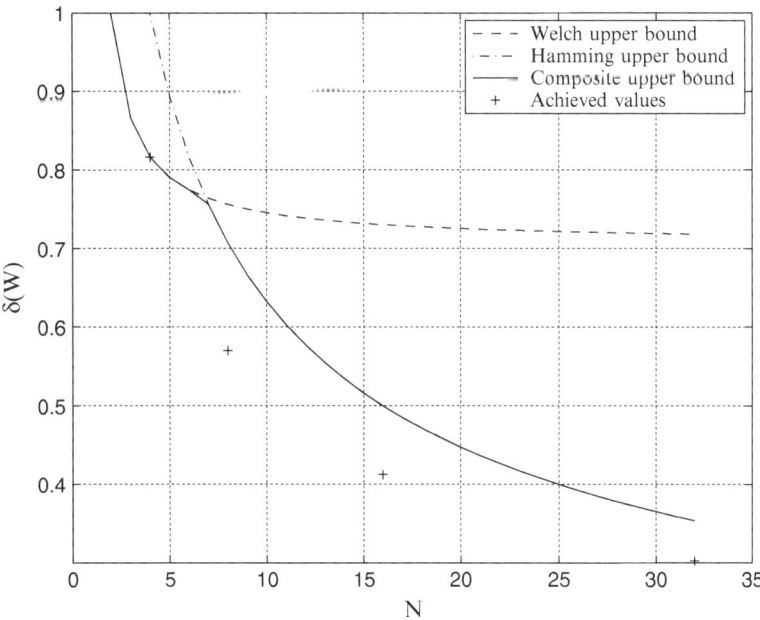

Fig. 4.4 Upper bounds and maximum achieved values of $\delta(\mathscr{W})$ for $N_t = 2$

The design of optimal Grassmannian line packings is in general a difficult problem except when $N \leq N_t$ which is not of our interest [16, 84, 86]. Several optimal and near optimal codebook constructions have been proposed in the literature such as difference sets [93] and alternating projection [89]. In order to reduce the storage requirements and search complexity, different methods have been proposed to design codebooks with finite alphabet entries [36, 63, 76]. For example in [76], quadrature amplitude modulation (QAM) and phase-shift keying (PSK) constellation codebooks have been proposed. Another solution is to build codebooks based on mutually unbiased bases (MUB) or using Kerdock codes [36, 63].

In the next section we will present the codebook construction based on the Fourier matrix, on MUB and on vector quantization.

4.2.2.3 Codebook Construction Based on the Fourier Matrix

The codebook design based on the Fourier matrix was first introduced in [33] in the context of noncoherent multiple antenna channel [97] and in [53] for equal gain codebook. Hochwald et al. have shown that unitary MIMO space-time constellations achieve the capacity of multiple antennas systems when the CSI is not available at the transmitter and receiver. It can be shown that the problem of designing unitary MIMO space-time codes is equivalent to the problem of the design of an optimal

Table 4.1 Example of codebook constructions based on the Fourier matrix

N_t	N	$\delta(\mathscr{W})$	Fourier matrix row indices
2	4	0.7071	$\{1, 2\}$
2	8	0.3826	$\{5, 8\}$
2	16	0.1952	$\{5, 14\}$
2	32	0.0980	$\{15, 32\}$
2	64	0.0490	$\{15, 48\}$
2	128	0.0245	$\{91, 112\}$
2	256	0.0123	$\{113, 224\}$
4	8	0.8660	$\{1, 4, 7, 8\}$
4	16	0.8134	$\{2, 3, 5, 14\}$
4	32	0.7071	$\{16, 21, 25, 28\}$
4	64	0.5141	$\{17, 20, 45, 61\}$
4	128	0.4989	$\{28, 73, 107, 116\}$
4	256	0.4088	$\{69, 127, 136, 140\}$

Grassmannian packing. They have proposed to build a codebook by selecting N_t rows from the Fourier Matrix. The codebook $\mathscr{W}$ is defined as follows:

$$\mathscr{W} = [\mathbf{w}_1, \mathbf{w}_2, \ldots, \mathbf{w}_N] = \frac{1}{\sqrt{N_t}} \begin{bmatrix} 1 & e^{j\frac{2\pi}{N}u_1} & \ldots & e^{j\frac{2\pi}{N}u_1(N-1)} \\ 1 & e^{j\frac{2\pi}{N}u_2} & \ldots & e^{j\frac{2\pi}{N}u_2(N-1)} \\ \vdots & \vdots & \vdots & \vdots \\ 1 & e^{j\frac{2\pi}{N}u_{N_t}} & \ldots & e^{j\frac{2\pi}{N}u_{N_t}(N-1)} \end{bmatrix} \tag{4.36}$$

where $u_m \in \{0, 1, \ldots, N-1\}$, $1 \leq m \leq N_t$ and $u_m \neq u_{m'} \ \forall m \neq m'$

The cross correlation between codeword $\mathbf{w}_i$ and $\mathbf{w}_{i'}$ is given by

$$|\mathbf{w}_i^H \mathbf{w}_{i'}| = \frac{1}{N_t} \left| \sum_{m=1}^{N_t} e^{j\frac{2\pi}{N}u_m(i-i')} \right| \tag{4.37}$$

We can now find the indexes $u_1, \ldots, u_{N_t}$ achieving the lowest possible correlations

$$\min_{0 \leq u_1, \ldots, u_{N_t} \leq N-1} \max_{i=1, \ldots, N} \frac{1}{N_t} \left| \sum_{m=1}^{N_t} e^{j\frac{2\pi}{N}u_m(i-1)} \right| \tag{4.38}$$

The correlation structure is circulant since the correlation between $\mathbf{w}_i$ and $\mathbf{w}_{i'}$ depends only on $(i - i') \mod N$. Consequently, we can perform the maximization only for $i = 1, \ldots, N$.

Based on this criterion, a list of the Fourier matrix row indices and the associated value $\delta(\mathscr{W})$ is given in Table 4.1 for $N_t = 2$ and 4.

4.2.2.4 Codebook Construction Based on Mutually Unbiased Bases

MUB are a useful tool for quantum information theory for the problem of quantum state determination but also for quantum error correction codes and cryptography [46]. Each MUB in the space $\mathbb{C}^{N_t}$ consists of N_t orthogonal unit vectors which can be thought of as a unitary $N_t \times N_t$ matrix. A MUB set is a set of two or more orthonormal bases such that the square magnitude of the inner product between the columns of different orthonormal bases is equal to the inverse of the basis dimension. Let $\mathbf{U} = [\mathbf{u}_1, \ldots, \mathbf{u}_{N_t}]$ and $\mathbf{V} = [\mathbf{v}_1, \ldots, \mathbf{v}_{N_t}]$ be the two orthonormal bases of a MUB, then we have

$$|\mathbf{u}_i^H \mathbf{v}_j|^2 = \frac{1}{N_t} \tag{4.39}$$

A codebook $\mathscr{W} = [\mathbf{w}_1, \ldots, \mathbf{w}_N]$ can be obtained from a set of MUBs by selecting the different columns of the orthonormal basis as the codewords $\mathbf{w}_i$. The chordal distance between two codewords $\mathbf{w}_i$ and $\mathbf{w}_j$ is

$$d^2(\mathbf{w}_i, \mathbf{w}_j) = \begin{cases} 1 & \text{if } \mathbf{w}_i \text{ and } \mathbf{w}_j \text{ are chosen from the same bases} \\ 1 - \frac{1}{N_t} & \text{if } \mathbf{w}_i \text{ and } \mathbf{w}_j \text{ are chosen from different bases} \end{cases} \tag{4.40}$$

We can compute the average chordal distance of the codebook as

$$d^2(\mathscr{W}) = \frac{1}{N(N-1)} \sum_{i=1}^{N} \sum_{j \neq i}^{N} d^2(\mathbf{w}_i, \mathbf{w}_j) = \frac{(N_t - 1)N}{N_t(N-1)} \tag{4.41}$$

Interestingly, this codebook construction is near optimal since it achieves the Welch bound given in Eq. (4.27) on average. The construction of a MUB set for a given dimension N_t is not straightforward. The maximal number of MUB N_m which can be constructed satisfies $N_m \leq N_t + 1$. When N_t is an integer power of a prime number, we have $N_m = N_t + 1$ but it is not known for arbitrary N_t. The standard form of the MUB set is given by $\{\mathbf{I}_{N_t} \mathbf{M}_1 \ldots \mathbf{M}_{N_t}\}$. The first basis is mapped to the identity of the space $\mathbb{C}^{N_t}$. From Eq. (4.39) it is seen that the remaining matrices $\mathbf{M}_1 \ldots \mathbf{M}_{N_t}$ must be complex Hadamard matrices.

For $N_t = 2$, the MUB set is composed of three orthonormal matrices which can be obtained starting from the Hadamard matrix of dimension two:

$$\mathbf{M}_0 = \mathbf{I}_2 \quad \mathbf{M}_1 = \frac{1}{\sqrt{2}} \begin{bmatrix} 1 & 1 \\ 1 & -1 \end{bmatrix} \quad \mathbf{M}_2 = \sqrt{2} \begin{bmatrix} 1 & 1 \\ +j & -j \end{bmatrix}$$

The identity matrix $\mathbf{I}_{N_t}$ corresponds to the antenna selection cases where only one antenna is selected among N_t antennas.

For $N_t = 3$, the MUB set is composed of four orthonormal matrices. There is only one de-phased complex Hadamard matrix given by the discrete Fourier matrix

$$\mathbf{M}_1 = \frac{1}{\sqrt{3}} \begin{bmatrix} 1 & 1 & 1 \\ 1 & e^{2j\pi/3} & e^{4j\pi/3} \\ 1 & e^{4j\pi/3} & e^{2j\pi/3} \end{bmatrix}$$

A search of the vectors which are mutually unbiased with respect to $\mathbf{M}_1$ leads to the orthonormal matrix $\mathbf{M}_2 = \mathbf{D}\mathbf{M}_1$ and $\mathbf{M}_3 = \mathbf{D}^2\mathbf{M}_1$ where $\mathbf{D} = \mathrm{diag}(1, e^{2j\pi/3}, e^{2j\pi/3})$ is a diagonal unitary matrix. The 4 orthonormal matrix are given by

$$\mathbf{M}_0 = \mathbf{I}_3 \quad \mathbf{M}_1 = \sqrt{3} \begin{bmatrix} 1 & 1 & 1 \\ 1 & e^{2j\pi/3} & e^{4j\pi/3} \\ 1 & e^{4j\pi/3} & e^{2j\pi/3} \end{bmatrix} \tag{4.42}$$

$$\mathbf{M}_2 = \sqrt{3} \begin{bmatrix} 1 & 1 & 1 \\ e^{2j\pi/3} & e^{4j\pi/3} & 1 \\ e^{2j\pi/3} & 1 & e^{4j\pi/3} \end{bmatrix} \quad \mathbf{M}_3 = \sqrt{3} \begin{bmatrix} 1 & 1 & 1 \\ e^{4j\pi/3} & 1 & e^{2j\pi/3} \\ e^{4j\pi/3} & e^{2j\pi/3} & 1 \end{bmatrix} \tag{4.43}$$

For $N_t = 4$, a MUB set composed of five orthonormal matrices can be used [8,46]:

$$\mathbf{M}_0 = \mathbf{I}_4 \quad \mathbf{M}_1 = \frac{1}{2} \begin{bmatrix} 1 & 1 & 1 & 1 \\ 1 & 1 & -1 & -1 \\ 1 & -1 & -1 & 1 \\ 1 & -1 & 1 & -1 \end{bmatrix} \quad \mathbf{M}_2 = \frac{1}{2} \begin{bmatrix} 1 & 1 & 1 & 1 \\ -1 & -1 & 1 & 1 \\ -j & j & j & -j \\ -j & j & -j & j \end{bmatrix}$$

$$\mathbf{M}_3 = \frac{1}{2} \begin{bmatrix} 1 & 1 & 1 & 1 \\ -j & -j & j & j \\ -j & j & j & -j \\ -1 & 1 & -1 & 1 \end{bmatrix} \quad \mathbf{M}_4 = \frac{1}{2} \begin{bmatrix} 1 & 1 & 1 & 1 \\ -j & -j & j & j \\ -1 & 1 & -1 & 1 \\ -j & j & j & -j \end{bmatrix}$$

The same set can be obtained using $\mathbb{Z}^4$ Kerdock codes for $N_t = 4$. Those codes have been proposed by Kerdock [44] for low-rate error correction and for limited feedback precoded MIMO systems by Inoue and Heath [36]. Interestingly, the codebooks based on MUB have the property that the elements belong to the finite alphabet $\{-1 + 1 - j + j\}$ for $N_t = 2$ and 4 and to $\{1, e^{2j\pi/3}, e^{4j\pi/3}\}$ for $N_t = 3$.

4.2.2.5 Codebook Construction Based on Vector Quantization

The quantization of the CDI has been introduced by Narula et al. in [67] where vector quantization (VQ) techniques were proposed. VQ is a lossy data compression method. The VQ maps a real- or complex-valued vector called source vector into one of a finite number of vector realizations, known as codebook. A pure VQ technique will attempt to obtain a good approximation of the channel realization. However, since the aim is to maximize the capacity, the codebook design based on VQ should minimize the average distortion between the original vector and the quantized one with respect to the source distribution. For the MISO case, the codebook design can then be formally stated as

$$\min_{\mathcal{W}} \mathbb{E}_{\mathbf{h}} \left[1 - |\mathbf{g}^H \mathbf{w}_{opt}|^2 \right] \tag{4.44}$$

$$\text{subject to } |\mathbf{w}_i^H \mathbf{w}_i| = 1 \quad \forall i \tag{4.45}$$

and the average SNR degradation can be written as

$$\frac{\mathbb{E} \left[\|\mathbf{h}\|^2 - |\mathbf{h}^H \mathbf{w}_{opt}|^2 \right]}{\mathbb{E} \left[\|\mathbf{h}\|^2 \right]} = \mathbb{E}_{\mathbf{h}} \left[1 - |\mathbf{g}^H \mathbf{w}_{opt}|^2 \right] \tag{4.46}$$

which can be viewed as a VQ problem with

$$\text{source input}: \quad \mathbf{g} \sim \text{uniform}(\mathbb{O}_{N_t})$$

$$\text{codebook}: \quad \mathcal{W} = [\mathbf{w}_1 \mathbf{w}_2 \dots \mathbf{w}_N], \|\mathbf{w}_i\| = 1, \forall i \tag{4.47}$$

$$\text{distortion metric}: \quad d(\mathbf{g}, \mathbf{w}_i) = \sqrt{1 - |\mathbf{g}^H \mathbf{w}_i|^2}, \forall \mathbf{g}, \forall i \tag{4.48}$$

The source vector, assuming i.i.d channel vector $\mathbf{h}$, is the (N_t)-dimensional normalized channel vector $\mathbf{g}$ uniformly distributed on $\mathbb{O}_{N_t}$.

The problem formulation in Eq. (4.48) is known as a spherical vector quantization (SVQ) problem [94]. Unique to this SVQ problem is the distortion metric, which is the root square of a projective distance from the source input vector to the codeword vector. Compared to VQ in source coding literature where the distortion function adopted for codebook designs is Euclidian, for the quantization of the CDI, the distortion function is non-Euclidean.

The design of such codebooks has been considered in several research studies which lead to some algorithms such as the Lloyd algorithm. The Lloyd algorithm was suggested for the design of the beamforming vector codebook in [67]. The authors in [54, 65, 74] have shown that the codebook should be constructed by minimizing the maximum inner product between any two beamforming vectors in the codebook.

Let S_n be the encoding region associated with codeword $\mathbf{w}_n$ and the partition of the space $\mathscr{P} = \{S_1, \ldots, S_N\}$.

The codebook $\mathscr{W}$ and the partition of the space $\mathscr{P}$ must satisfy two necessary conditions [23]. Firstly, the so-called centroid condition requires that for each region, the optimal codeword should be chosen to minimize the distortion measure averaged over that region. Secondly, necessary to the optimality of the channel space partition is the nearest neighbor rule which dictates that all input vectors closer to the codeword $\mathbf{w}_i$ than to any other codeword should be assigned to region S_i.

$$S_i = \{\mathbf{g} | d(\mathbf{g}, \mathbf{w}_i) \le d(\mathbf{g}, \mathbf{w}_j), \quad \forall j \ne i\} \tag{4.49}$$

The Lloyd algorithm is described as follows:

1. *Step1: initialization phase*

   - Generate a training sequence consisting of source vectors $\mathbf{h}$ of size $N_t \times 1$. The coefficients h_j are i.i.d. complex Gaussian random variables with zero mean and unit variance.

2. *Step2: nearest neighbor rule*

   - All input vectors $\mathbf{h}$ closer to the codeword $\mathbf{w}_i$ than to any other codeword, be assigned to the neighborhood of $\mathbf{w}_i$ or region S_i

     - $\mathbf{g} \in S_i$ if and only if $d(\mathbf{g}, \mathbf{w}_i) \le d(\mathbf{g}, \mathbf{w}_j), \quad \forall j \ne i$.

3. *Step3: centroid condition*

   - Take the ith region S_i, for example, whose local correlation matrix $\Sigma_i := \mathbb{E}\left[\mathbf{h}\mathbf{h}^H | \mathbf{h} \in S_i\right]$. To avoid the expectation operation, we compute Σ_i using the Monte Carlo approach as follows:

$$\Sigma_i = \frac{1}{|S_i|} \sum_{\mathbf{h}_n \in S_i} \mathbf{h}_n \mathbf{h}_n^H \tag{4.50}$$

According to the centroid condition, the optimal beamforming vector $\mathbf{w}_i^{opt}$ should maximize $\mathbf{w}_i^H \Sigma_i \mathbf{w}_i$ subject to the unit norm constraint:

$$\mathbf{w}_i^{opt} = \arg \max_{|\mathbf{w}_i^H \mathbf{w}_i| = 1} \mathbf{w}_i^H \Sigma_i \mathbf{w}_i$$

$$= \mathbf{u}_i \tag{4.51}$$

where $\mathbf{u}_i$ is the left eigenvector corresponding to the largest eigenvalue of Σ_i. This operation ensures the construction of the codebook which maximizes the SNR at the receiver.

Table 4.2 Comparison of different codebooks minimum chordal distances

N_t	N	Combined upper bound	Lloyd search	Fourier based
2	4	0.8165	0.8165	0.7071
2	8	0.7071	0.5701	0.3826
2	16	0.4851	0.4122	0.1952
2	32	0.3536	0.3090	0.0980
2	64	0.25	0.2013	0.0490
2	128	0.1768	0.1368	0.0245
2	256	0.125	0.0891	0.0123
4	8	0.9258	0.9043	0.8660
4	16	0.8944	0.8778	0.8134
4	32	0.8799	0.7346	0.7071
4	64	0.8729	0.6454	0.5141
4	128	0.8694	0.5694	0.4989
4	256	0.7937	0.4782	0.4088

4. *Loop back to Step2 until convergence*

A comparison of the minimum chordal distances of different codebooks is given in Table 4.2. For the Lloyd search, we have performed the Lloyd algorithm 20 times with different initialization and the best codebook has been selected. We can see that in terms of minimum chordal distance, the codebooks based on VQ outperform the Fourier-based codebooks.

The Lloyd algorithm can also be applied for the MIMO with receive combining case by replacing the average distortion function $\mathbb{E}_{\mathbf{h}}\left[1 - |\mathbf{g}^H \mathbf{w}_{opt}|^2\right]$ by $\mathbb{E}_{\mathbf{H}}\left[\lambda_1 - ||\mathbf{H}\mathbf{w}_{opt}||^2\right]$ given in Eq. (4.13).

Other distortion functions can be used for the codebook construction. In [37], the authors have designed a codebook in order to minimize the symbol error probability for rectangular M-QAM using the Lloyd algorithm. A codebook design for spatially correlated channels has been proposed in [98].

4.2.2.6 Data Rate Derivation

We define the random variable Z as follows:

$$Z = \min_i d^2(\mathbf{g}, \mathbf{w}_i)$$
$$= \min_i(1 - |\mathbf{g}^H \mathbf{w}_i|^2) \qquad (4.52)$$

Z is a random variable within the interval $[0, 1]$. Denote $p(z)$ and $F(z)$ as the probability density function (pdf) and the cumulated distribution function (cdf) of Z.

Fig. 4.5 Spherical cap on the surface of unitary radius hypersphere

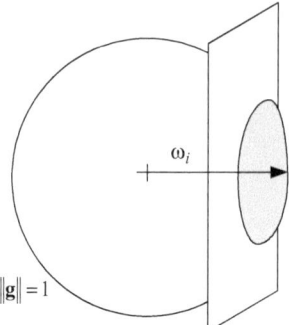

We first compute the cdf of Z.

For $1 \leq i \leq N$, we consider the regions $\mathcal{R}_z(\mathbf{w}_i)$ with center $\mathbf{w}_i$ and square radius z given by

$$\mathcal{R}_z(\mathbf{w}_i) = \{\mathbf{g} \in \mathbb{O}^{N_t} : d^2(\mathbf{g}, \mathbf{w}_i) \leq z\} \tag{4.53}$$

where $\mathbb{O}^{N_t}$ be the unit sphere lying in $\mathbb{C}^{N_t}$ and centered at the origin.

These regions are spherical cap on the surface of the unitary radius hypersphere as shown on Fig. 4.5. The surface areas $A(\mathcal{R}_z(\mathbf{w}_i))$ and A_1 are described by Eqs. (4.21) and (4.23) respectively. The cdf of Z is given by

$$\begin{aligned}
F(z) &= Pr\,(Z \leq z) \\
&= Pr\,(\min_i d^2(\mathbf{g}, \mathbf{w}_i) \leq z) \\
&= Pr\,(d^2(\mathbf{g}, \mathbf{w}_1) \leq z \text{ or } d^2(\mathbf{g}, \mathbf{w}_2) \leq z \ldots \text{ or } d^2(\mathbf{g}, \mathbf{w}_N) \leq z) \\
&= \frac{A\left(\bigcup_{i=1}^{N} \mathcal{R}_z(\mathbf{w}_i)\right)}{A_1}
\end{aligned} \tag{4.54}$$

Assuming that the surfaces are not overlapping, we have

$$\begin{aligned}
A\left(\bigcup_{i=1}^{N} \mathcal{R}_z(\mathbf{w}_i)\right) &= \sum_{i=1}^{N} A(\mathcal{R}_z(\mathbf{w}_i)) \\
&= N\left(\frac{2\pi^{N_t} z^{N_t-1}}{(N_t - 1)!}\right)
\end{aligned} \tag{4.55}$$

Consequently, we can introduce an upper bound on $F(z)$ for the quantized feedback scheme assuming that the regions associated to each codeword do not overlap [65, 101] given by

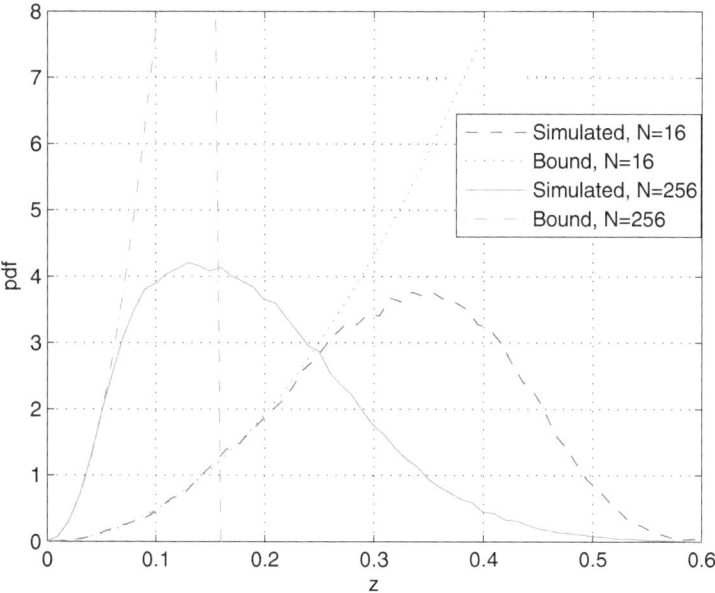

Fig. 4.6 Probability density function $p(z)$ and $\tilde{p}(z)$ for $N_t = 4$ and $N = 16, 256$ for the quantized feedback

$$F(z) \leq \tilde{F}(z) \begin{cases} Nz^{N_t-1} & \text{if} \quad 0 \leq z < N^{-\frac{1}{N_t-1}} \\ 1 & \text{if} \quad z \geq N^{-\frac{1}{N_t-1}} \end{cases} \tag{4.56}$$

Then, the associated power density function (pdf) $\tilde{p}(z)$ is as follows:

$$\tilde{p}(z) = \begin{cases} N(N_t-1)z^{N_t-2} & \text{if} \quad 0 \leq z < N^{-\frac{1}{N_t-1}} \\ 0 & \text{if} \quad z \geq N^{-\frac{1}{N_t-1}} \end{cases} \tag{4.57}$$

This pdf is given in Fig. 4.6 for $N_t = 4$ and $N = 16, 256$. In this figure, we also present the pdf obtained using the best packing codebook given in [52] for $N = 16$ and a codebook based on fast Fourier transform matrices for $N = 256$ using a randomly incomplete search [33]. We can observe that there is a noticeable difference between the bound and the simulated pdf.

Like previously, we can compute the expectation of $\tilde{Z}$:

$$E[\tilde{Z}] = \frac{N_t - 1}{N_t} 2^{-\frac{B}{N_t-1}} \tag{4.58}$$

From these results, we can obtain an approximation of the achievable data rate of the beamforming scheme. For point-to-point MISO single user, the achievable data rate is approximated as follows:

$$R_{quantCSIT} = \mathbb{E}_{\mathbf{h}}\left[B_w \log_2\left(1 + \frac{P}{N_0 B_w}||\mathbf{h}||^2|\mathbf{g}^H \mathbf{w}_{opt}|^2\right)\right]$$

$$= \mathbb{E}_{\mathbf{h}}\left[B_w \log_2\left(1 + \frac{P}{N_0 B_w}||\mathbf{h}||^2(1 - d^2(\mathbf{g}, \mathbf{w}_{opt}))\right)\right]$$

$$\approx \mathbb{E}_{\mathbf{h}}\left[B_w \log_2\left(1 + \frac{P}{N_0 B_w}||\mathbf{h}||^2(1 - \frac{N_t - 1}{N_t}2^{-\frac{B}{N_t-1}})\right)\right]$$

In the literature, different criteria have been considered for the performance evaluation such as ergodic capacity, outage probability or SNR. Some authors have derived a relation between the number of required bits to quantized the precoding vector $\mathbf{w}$ and the degradation compared to the full CSIT case when random vector quantization (RVQ) is used.

4.2.2.7   Random Vector Quantization

RVQ has been used to analyze the achievable average rate of point-to-point MIMO channels in [81] since the optimal vector quantizer is not known in general.

In RVQ, the $N = 2^B$ quantization vectors are independently chosen from the isotropic distribution on the N_t-dimensional unit sphere [2]. It has been shown that RVQ is very useful for performance analysis and performs close to optimal quantization when $B \to \infty$.

In [2], the authors have computed the cdf of $\sin^2(\angle(\mathbf{g}, \mathbf{w}_{opt})$ as follows:

$$Pr\left(\sin^2(\angle(\mathbf{g}, \mathbf{w}_{opt}) < z)\right) = (1 - z^{N_t-1})2^B \qquad (4.59)$$

when using a random 2^B unit vector beamforming codebook. From this result, we can compute the expectation of this quantity over the channel realization $\mathbf{h}$ as follows [39]:

$$\mathbb{E}_{\mathbf{h},\mathscr{W}}[d^2(\mathbf{g}, \mathbf{w}_{opt})] = \mathbb{E}_{\mathbf{h},\mathscr{W}}[\sin^2(\angle(\mathbf{g}, \mathbf{w}_{opt}))]$$

$$= \int_0^1 (1 - z^{N_t-1})2^B dz$$

$$= \frac{1}{N_t - 1}\beta(2^B + 1, \frac{1}{N_t - 1})$$

$$= 2^B.\beta(2^B, \frac{N_t}{N_t - 1}) \qquad (4.60)$$

where $\beta(.,.)$ denotes the beta function which is defined using the gamma function as $\beta(x, y) = \frac{\Gamma(x)\Gamma(y)}{\Gamma(x+y)}$

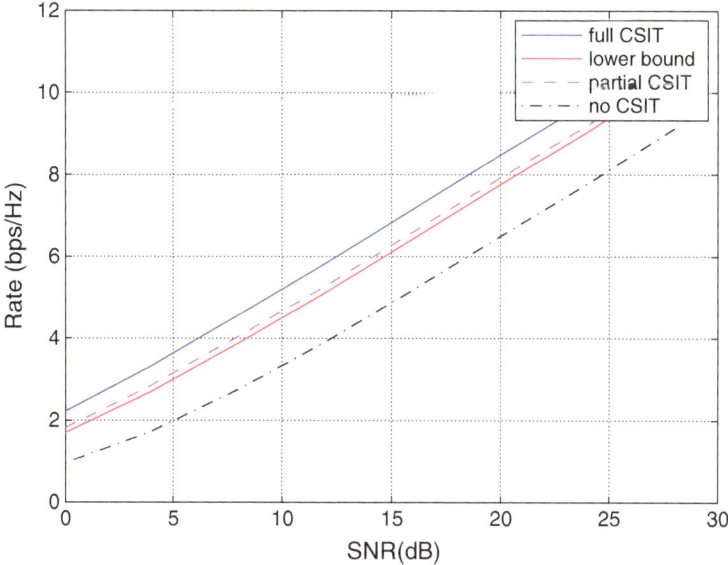

Fig. 4.7 Rate comparison for point-to-point MISO single user communication for $N_t = 4$ and $N = 16$ for the quantized feedback

This expectation can be upper bounded as [2, 39]:

$$\mathbb{E}_{\mathbf{h},\mathscr{W}}[d^2(\mathbf{g}, \mathbf{w}_{opt})] < 2^{-\frac{B}{N_t-1}} \tag{4.61}$$

Consequently, we can evaluate the point-to-point MISO single user achievable rate with RVQ as follows:

$$
\begin{aligned}
R_{quantCSIT} &= \mathbb{E}_{\mathbf{h}}\left[B_w \log_2\left(1 + \frac{P}{N_0 B_w}||\mathbf{h}||^2(1 - d^2(\mathbf{g}, \mathbf{w}_{opt}))\right)\right] \\
&\approx \mathbb{E}_{\mathbf{h}}\left[B_w \log_2\left(1 + \frac{P}{N_0 B_w}||\mathbf{h}||^2(1 - 2^{-\frac{B}{N_t-1}})\right)\right]
\end{aligned}
$$

Compared to the full CSIT capacity and at high SNR, the SNR degradation can be approximated as

$$\Delta R_{quant} = 10 \log_{10}(1 - 2^{-\frac{B}{N_t-1}}) \quad \text{dB.} \tag{4.62}$$

In Fig 4.7, we compare the rate for $N_t = 4$ and $N = 16$. In this figure, we present the capacity of full CSIT, no CSIT, the lower bound obtained using RVQ and the partial rate computed from $E[\tilde{Z}]$. For this configuration, we have an SNR degradation of 2.2 dB compared to the full CSIT case.

4.2.3   Analog Feedback

Another solution is to directly transmit the unquantized precoding vector or the unquantized channel vector. It can be shown that this solution achieves the minimum mean square error (MMSE) distortion in the Shannon sense. Analog feedback has been described in [21] and studied for multiuser wireless systems in [60]. In this section we consider two different analog feedback links. Since we focus on the impact of the feedback link on the performance, we will consider that the channel vector $\mathbf{h}$ has been perfectly estimated by the receiver.

Since analog feedback uses analog modulation, we avoid the problem associated with the quantization of the channel. However, compared to digital modulation, the dynamic range is much larger. In this section, we will not consider the possible RF transmitter problems related to this large dynamic range and we will focus on two different analog feedback schemes:

- Analog transmission of the normalized channel vector $\mathbf{g}$
- Analog transmission of the channel vector $\mathbf{h}$

We will assume that the receiver has performed a perfect estimation of the channel vector $\mathbf{h}$.

When the normalized channel vector $\mathbf{g}$ is transmitted over the noisy feedback channel, the received vector is given by

$$\mathbf{w}' = \mathbf{g} + \boldsymbol{\epsilon} \tag{4.63}$$

After normalization we have

$$\mathbf{w} = \frac{\mathbf{w}'}{||\mathbf{w}'||} \tag{4.64}$$

Each element of the noisy vector $\boldsymbol{\epsilon}$ is a zero-mean complex Gaussian noise with variance $\sigma_{UL}^2 = \frac{(N_0)_{UL}}{2}$.

In the second version where the channel vector $\mathbf{h}$ is transmitted over the noisy feedback channel, the received vector is given by

$$\mathbf{w}' = \mathbf{h} + \boldsymbol{\epsilon} \tag{4.65}$$

After normalization, we have

$$\mathbf{w} = \frac{\mathbf{h} + \boldsymbol{\epsilon}}{||\mathbf{h} + \boldsymbol{\epsilon}||} \tag{4.66}$$

In both cases, the precoding vector is applied to the transmitted vector. The received signal y is given by

$$y = \mathbf{h}^H \mathbf{w} s + n \tag{4.67}$$

4.2.4 *Performance Analysis*

4.2.4.1 SER Approximation for Quantized Feedback

The instantaneous SNR can also be written as

$$\gamma = \frac{P}{N_0 B_w} ||\mathbf{h}||^2 (1 - Z) \tag{4.68}$$

where Z is the random variable introduced previously.

We will now evaluate the symbol error rate (SER) from $p(z)$.

For the Rayleigh fading channel, the SER can be evaluated using Craig's formula [83]:

$$SER(\gamma) = \frac{1}{\pi} \int_0^{\frac{(M-1)\pi}{M}} \exp\left(-\frac{g_{PSK}\gamma}{\sin^2\theta}\right) d\theta \tag{4.69}$$

where M is the constellation size and $g_{PSK} = \sin^2(\pi/M)$ is the constellation dependent term.

The average SER can be calculated by averaging over all the possible instantaneous SNR γ [101]:

$$\overline{SER} = \int_{\gamma=0}^{\infty} SER(\gamma) p(\gamma) d\gamma$$

$$= \frac{1}{\pi} \int_0^{\frac{(M-1)\pi}{M}} \int_0^1 \left(1 + \frac{g_{PSK}(1-z)\frac{E_s}{N_0}}{\sin^2\theta}\right)^{-N_t} p(z) dz d\theta \tag{4.70}$$

We can obtain a lower bound on the average SER for quantized feedback channel by replacing $p(z)$ by Eq. (4.57) in Eq. (4.70) [101]. We have

$$\overline{SER}_{lb} = \frac{1}{\pi} \int_0^{\frac{(M-1)\pi}{M}} A(\theta) d\theta \tag{4.71}$$

with

$$A(\theta) = \left(1 + \frac{g_{PSK}\bar{\gamma}}{\sin^2\theta}\right)^{-1} \left[1 + \left[1 - \left(\frac{1}{N}\right)^{\frac{1}{N_t-1}}\right] \frac{g_{PSK}\bar{\gamma}}{\sin^2\theta}\right]^{1-N_t} \tag{4.72}$$

While the pdf is quite different we can see in Fig. 4.11 that for $N = 16$ the lower bound is very close to the average simulated SER obtained using the codebook presented previously (0.2 dB of difference only).

When the feedback channel is noisy, the decoded codeword index can be different from the original one. We introduce a random variable Y taking into account the possible transmission errors:

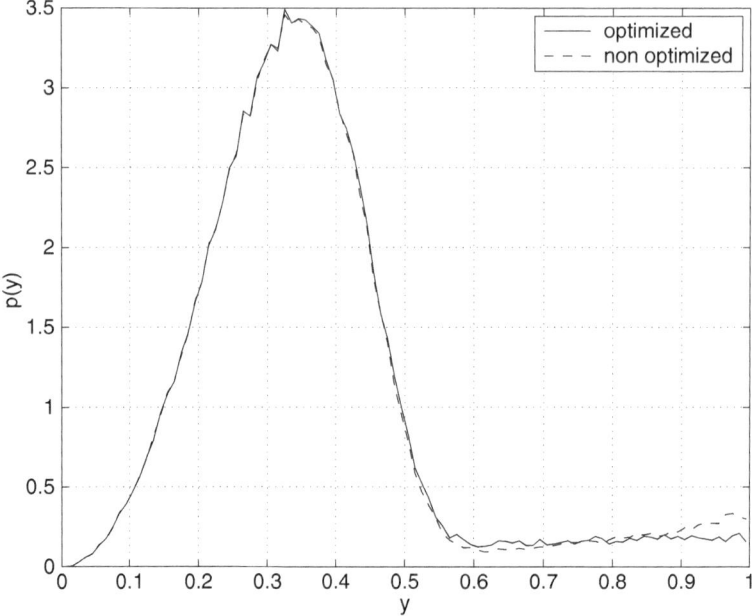

Fig. 4.8 Probability density function $p(y)$ for the quantized feedback

$$Y = 1 - |\mathbf{g}^H \mathbf{w}_i|^2 \tag{4.73}$$

where

$$\begin{cases} j = \arg\min_i (1 - |\mathbf{g}^H \mathbf{w}_i|^2) & \text{if} \quad \text{no transmission error} \\ j \neq \arg\min_i (1 - |\mathbf{g}^H \mathbf{w}_i|^2) & \text{if} \quad \text{transmission error} \end{cases} \tag{4.74}$$

The pdf of Y is obtained by the association of two pdf $p_C(y)$ and $p_E(y)$ corresponding, respectively, to the no transmission/transmission error case. $p_C(y)$ is the same pdf than the previous pdf $p(z)$. We have

$$p(y) = (1 - CER)p_C(y) + CERp_E(y) \tag{4.75}$$

The codeword error rate (CER) depends on the chosen modulation.

In Fig. 4.8, we give the pdf $p(y)$ obtained for $N_t = 4$, $N = 16$ codewords and $(E_s/N_0)_{UL} = 4\,\text{dB}$ using Monte Carlo simulations. The 4 bits have been encoded using 4 BPSK symbols. We can observe the impact of the noise on this pdf. As shown in Fig. 4.8, by optimizing the mapping between the codewords and the transmitted symbols, it is possible to slightly improve the performance.

The pdf $p_E(y)$ is difficult to evaluate analytically. However, for a given normalized channel vector $\mathbf{g}$, asymptotically the received vectors $\mathbf{w}_i$ are uniformly distributed over the unitary radius hypersphere. Consequently, we can find a SER lower bound by approximating $p_E(z)$ as a uniform pdf.

From Eq. (4.70), the SER is expressed as follows:

$$
\overline{SER} = \frac{1}{\pi} \int_0^{\frac{(M-1)\pi}{M}} \left[\int_0^1 \left(1 + \frac{g_{PSK}(1-y)\frac{P}{N_0 B_w}}{\sin^2 \theta} \right)^{-N_t} \right.
$$

$$
\times (1 - CER) p_C(y) dy
$$

$$
\left. + \int_0^1 \left(1 + \frac{g_{PSK}(1-y)\frac{P}{N_0 B_w}}{\sin^2 \theta} \right)^{-N_t} CER p_E(y) dy \right] d\theta \qquad (4.76)
$$

From this result and using the previous lower bound we obtain the modified SER lower bound as follows:

$$
\overline{SER}_{lb} = \frac{1}{\pi} \int_0^{\frac{(M-1)\pi}{M}} \left[A(\theta)(1 - CER) + \frac{1}{N_t - 1} CER \right] d\theta \qquad (4.77)
$$

4.2.4.2 SER Approximation for Analog Feedback

Let's first assume that the normalized channel vector $\mathbf{g}$ is transmitted over the noisy feedback channel. As previously, Z is a random variable within the interval $[0, 1]$. We have the following relation [49]:

$$
Z = 1 - |\mathbf{g}^H \mathbf{w}|^2
$$

$$
= 1 - \left| \mathbf{g}^H \frac{\mathbf{g} + \epsilon}{||\mathbf{g} + \epsilon||} \right|^2
$$

$$
= 1 - \left| \frac{1 + \mathbf{g}^H \epsilon}{||\mathbf{g} + \epsilon||} \right|^2
$$

$$
= 1 - \frac{1 + 2\Re(\mathbf{g}^H \epsilon) + |\mathbf{g}^H \epsilon|^2}{1 + 2\Re(\mathbf{g}^H \epsilon) + ||\epsilon||^2} \qquad (4.78)
$$

where $\Re(a)$ is the real part of a.

In the second version where the channel vector $\mathbf{h}$ is transmitted over the noisy feedback channel, we have the following relation:

$$
Z = 1 - |\mathbf{g}^H \mathbf{w}|^2
$$

$$
= 1 - \left| \mathbf{g}^H \frac{\mathbf{h} + \epsilon}{||\mathbf{h} + \epsilon||} \right|^2 \qquad (4.79)
$$

$$
= 1 - \frac{||\mathbf{h}||^2 + 2||\mathbf{h}|| \Re(\mathbf{g}^H \epsilon) + |\mathbf{g}^H \epsilon|^2}{||\mathbf{h}||^2 + 2||\mathbf{h}|| \Re(\mathbf{g}^H \epsilon) + ||\epsilon||^2} \qquad (4.80)
$$

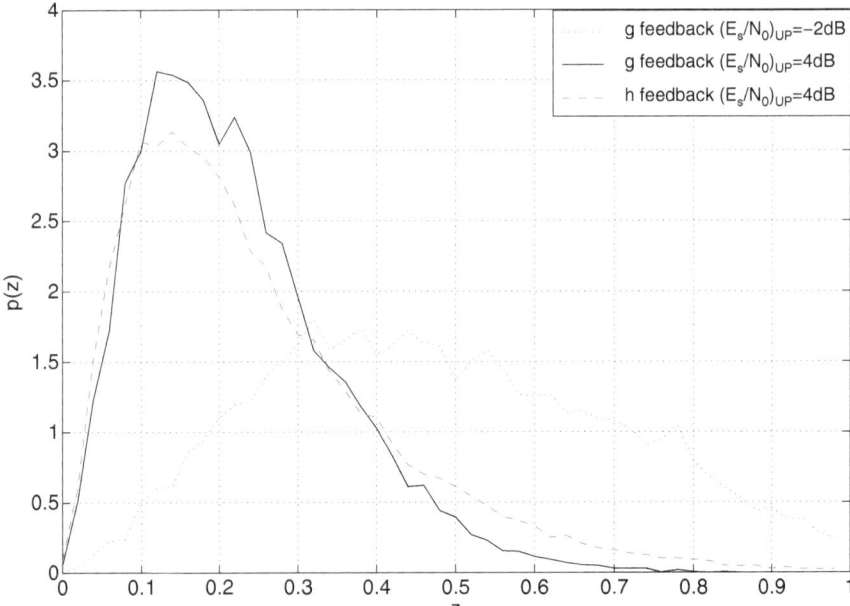

Fig. 4.9 Probability density function $p(z)$ for the analog feedback

Since it is difficult to compare the expressions (4.78) and (4.80), we will evaluate the pdf of Z using Monte Carlo simulations.

The pdf $p(z)$ are given in Fig. 4.9 with $N_t = 4$ for the analog feedback of **h** and $(E_s/N_0)_{UL} = 4\,\text{dB}$ ($E_s = 1$, $\sigma_{UL}^2 = 0.2$) and for the analog feedback of **w** and $(E_s/N_0)_{UL} = 4\,\text{dB}$ ($E_s = 1/N_t$, $\sigma_{UL}^2 = 0.05$) and $(E_s/N_0)_{UL} = -2\,\text{dB}$ ($\sigma_{UL}^2 = 0.2$).

From these results and using Eq. (4.70), we can predict that an analog feedback of the normalized channel vector **g** gives better performances than the analog feedback of the channel vector **h** since the tail of $p(z)$ is longer for the feedback of **h**. While the feedback of **h** achieves the same MMSE distortion as a scheme that optimally quantizes and encodes the CSI in the Shannon sense [21], this scheme is not optimal from the overall performance point of view.

This result can be explained since the constraint imposed to the norm of **g** allows us to eliminate some energy of the uplink noise compared to the other scheme. Consequently Z is on average smaller using the normalized channel vector feedback.

4.2.4.3 Simulation Results

In this section we consider the different feedback schemes for $N_t = 4$ and i.i.d. channels.

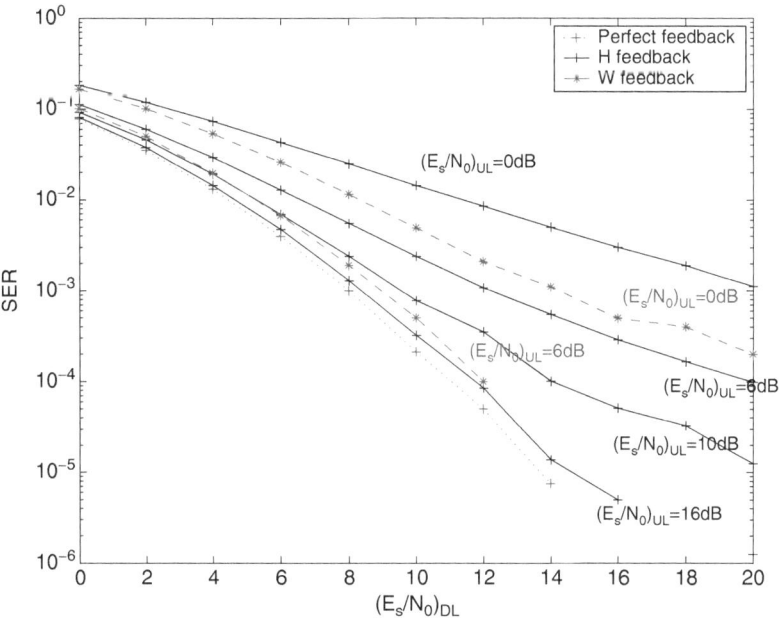

Fig. 4.10 $SER = f(E_S/N_0)_{DL}$ for the analog feedback schemes

In order to fairly compare all the different schemes, we have fixed the same average transmit energy per symbol. While for the **g** feedback scheme, the energy corresponding to the transmission of the vector **g** is a constant (since $\|\mathbf{g}\| = 1$), at the symbol level, the transmitted signal is Gaussian with the same variance as for the **h** feedback scheme.

In Fig. 4.10, we give the performance SER versus $(E_S/N_0)_{DL}$ of the two analog feedback schemes for different $(E_S/N_0)_{UL}$. As shown previously, the **g** feedback scheme outperforms the **h** feedback scheme.

In Fig. 4.11, we show the performance of the quantized feedback scheme for different $(E_S/N_0)_{UL}$. The number of codewords is $N = 16$ and the modulation is BPSK (four symbols are needed to transmit one codeword index). Compared to the analog feedback schemes, we can observe a floor effect on the SER depending on the uplink noise power. We also give the curves obtained using the lower bound Eq. (4.77) for different $(E_S/N_0)_{UL}$. We can see from the curves that the lower bounds are quite tight.

Of course, it is possible to decrease the floor effect by adding an error correcting code (for example using 4 QPSK symbols and a Reed-Muller (8,4) code to transmit one codeword index). Other signaling designs to compensate the errors have been proposed in [50].

In Fig. 4.12, we compare the performance of the analog and partial feedback schemes for $f(E_S/N_0)_{DL} = 6$ dB. For all the schemes, four symbols are needed to transmit one codeword index. As shown previously, the **g** feedback scheme outperforms the other feedback schemes. While the quantized precoding scheme

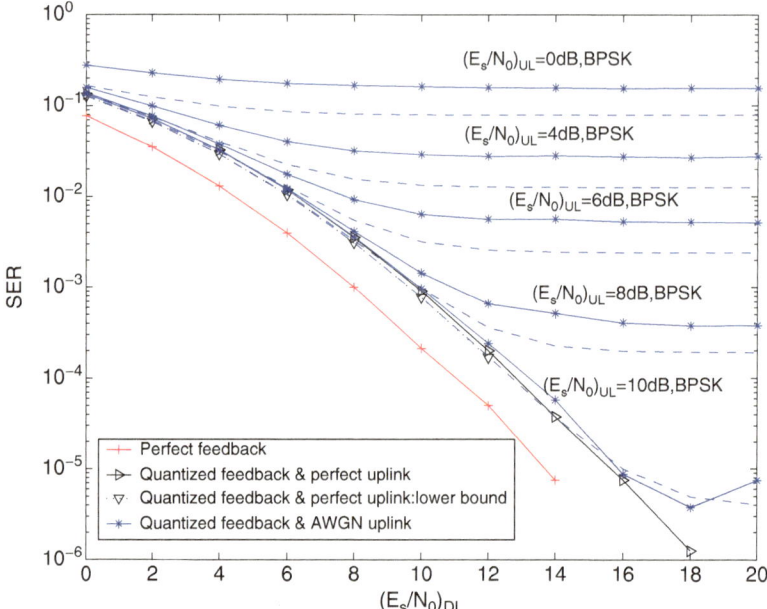

Fig. 4.11 $SER = f(E_S/N_0)_{DL}$ for the quantized feedback schemes

with $N = 256$ gives better performance than with $N = 16$, since the quantization error is lower, at low $f(E_S/N_0)_{UL}$ this scheme is more sensitive to transmission errors.

4.3   Linear Precoding for Spatial Multiplexing

We have seen in Chap. 2 that spatial multiplexing is a practical space-time modulation technique to exploit the high capacity and high quality of MIMO wireless channels. However, spatial multiplexing is sensitive to rank deficiencies in the MIMO channel. If some channel knowledge is available at the transmitter, linear precoding can be used to improve significantly the performance of the system. Linear precoding extends beamforming precoding by premultiplying the different transmitted data streams with a precoding matrix chosen based on the eigenstructure of the MIMO channel matrix. Linear precoding has been studied assuming the knowledge of the full CSIT in [78, 82] .

Perfect CSIT is not always available for example when the channel varies too rapidly. By exploiting channel statistics, such as the first-order statistics (channel mean value) [90, 99] or the second-order statistics of the channel (channel covariance matrix) [79, 90, 100], it is still possible to significantly improve the

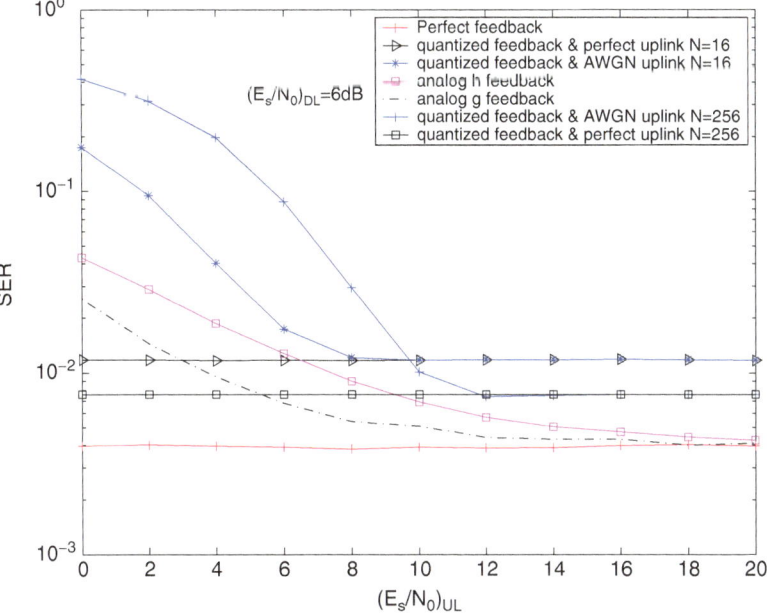

Fig. 4.12 Comparison of all feedback schemes with $SER = f(E_S/N_0)_{UL}$

performance of the communication scheme. Indeed, compared to the instantaneous channel response, the variation of the second-order channel statistics is slow and can be more easily fed back to the transmitter through a low data rate feedback link. When some channel statistical information is known at the transmitter we can design the precoder using capacity-based or error probability-based criteria. As shown in Chap. 2, from Eq. (2.94), the covariance matrix of the transmitted signal **x** must be determined in order to maximize the ergodic capacity. The problem is stated as follows:

$$\max_{\mathbf{Q}:\mathrm{tr}(\mathbf{Q})=P} \log_2\left(\det\left(\mathbf{I} + \frac{1}{N_0 B_W}\mathbf{HQH}^H\right)\right) \tag{4.81}$$

where $\mathbf{Q} = \mathbb{E}(\mathbf{xx}^H)$ is the input covariance matrix.

We have shown in Chap. 2 that for the case of no feedback, the optimal strategy is the diversity strategy where the power is equally distributed:

$$\mathbf{Q} = \sqrt{\frac{P}{N_t}}\mathbf{I}_{N_t} \tag{4.82}$$

In Rayleigh fading channels with transmit correlation only where the channel covariance matrix $\mathbf{R}_{tx} = \mathbb{E}(\mathbf{H}^H\mathbf{H}) = \mathbf{U}_t \Delta_t \mathbf{U}_t^H$ is known at the transmitter, the

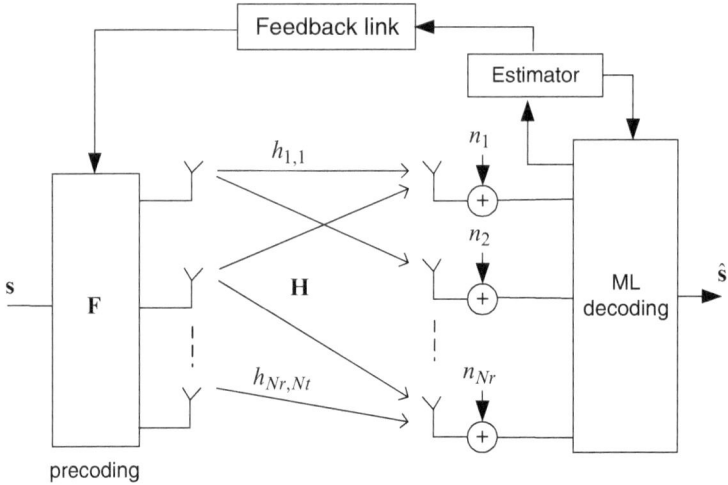

Fig. 4.13 Block diagram of MIMO scheme with linear precoding

optimal covariance matrix $\mathbf{Q} = \mathbf{U}_Q \Delta_Q \mathbf{U}_Q^H$ is obtained by selecting the eigenvectors of $\mathbf{Q}$ equal to the eigenvectors of the channel covariance matrix $\mathbf{R}_{tx}$ [38,41,64,90]:

$$\mathbf{U}_Q = \mathbf{U}_t \tag{4.83}$$

4.3.1 System Model

We consider a point-to-point single user communication system employing linear precoding for spatial multiplexing with N_t transmit antennas and N_r receive antennas. This communication system is presented in Fig. 4.13.

The input data stream is demultiplexed into N_s different streams. The baseband input–output discrete relationship for a given complex transmitted symbol $\mathbf{s}$ is now presented by

$$\mathbf{y} = \mathbf{H}\mathbf{x} + \mathbf{n} \tag{4.84}$$

with

$$\mathbf{x} = \mathbf{F}\mathbf{s} \tag{4.85}$$

where $\mathbf{s}$ is the symbol vector of size $N_s \times 1$ of power P, $\mathbf{H}$ is the $N_r \times N_t$ channel matrix, and $\mathbf{F}$ is the unitary precoding matrix vector of size $N_t \times N_s$ with $N_s \le N_t$. The entries of the noise vector $\mathbf{n}$ are circularly symmetric i.i.d. Gaussian.

We model the entries of $\mathbf{H}$ as i.i.d. zero-mean complex Gaussian random variable with variance one.

In this chapter, since we consider only the single user case, the precoding matrix $\mathbf{F}$ is always the same matrix as the one computed by the single receiver $\mathbf{W}$ as

$$\mathbf{F} = \mathbf{W} \tag{4.86}$$

Consequently, in the rest of the chapter, we will only refer to $\mathbf{W}$ as the precoding matrix. We will see in Chap. 6, that in the multiuser case, there is usually a significant difference between the matrices $\mathbf{W}$ computed by the receiver and the precoding matrix $\mathbf{F}$. In the long term evolution (LTE) standard, these two cases are called codebook-based ($\mathbf{F} = \mathbf{W}$) and non-codebook-based ($\mathbf{F} \neq \mathbf{W}$) precoding schemes.

Assuming perfect knowledge of the product $\mathbf{HW}$ at the receiver, the vector decoder produces the symbol estimate $\hat{\mathbf{s}}$. Depending on the criterion such as maximizing the mutual information or minimizing the error rate for a given receiver, different optimal precoding matrices $\mathbf{W}$ can be obtained.

The receiver will choose the unitary precoding matrix $\mathbf{W}$ from a finite set of possible precoding matrices $\mathcal{W} = \{\mathbf{W}_1 \ \mathbf{W}_2 \ \ldots \ \mathbf{W}_N\}$.

The proposed codebook will satisfy $\mathcal{W} \in U(N_t, N_s)$ where $U(N_t, N_s)$ is the set of the $N_t \times N_s$ matrices with orthonormal columns. In other words, the N_s columns of each unitary precoding matrix $\mathbf{W}$ are unitary and orthogonal.

In the next section, we will consider the case where there exists a low-rate feedback link and the precoding vector should be chosen among a given set of beamforming vectors.

4.3.2 Codebook Design

4.3.2.1 Criterion and Distortion Function

Before focusing on the codebook design, we will first derive the optimal precoding matrix $\mathbf{W}$ for different criteria [56, 82].

Criterion 1: Minimizing the Symbol Error Rate for an ML Receiver

The ML receiver compares all possible combinations of symbols which could have been transmitted and results in the estimate

$$\hat{\mathbf{s}} = \arg \min_{\mathbf{s} \in \Lambda_s^{N_s}} \|\mathbf{y} - \mathbf{HWs}\|^2. \tag{4.87}$$

Since the performance of ML decoding is related to the receive minimum distance between each pair of symbols $d_{min} = \min_{\mathbf{s}_1 \neq \mathbf{s}_2} \|\mathbf{HW}(\mathbf{s}_1 - \mathbf{s}_2)\|$, we can derive the following criterion for the codebook design

$$\mathbf{W}_{opt} = \arg \max_{\mathbf{W}_i \in \mathscr{W}} d_{min} \qquad (4.88)$$

Unfortunately, there is no closed-form expression for this selection criterion since the computation of the minimum distance for a given channel realization requires a search over all the symbol pairs.

Criterion 2: Minimizing the Symbol Error Rate for a ZF Receiver

The ZF receiver applies a $N_s \times N_r$ matrix $\mathbf{G} = (\mathbf{HW})^H$. The SNR per stream is given by

$$\text{SNR}_s^{ZF} = \frac{P}{N_s N_0 B_W \, [\mathbf{GG}^H]_{s,s}^{-1}} \qquad (4.89)$$

In order to minimize the average symbol vector error probability, the minimum substream SNR must be maximized. Since a criterion based on the minimum SNR is difficult to derive, we use the following lower bound:

$$\text{SNR}_{min}^{ZF} = \min_{1 \le s \le N_s} \text{SNR}_s^{ZF}$$

$$\ge \lambda_{min}^2 \{\mathbf{HW}\} \frac{P}{N_s N_0 B_W} \qquad (4.90)$$

where $\lambda_{min}\{\mathbf{A}\}$ is the minimum singular value of $\mathbf{A}$. Using the above lower bound, the ZF selection criterion consists in picking $\mathbf{W}$ that maximizes the minimum singular value of $\mathbf{HW}$ and is stated as

$$\mathbf{W}_{opt} = \arg \max_{\mathbf{W}_i \in \mathscr{W}} \lambda_{min}\{\mathbf{HW}_i\} \qquad (4.91)$$

Criterion 3: Minimizing the Symbol Error Rate for an MMSE Receiver

A classical criterion to design a linear receiver, for given precoding matrix $\mathbf{W}$ and channel matrix $\mathbf{H}$, is to minimize the mean square error (MSE) matrix that is given by

$$\overline{MSE}(\mathbf{W}, \mathbf{G}) = \mathbb{E}[(\mathbf{Gy} - \mathbf{s})(\mathbf{Gy} - \mathbf{s})^*] \qquad (4.92)$$

The $N_s \times N_r$ matrix $\mathbf{G}$ that minimizes the MSE is the same as the MMSE (Wiener) receiver and is given by

$$\mathbf{G} = \left[\mathbf{W}^H \mathbf{H}^H \mathbf{HW} + (N_s N_0 B_W / P)\mathbf{I}_{N_s}\right]^{-1} \mathbf{W}^H \mathbf{H}^H \qquad (4.93)$$

The SNR per stream is given by

$$\text{SNR}_s^{MMSE} = \frac{P}{N_s N_0 B_W \, [\mathbf{W}^H \mathbf{H}^H \mathbf{HW} + (N_s N_0 B_W)\mathbf{I}_{N_s}]_{s,s}^{-1}} - 1 \qquad (4.94)$$

A lower bound of the MSE is given by

$$\overline{MSE}(\mathbf{W}) = \frac{P}{N_s N_0 B_W} \left(\mathbf{I}_{N_s} + \frac{P}{N_s N_0 B_W} \mathbf{W}^H \mathbf{H}^H \mathbf{H} \mathbf{W} \right)^{-1} \qquad (4.95)$$

The MMSE selection criterion consists in minimizing the trace of $\overline{MSE}(\mathbf{W})$ and is stated as

$$\mathbf{W}_{opt} = \arg \min_{\mathbf{W}_i \in \mathscr{W}} \mathrm{tr}(\overline{MSE}(\mathbf{W}_i)) \qquad (4.96)$$

Criterion 4: Maximizing the Mutual Information

For a given precoding matrix $\mathbf{W}$ and channel matrix $\mathbf{H}$, the mutual information between $\mathbf{s}$ and $\mathbf{y}$ is

$$I(\mathbf{W}) = \log_2 \det \left(\mathbf{I}_{N_s} + \frac{P}{N_s N_0 B_W} \mathbf{W}^H \mathbf{H}^H \mathbf{H} \mathbf{W} \right) \qquad (4.97)$$

The capacity selection criterion can be stated as

$$\mathbf{W}_{opt} = \arg \max_{\mathbf{W}_i \in \mathscr{W}} I(\mathbf{W}_i) \qquad (4.98)$$

Optimal Precoding Matrix for Full CSIT

For i.i.d Rayleigh channel matrix $\mathbf{H}$, if the full CSIT is available, it can be shown that the optimal precoding matrix $\mathbf{W}_{opt}$ for criteria 2, 3 and 4 is composed by the first N_s columns of the unitary matrix $\mathbf{V}$ [56, 82]:

$$\mathbf{W}_{full} = \bar{\mathbf{V}}$$

$$= \mathbf{V} \begin{bmatrix} \mathbf{I}_{N_s} \\ \mathbf{0}_{(N_t - N_s) \times (N_s)} \end{bmatrix} \qquad (4.99)$$

where MIMO channel matrix $\mathbf{H}$ is decomposed as $\mathbf{H} = \mathbf{U} \Sigma \mathbf{V}^H$ and where $\mathbf{V} = [\mathbf{v}_1 \ldots \mathbf{v}_{N_t}]$ and $\Sigma = \mathrm{diag}(\sqrt{\lambda_1}, \sqrt{\lambda_2}, \ldots, \sqrt{\lambda_r}, 0, \ldots, 0)$.

From the expression of the MIMO capacity given in Chap. 2 and assuming that the power is equally shared between the streams, the achievable data rate of the point-to-point single user MIMO communication link using the optimal precoding matrix with full CSIT is given by

$$R_{fullCSIT} = \mathbb{E}_{\mathbf{H}} \left[B_w \log_2 \det \left(1 + \frac{P}{N_s N_0 B_w} \bar{\mathbf{V}}^H \mathbf{H}^H \mathbf{H} \bar{\mathbf{V}} \right) \right]$$

$$= \mathbb{E}_{\mathbf{H}} \left[\sum_{i=1}^{N_s} B_w \log_2 \left(1 + \frac{P}{N_s N_0 B_w} \lambda_i \right) \right] \qquad (4.100)$$

4.3.2.2 Grassmannian Subspace Packing

We will first introduce some geometric background about Stiefel and Grassmann manifolds [7, 18, 97]. While Stiefel and Grassmann manifolds can be described using real or complex matrices, for this application, we will concentrate on complex matrices.

The orthogonal group $O(n)$ is defined as the set of the $n \times n$ orthogonal matrices:

$$O(n) = \{\mathbf{Q} \in \mathbb{C}^{n \times n} : \mathbf{Q}^H \mathbf{Q} = \mathbf{I}_n\} \tag{4.101}$$

The Stiefel manifold $U(n, p)$ for $n \geq p$ is defined as the set of all "tall-skinny" matrices of size $n \times p$ whose column vectors are orthogonal unit vectors:

$$U(n, p) = \{\mathbf{Q} \in \mathbb{C}^{n \times p} : \mathbf{Q}^H \mathbf{Q} = \mathbf{I}_p\} \tag{4.102}$$

The Stiefel manifold $U(n, p)$ can be viewed as a submanifold of $\mathbb{C}^{n \times p}$ of real dimension $2np - p^2$. The Stiefel manifold may also be defined as a quotient space arising from the orthogonal group as $U(n, p) = O(n)/O(n - p)$.

The Grassmann manifold $G(n, p)$ is the set of all p-dimensional subspaces spanned by matrices in Stiefel manifold $U(n, p)$. The Grassmann manifold may also be defined as a quotient space arising from the Stiefel manifold as $G(n, p) = U(n, p)/O(p)$.

Since the precoding matrices can be any orthonormal matrices of size $N_t \times N_s$, the Grassmann manifold is a natural description of the domain associated to these matrices.

Let $\mathbf{W}_i, \mathbf{W}_j \in U(N_t, N_s)$, then the column spaces $\mathscr{P}_{\mathbf{W}_i}$ and $\mathscr{P}_{\mathbf{W}_j}$ of $\mathbf{W}_i$ and $\mathbf{W}_j$, respectively, are in the Grassmann manifold $G(N_t, N_s)$. The codebook $\mathscr{W}$ which consists of a finite number of matrices chosen from $U(N_t, N_s)$ thus represents a set or a packing, of subspaces in the Grassmann manifold $G(N_t, N_s)$. The design of the codebook $\mathscr{W}$ composed of N matrices that maximize the minimum subspace distance is known as Grassmannian subspace packing.

We first introduce the main distances between two subspaces $\mathscr{P}_{\mathbf{W}_i}$ and $\mathscr{P}_{\mathbf{W}_j} \in G(N_t, N_s)$.

The geodesic distance between two subspaces $\mathscr{P}_{\mathbf{W}_i}$ and $\mathscr{P}_{\mathbf{W}_j}$ spanned by the columns of the matrices $\mathbf{W}_i$ and $\mathbf{W}_j$ is

$$d_g^2(\mathbf{W}_i, \mathbf{W}_j) = \sum_{s=1}^{N_s} \phi_s^2$$

where $\phi_1, \phi_2, \ldots, \phi_{N_s}$ are the principal angles between the 2 subspaces $\mathscr{P}_{\mathbf{W}_1}$ and $\mathscr{P}_{\mathbf{W}_2}$

The chordal distance between two subspaces $\mathscr{P}_{\mathbf{W}_i}$ and $\mathscr{P}_{\mathbf{W}_j}$ is given by

$$d^2(\mathbf{W}_i, \mathbf{W}_j) = \frac{1}{2}||\mathbf{W}_i\mathbf{W}_i^H - \mathbf{W}_j\mathbf{W}_j^H||_F^2$$

$$= \sum_{s=1}^{N_s} \sin^2(\phi_s)$$

$$= N_s - \sum_{s=1}^{N_s} \lambda_s^2\{\mathbf{W}_i^H\mathbf{W}_j\} \qquad (4.103)$$

$||\mathbf{A}||_F$ is the Frobenius norm of the matrix $\mathbf{A}$. The chordal distance generalizes the distance between points on the unit sphere through an isometric embedding from $G(N_t, N_s)$ to the unit sphere.

The chordal distance can be reexpressed as

$$d^2(\mathbf{W}_i, \mathbf{W}_j) = \mathrm{tr}(\mathbf{I}_{N_s} - \mathbf{W}_j^H\mathbf{W}_i\mathbf{W}_i^H\mathbf{W}_j) \qquad (4.104)$$

The Fubini-Study distance between two subspaces $\mathscr{P}_{\mathbf{W}_i}$ and $\mathscr{P}_{\mathbf{W}_j}$ is given by

$$d_f^2(\mathbf{W}_i, \mathbf{W}_j) = \mathrm{arccos}^2 \sum_{s=1}^{N_s} \cos(\phi_s)$$

$$= \mathrm{arccos}^2|\det(\mathbf{W}_i^H\mathbf{W}_j)| \qquad (4.105)$$

The projection two-norm distance between two subspaces $\mathscr{P}_{\mathbf{W}_i}$ and $\mathscr{P}_{\mathbf{W}_j}$ associated to the matrices $\mathbf{W}_i$ and $\mathbf{W}_j$ is

$$d_p^2(\mathbf{W}_i, \mathbf{W}_j) = ||\mathbf{W}_i\mathbf{W}_i^H - \mathbf{W}_j\mathbf{W}_j^H||^2$$

$$= 1 - \lambda_{min}^2\{\mathbf{W}_i^H\mathbf{W}_j\} \qquad (4.106)$$

The projection two-norm distance is maximized by minimizing the smallest eigenvalue of $\mathbf{W}_i^H\mathbf{W}_j$ while the chordal distance is maximized by minimizing the sum of the eigenvalues of $\mathbf{W}_i^H\mathbf{W}_j$.

In [56], the authors have shown that criteria 1, 2 and 3 relate to maximizing the average singular value of the effective channel $\mathbf{HW}$. The average distortion function $G(\mathbf{W})$ can be written as

$$G(\mathbf{W}) = \mathbb{E}_{\mathbf{H}}\left[\lambda_{min}^2\{\mathbf{HW}_{opt}\} - \max_{\mathbf{W}_i \in \mathscr{W}} \lambda_{min}^2\{\mathbf{HW}_i\}\right] \qquad (4.107)$$

In order to minimize $G(\mathbf{W})$, Love and Heath [56] have shown that $\mathscr{W}_{opt}$ should be designed to maximize the minimum projection two-norm distance between any pair of matrices of the codebook. On the other hand, for criterion 4, it can be shown that $\mathscr{W}_{opt}$ should be designed to maximize the minimum of the Fubini-Study distance between any pair of matrices. In [102], the authors have proposed BER-based criterion. For this criterion, $\mathcal{W}_{opt}$ is designed in order to maximize the minimum chordal distance between any pair of matrices.

One method to build a good codebook is to use the Fourier matrices like for the beamforming case [33].

Up to now we have considered that the number of streams N_s is fixed. Some substantial performance gains over constant stream strategy can be achieved by adapting N_s in function of the channel matrix for a fixed data rate using the so-called multimode precoding [57].

4.3.2.3 Codebook Construction Based on Vector Quantization

The quantization of the CDI based on vector quantization (VQ) introduced previously for the beamforming scheme can be extended to build the codebook for the linear precoding scheme.

The codebook construction is linked to the vector quantization problem as follows. Suppose that we have a random matrix $\mathbf{V}$ which is isotropically distributed. We now want to quantize this matrix to a finite number of codewords $\mathcal{W} = \{\mathbf{W}_1 \ \mathbf{W}_2 \ \dots \ \mathbf{W}_N\}$ using the chordal distance as subspace distance.

The problem can be formally stated as the minimization of an average distortion as

$$\min \mathbb{E}_{\mathbf{H}}\left[d^2(\mathbf{V}, \mathbf{W}_{opt})\right] \tag{4.108}$$

$$\text{subject to } \mathbf{W}_i^H \mathbf{W}_i = \mathbf{I}_{N_s} \quad \forall i \tag{4.109}$$

which can be viewed as a VQ problem with

source input : $\mathbf{V} \sim \text{uniform}(U(N_t, N_s))$

codebook : $\mathcal{W} = \{\mathbf{W}_1 \ \mathbf{W}_2 \ \dots \ \mathbf{W}_N\}, \quad \mathbf{W}_i^H \mathbf{W}_i = \mathbf{I}_{N_s} \quad \forall i$ (4.110)

distortion metric : $d^2(\mathbf{V}, \mathbf{W}_i) = \dfrac{1}{2}||\mathbf{V}\mathbf{V}^H - \mathbf{W}_i\mathbf{W}_i^H||_F^2, \quad \forall \mathbf{V}, i$

The source vector, in this case, is the $(N_t \times N_s)$-dimensional matrix $\mathbf{V}$ uniformly distributed on $U(N_t, N_s)$.

The design of such a codebook using the Lloyd algorithm has been proposed in [102].

The Lloyd algorithm is described as follows:

Let S_n be the encoding region associated with codeword $\mathbf{W}_n$ and the partition of the space $\mathcal{P} = \{S_1, \dots, S_N\}$.

Like in the beamforming case, the codebook $\mathcal{W}$ and the partition of the space $\mathcal{P}$ must satisfy the centroid condition and the nearest neighbor rule.

$$S_i = \{\mathbf{V} | d(\mathbf{V}, \mathbf{W}_i) \le d(\mathbf{V}, \mathbf{W}_j), \quad \forall j \ne i\} \tag{4.111}$$

1. *Step1: initialization phase*

   • Generate a training sequence consisting of source matrices $\mathbf{V}$ which are isotropically distributed on $U(N_t, N_s)$.

2. *Step2: nearest neighbor rule*

   • All input matrices $\mathbf{V}$ closer to the codeword $\mathbf{W}_i$ than to any other codeword, be assigned to the neighborhood of $\mathbf{W}_i$ or region S_i

     – $\mathbf{V} \in S_i$ if and only if $d(\mathbf{V}, \mathbf{W}_i) \le d(\mathbf{V}, \mathbf{W}_j), \quad \forall j \ne i$

3. *Step3: centroid condition*

   • Take the ith region S_i, for example, whose local correlation matrix $\Sigma_i := \mathbb{E}\left[\mathbf{V}\mathbf{V}^H | \mathbf{V} \in S_i\right]$. To avoid the expectation operation, we compute Σ_i using Monte Carlo approach as follows:

$$\Sigma_i = \frac{1}{|S_i|} \sum_{\mathbf{V}_n \in S_i} \mathbf{V}_n \mathbf{V}_n^H \tag{4.112}$$

   According to the centroid condition, the optimal beamforming vector $\mathbf{W}_i^{opt}$ should maximize $\mathbf{W}_i^H \Sigma_i \mathbf{W}_i$ subject to the unit norm constraint.

$$\mathbf{W}_i^{opt} = \underset{\mathbf{W}_i^H \mathbf{W}_i = \mathbf{I}_{N_s}}{\arg\max} \ \mathbf{W}_i^H \Sigma_i \mathbf{W}_i$$

$$= \mathbf{V}_i \tag{4.113}$$

   where $\mathbf{V}_i$ is built from the N_s eigenvectors corresponding to the largest eigenvalue of Σ_i.

4. *Loop back to Step2 until convergence*

To generate each matrix of the training sequence in Step1, we draw a $N_s \times N_t$ matrix whose entries are i.i.d. complex, Gaussian random variables and perform a QR decomposition. The first N_s columns of the unitary part of the QR decomposition form the matrix.

In Fig. 4.14, we evaluate the BER performance of linear precoding for a MIMO system composed of $N_t = 4$ transmit antennas and $N_r = 2$ receive antennas and using 4-QAM modulation and different codebooks designed using the Lloyd algorithm described in Sect. 4.3.2.3. As a reference, we show the BER performance using full CSIT linear precoder. The four-bit, six-bit and eight-bit codebooks achieve 0.05 dB, 0.4 dB and 1.2 dB loss with respect to the full CSIT precoder at $BER = 10^{-3}$.

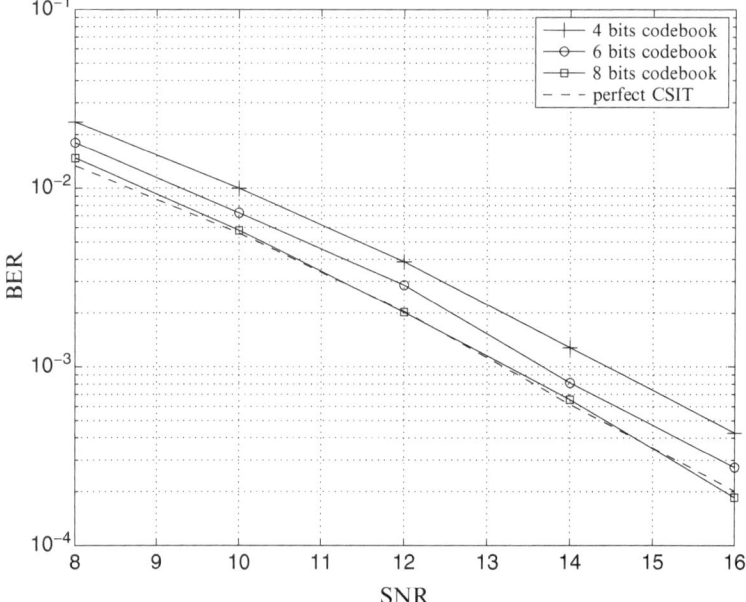

Fig. 4.14 BER comparison of different linear precoding schemes for $N_t = 4$ and $N_r = 2$

4.4   Improved Space-Time Codes

4.4.1   Limited Feedback for Orthogonal Space-Time Block Codes

In Chap. 2, we have introduced different classes of STBCs including OSTBCs. While OSTBCs are an open loop scheme, they achieve full diversity. However, an additional array gain can be obtained using OSTBCs by exploiting channel statistics at the transmitter [99, 100] or when partial CSIT is available [40, 48, 55]. In order to achieve both full diversity and array gain, a precoding matrix is applied after the STBC encoder. In this section, we will describe this class of limited feedback scheme for OSTBCs.

Let's consider an N_t transmit antenna and N_r receive antenna MIMO system with limited feedback link. At the transmitter, information symbols s_i belonging to the constellation set Λ_s are parsed into blocks $\mathbf{s} = [s_1 \quad s_2 \quad \ldots \quad s_Q]^T$ of size $Q \times 1$ in the symbol vector where Q is the number of information symbols. The OSTBC encoder performs the mapping of the vector $\mathbf{s}$ to the matrix code $\mathbf{C}$ of size $M \times T$ with $M < N_t$. Since the STBC is orthogonal we have $\mathbf{C}\mathbf{C}^H = \left(|s_1|^2 + \cdots + |s_Q|^2\right)\mathbf{I}_M$.

Before transmission, the codeword is premultiplied by the $N_t \times M$ precoding matrix $\mathbf{W}$. The received signal can be written as follows:

$$\mathbf{Y} = \mathbf{HWC} + \mathbf{N} \tag{4.114}$$

where $\mathbf{Y}$ and $\mathbf{N}$ are the received and noise matrices, respectively, with the dimension of $N_r \times T$.

From Eq. (2.109), the conditional PEP $P(\mathbf{C} \to \mathbf{C}'|\mathbf{H})$ can be upper bounded as

$$P(\mathbf{C} \to \mathbf{C}'|\mathbf{H}) \le \exp\left(-\frac{P}{4N_0 B_W N_t}||\mathbf{HW}(\mathbf{C} - \mathbf{C}'||_F^2\right) \tag{4.115}$$

Minimizing the bound in Eq. (4.115) is equivalent to choosing the precoding matrix $\mathbf{W}$ to maximize the minimum distance

$$d_{min} = \min_{\mathbf{C} \ne \mathbf{C}'} ||\mathbf{HW}(\mathbf{C} - \mathbf{C}')||_F \tag{4.116}$$

Due to the orthogonality of the codewords, the following selection criterion can be deduced.

Selection Criterion: Maximizing the Frobenius Norm

$$\mathbf{W} = \arg\max_{\mathbf{W} \in \mathscr{W}} ||\mathbf{HW}||_F \tag{4.117}$$

Optimal Precoding Matrix for Full CSIT

For i.i.d Rayleigh channel matrix $\mathbf{H}$, if the full CSIT is available, it can be shown that the optimal precoding matrix $\mathbf{W}_{opt}$ for the maximization of the Frobenius norm is composed of the first M columns of the unitary matrix $\mathbf{V}$ [55]:

$$\mathbf{W}_{full} = \bar{\mathbf{V}}$$

$$= \mathbf{V}\begin{bmatrix} \mathbf{I}_M \\ \mathbf{0}_{(N_t-M)\times(M)} \end{bmatrix} \tag{4.118}$$

where the MIMO channel matrix $\mathbf{H}$ is decomposed as $\mathbf{H} = \mathbf{U}\Sigma\mathbf{V}^H$.

Like in the linear precoding section, the codebook should be designed in order to minimize a distortion measure. Since we would like to obtain the closest performance to the optimal precoding strategy, the design of the codebook can be performed by minimizing the measure of distortion related to the average loss in received channel power and is given by

$$G(\mathbf{W}) = \mathbb{E}_\mathbf{H}\left[\min_{\mathbf{W} \in \mathscr{W}}(||\mathbf{HW}_{full}||_F^2 - ||\mathbf{HW}||_F^2)\right] \tag{4.119}$$

An upper bound of this distortion function can be derived [55]:

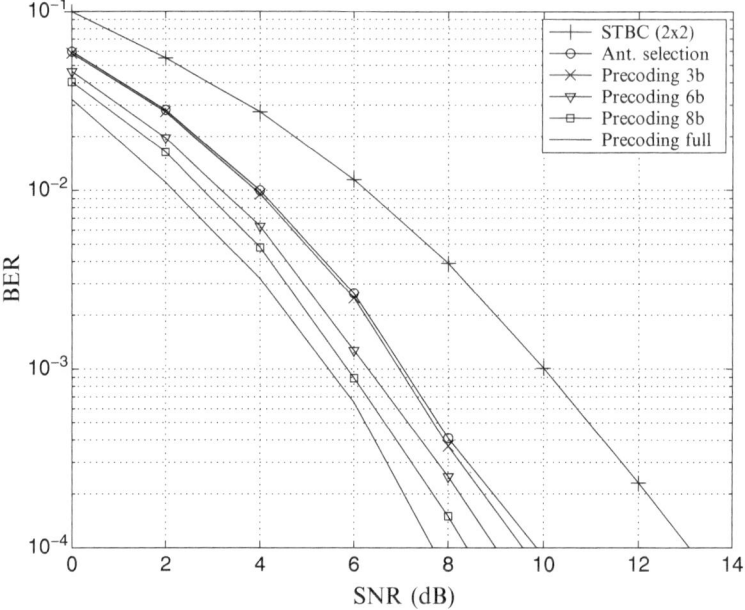

Fig. 4.15 BER comparison of different orthogonal STBC precoding schemes for $N_t = 4$ and $N_r = 2$

$$G_{up}(\mathbf{W}) = \mathbb{E}_{\mathbf{H}}(\lambda_1)\mathbb{E}_{\mathbf{H}}\left[\min_{\mathbf{W}\in\mathscr{W}}\frac{1}{2}||\bar{\mathbf{V}}\bar{\mathbf{V}}^H - \mathbf{W}\mathbf{W}^H||_F^2\right] \qquad (4.120)$$

The right factor represents the effect of the codebook design and the left factor represents the channel impact. Consequently, the codebook should be designed such that the minimum chordal distance

$$\min_{1\le j\le l\le N} d(\mathbf{W}_j, \mathbf{W}_l)$$

should be maximized.

In Fig. 4.15 we evaluate the BER performance of different precoders associated to the Alamouti STBC for a system composed of $N_t = 4$ transmit antennas and $N_r = 2$ receive antennas using 4-QAM modulation. As a reference, we show the performance achieved using the 2×2 Alamouti code without precoding and the performance using a full CSIT linear precoder. The antenna selection scheme requiring $\lceil \log_2 \binom{4}{2} \rceil = 3$ bits of feedback achieves a 3 dB gain with respect to the Alamouti scheme. The three-bit, six-bit and eight-bit codebooks provide 0.2 dB, 0.7 dB and 1.2 dB gain, respectively, over the antenna selection scheme at $BER = 10^{-3}$. The codebooks have been designed according to Eq. (4.121) using the Lloyd algorithm described in Sect. 4.3.2.3.

4.4.2 Limited Feedback for Non-orthogonal Space-Time Block Codes

We have seen in Chap. 2 that full rate quasi-orthogonal space-time codes (QO-STBC) sacrifice orthogonality to retain full rate. Due to the self-interference of these codes, they do not fully exploit the available diversity. When full or partial CSIT is available, it was shown in [75, 88] that QO-STBC can achieve both full diversity and array gain. However, due to the non-orthogonality of the QO-STBC, how to derive the optimal precoding scheme is not straightforward.

The derivation of the optimal precoder for the association of linear precoding with STBC codes introduced in [40] has been extended to the QO-STBCs codes by Liu and Jafarkhani in [51]. The proposed QO-STBC linear precoder can be seen as a four-directional eigen-beamformer that works for systems with four transmit antennas.

In order to cancel the nonzero element α in the interference matrix $\mathbf{J}$ given in Eq. (2.127), another solution consists in rotating the signals from the third and fourth transmit by two phasors ϕ and θ, then the self-interference term α becomes

$$\alpha' = 2\Re(h_{11}^* h_{14} e^{j\phi} - h_{12}^* h_{13} e^{j\theta}) \tag{4.121}$$

By choosing ϕ and θ, it is possible to cancel the self-interference ($\alpha' = 0$). Another advantage of this approach is that a full diversity and coding gain can be achieved with a simple linear receiver.

4.5 Antenna Selection

In Chap. 2, we have shown that MIMO systems can increase the reliability by exploiting the spatial diversity using space-time coding scheme or the data rate through spatial multiplexing. We have seen that depending on the number of transmit and receive antennas of the MIMO system, there is a trade-off between the diversity gain and the multiplexing gain. Since the main drawback of MIMO systems is its complexity and cost, there has been an increasing interest in applying antenna selection at the transmitter and the receiver in order to limit the number of required RF chains.

When a limited feedback is allowed, a simple solution to exploit transmit diversity consists to simply select a set of transmit antennas among the set of antennas [80, 85, 92]. Antenna selection can be seen as the simplest form of beamforming or linear precoding and has been first considered in the past in the context of both transmit and receive diversity [92]. Selection for multiple transmit and multiple receive antenna systems was first presented in [25] using a criterion based on Shannon capacity and has been studied in [26,28,62,80]. Antenna selection has been adopted for the uplink of the next generation LTE standard [61].

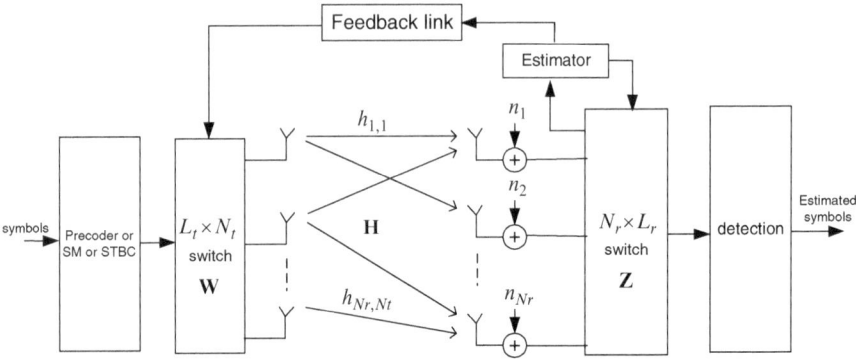

Fig. 4.16 Block diagram of MIMO scheme with antenna selection

4.5.1   System Model

We consider a point-to-point single user MIMO system with N_t transmit antennas and N_r receive antennas where L_t transmit and L_r receive antennas signal are selected as shown in Fig. 4.16.

We define the matrix $\tilde{\mathbf{H}}$ of size $L_r \times L_t$ as the equivalent channel matrix including the $N_t \times L_t$ transmit selection antenna matrix $\mathbf{W}$, the MIMO channel matrix $\mathbf{H}$ of size $N_r \times N_t$ and the $L_r \times N_r$ receive selection antenna matrix $\mathbf{Z}$ as follows:

$$\tilde{\mathbf{H}} = \mathbf{ZHW} \tag{4.122}$$

The matrix $\tilde{\mathbf{H}}$ is created by striking $N_t - L_t$ columns and $N_r - L_r$ lines from the MIMO channel matrix $\mathbf{H}$. By successively sounding all the channel elements using training signal, the receiver can estimate $\mathbf{H}$. In the case where the multiple antennas are used for spatial multiplexing, the scheme is called "hybrid selection/MIMO" (H-S/MIMO) [62]. Compared to Eq. (2.130), we have the following input–output relation:

$$
\begin{aligned}
\mathbf{y} &= \mathbf{ZHWs} + \mathbf{n} \\
&= \tilde{\mathbf{H}}\mathbf{s} + \mathbf{n}
\end{aligned} \tag{4.123}
$$

where $\mathbf{s}$ is the transmit vector of size $L_t \times 1$ and $\mathbf{y}$ is the receive vector of size $L_r \times 1$.

Assuming equal power transmission from antennas, the selected matrix $\tilde{\mathbf{H}}$ is the one which maximizes the following information-theoretic expressions:

$$R_{H-S/MIMO} = \mathbb{E}_{\mathbf{H}}\left[\max_{\tilde{\mathbf{H}} \in \mathscr{A}} B_W \log_2\left(\det\left(\mathbf{I}_{L_r} + \frac{P}{N_t N_0 B_W} \tilde{\mathbf{H}}\tilde{\mathbf{H}}^H \right) \right) \right] \tag{4.124}$$

The exhaustive search is performed over the set $\mathscr{A}$ of size $\binom{N_t}{L_t}\binom{N_r}{L_r}$ of all possible matrices $\tilde{\mathbf{H}}$. A simplified selection rule proposed in [28] consists in decoupling the transmit and receive selection. Further simplifications have been proposed by performing a recursive maximization of the capacity. While receive selection does not require any feedback link since the selection is performed by the receiver, transmit selection implies signalization using a feedback link. Given N_t and L_t, the codebook $\mathscr{W} = \{\mathbf{W}_1, \ldots, \mathbf{W}_N\}$ for the transmit antenna selection is composed of $\binom{N_t}{L_t}$ matrices. For example, if $N_t = 3$ and $L_t = 1$ and $L_t = 2$, we have, respectively,

$$\mathscr{W}_1 = \left\{ \begin{bmatrix} 1 \\ 0 \\ 0 \end{bmatrix} \begin{bmatrix} 0 \\ 1 \\ 0 \end{bmatrix} \begin{bmatrix} 0 \\ 0 \\ 1 \end{bmatrix} \right\} \mathscr{W}_2 = \left\{ \begin{bmatrix} 1 & 0 \\ 0 & 1 \\ 0 & 0 \end{bmatrix} \begin{bmatrix} 1 & 0 \\ 0 & 0 \\ 0 & 1 \end{bmatrix} \begin{bmatrix} 0 & 0 \\ 0 & 1 \\ 1 & 0 \end{bmatrix} \right\} \tag{4.125}$$

In [31], Heath and Love have extended the transmit antenna selection for spatial multiplexing systems by dynamically adjusting the number of streams and the mapping of streams to antennas for a fixed total data rate. This scheme is referred as multimode antenna selection and provides an additional array gain compared to a fixed number of streams. Like for linear precoding, different criteria for selecting the number of streams and the mapping of streams to transmit antennas have been proposed such as minimizing the average vector error rate or maximizing the SNR. In the general case where the range of L_t can vary from 1 to N_t, the total number of codewords in the codebook is

$$\sum_{m=1}^{N_t} \binom{N_t}{m} = 2^{N_t} - 1 \tag{4.126}$$

which can be implemented using N_t bits of feedback. This approach can be compared to the solution using the Grassmannian codebook introduced in the previous section. Compared to the previous antenna selection methods, the aim of the multimode antenna selection is not to reduce the number of transmission chains but to improve the performance of the forward communication link thanks to the limited feedback link. In this case there is no antenna selection at the receiver so we have $L_r = N_r$. Techniques for selecting the mode using limited feedback are given in [47]. Since the number of bits per channel use R is fixed, the modulation will be chosen depending on the number of streams. Assuming that at the receiver a $L_t \times L_r$ ZF linear matrix $\mathbf{G} = \tilde{\mathbf{H}}^\dagger$ is applied to $\mathbf{y}$, the effective SNR for the sth stream after the linear processing is given by

$$SNR_s = \frac{P}{N_0 B_W} \frac{1}{[\tilde{\mathbf{H}}^H \tilde{\mathbf{H}}]_{s,s}^{-1}} \tag{4.127}$$

where $[\mathbf{A}]_{s,s}^{-1}$ is the entry (s, s) of $\mathbf{A}^{-1}$. Since the total data rate is fixed, the optimal criterion is to choose the number of streams and the selected antennas in order to minimize the overall probability of error. For an M-QAM modulation the bit error probability conditioned on $\tilde{\mathbf{H}}$ can be approximated as follows:

$$BER_s(\tilde{\mathbf{H}}) \approx \frac{2}{\log_2 M} \text{erfc}\left(\sqrt{\frac{3}{2(M-1)} SNR_s}\right) \qquad (4.128)$$

Instead of evaluating these quantities for each of the $2^{N_t} - 1$ transmit configurations, in [31], the authors propose to select first the number of streams L_t and then the selected antennas.

A single-stream signaling also called hybrid selection/maximum ratio combining (H-S/MRC) consists in fixing the transmit precoding vector $\mathbf{w} = [1/\sqrt{L_t}, \ldots, 1/\sqrt{L_t}]^T$ before the transmit switch and performing MRC at the receiver. The achievable rate of this scheme is given as [6]

$$R_{H-S/MRC} = \mathbb{E}_{\mathbf{H}}\left[\max_{\tilde{\mathbf{H}} \in \mathscr{A}} B_W \log_2\left(1 + \frac{P}{N_0 B_W}|\tilde{\mathbf{H}}\mathbf{w}|^2\right)\right] \qquad (4.129)$$

For small SNR, using Taylor series approximations the above achievable rates can be approximated as [6]

$$R_{H-S/MIMO} \approx \mathbb{E}_{\mathbf{H}}\left[\frac{P}{L_t N_0 \ln(2)}\left(\sum_{i=1}^{L_r}\sum_{j=1}^{L_t}|\tilde{H}_{i,j}|^2\right)\right] \qquad (4.130)$$

$$R_{H-S/MRC} \approx \mathbb{E}_{\mathbf{H}}\left[\frac{P}{L_t N_0 \ln(2)}\left(\sum_{i=1}^{L_r}\left|\sum_{j=1}^{L_t}\tilde{H}_{i,j}\right|^2\right)\right] \qquad (4.131)$$

where $\tilde{H}_{i,j}$ are the elements of $\tilde{\mathbf{H}}$.

The difference between these two approximations relies in the cross terms that appear in Eq. (4.131) and are missing in Eq. (4.130). Since the cross terms are positive on average, the achievable rate $R_{H-S/MRC}$ is higher than $R_{H-S/MIMO}$.

Like for the MIMO scheme, the achievable rate of $R_{H-S/MIMO}$ is proportional to $\min(L_t, L_r)$. However, antenna selection provides an additional gain. For example, when $N_r = N_t = 8$ and $L_r = L_t = 2$, a 7 dB gain is obtained [6] compared to the MIMO scheme.

Antenna selection can also be used with STBC codes. In this case, a STBC with L_t outputs is sent over N_t antennas. The most popular form of space-time coding, orthogonal space-time block codes perfectly fits with antenna selection, particularly the Alamouti space-time block code with rate one for the case $L_t = 2$. Antenna subset selection with orthogonal space-time block coding was proposed in [1, 27] where it was demonstrated that the transmitter should choose the antenna set that maximizes the channel Frobenius norm leading to both full diversity and coding gain. Analysis of the error probability of OSTBCs with antenna subset selection has been analysed in [29, 43, 58]. The capacity was calculated in [70]. In that case, the equivalent channel gain is given by

$$H_{eq} = \sqrt{\frac{1}{L_t} \sum_{i=1}^{L_r} \sum_{j=1}^{L_t} |\tilde{H}_{i,j}|^2}$$

$$= \sqrt{\frac{1}{L_t} ||\tilde{\mathbf{H}}||_F^2} \tag{4.132}$$

and the SNR γ is given by

$$\gamma = \frac{P}{L_t N_0 B_W} ||\tilde{\mathbf{H}}||_F^2 \tag{4.133}$$

Consequently, for this scheme, the selection of the matrix channel $\tilde{\mathbf{H}} \in \mathscr{A}$ is performed by maximizing the SNR γ. Like previously, the optimal selection phase is rather complex and simplifications have been proposed [70].

We have seen in Chap. 2 that the conditional PEP $P(\mathbf{C} \rightarrow \mathbf{C}'|\mathbf{H})$ can be upper bounded as

$$P(\mathbf{C} \rightarrow \mathbf{C}'|\mathbf{H}) \leq \exp\left(-\frac{P}{4N_0 B_W N_t} d^2(\mathbf{C}, \mathbf{C}')\right) \tag{4.134}$$

From Eq. (4.134), it was shown in [27] that at high SNR, the average PEP of antenna subset selection employing space-time coding can be obtained as follows:

$$P(\mathbf{C} \rightarrow \mathbf{C}') \leq \frac{\gamma_c^{-L_t L_r}}{\det(\mathbf{R}_r)^{L_t} \det(\mathbf{R}_t)^{L_r} \det(\mathbf{D}\mathbf{D}^H)^{L_r}} \tag{4.135}$$

assuming the Kronecker channel model introduced in Chap. 2 where the correlation matrix $\mathbf{R} = \mathbf{R}_r \otimes \mathbf{R}_r$, $\mathbf{D}$ is the difference matrix defined in Eq. (2.108) and $\gamma_c = \frac{P}{4N_0 B_W L_t}$. If the underlying space-time code has full rank, the achievable diversity order is $N_r \times N_t$.

In [1], a method to extend any space-time code constructed for L_t transmit antennas to N_t transmit antennas through group-coherent codes (GCC) has been proposed. GCC use very limited feedback from the receiver (as low as one bit) and preserve low decoding complexity while achieving full diversity and full data rate. The association of quasi-orthogonal STBC (QO-STBC) with antenna selection has also been considered in [88] to alleviate the self-interference and restore full diversity.

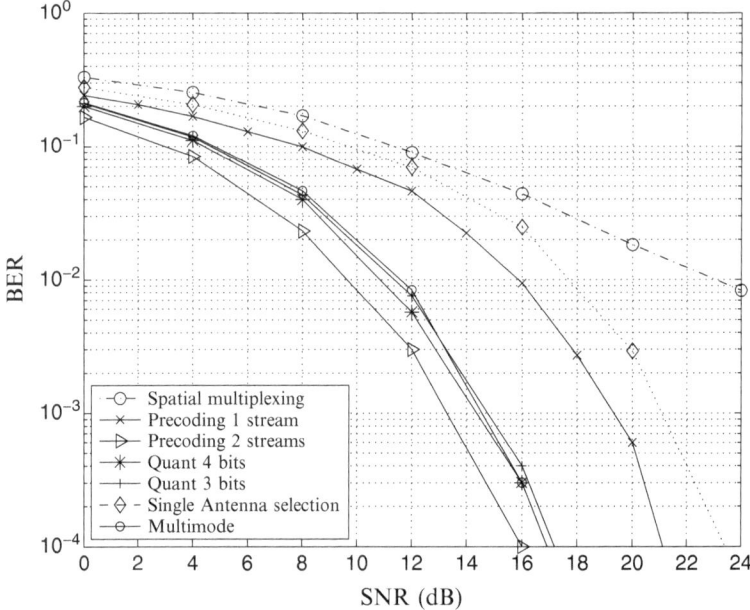

Fig. 4.17 Comparison of different antenna selection and precoding strategies for $N_r = N_t = 4$

If full CSIT is available at the transmitter, as seen in Sect. 4.2, beamforming can be performed before antenna selection. This scheme is referred as hybrid selection/maximum ratio transmission (H-S/MRT) in the antenna selection literature. From Eq. (4.10), the achievable rate is given by

$$R_{H-S/MRT} = \mathbb{E}_{\mathbf{H}} \left[\max_{\tilde{\mathbf{H}} \in \mathscr{A}} B_W \log_2 \left(1 + \frac{P}{N_0 B_W} \tilde{\lambda}_1 \right) \right] \qquad (4.136)$$

when the precoding vector $\mathbf{w}_{full}$ is equal to $\mathbf{v}_1$ the dominant right singular vector corresponding to the maximum eigenvalue $\sqrt{\tilde{\lambda}_1}$ of the channel matrix $\tilde{\mathbf{H}}$.

In Fig. 4.17, we compare different antenna selection and precoding strategies for a 4×4 system using ZF decoding with $R = 8$ bits per channel use. Depending on the mode, we will use 256-QAM, 16-QAM or QPSK modulation. As a reference, we show the spatial multiplexing performance without any CSIT. While the single transmit antenna selection scheme achieves a 5 dB gain compared to spatial multiplexing, a significant improvement can be obtained using the multimode strategy. The single antenna selection and the multimode strategy require, respectively, $B = 2$ and $B = 4$ feedback bits. For comparison, we also provide the performance of linear precoding scheme with full CSIT using one stream with 256-QAM modulation and two streams with 16-QAM modulation. We also show the performance results of the linear precoding with two streams obtained when a 3-bit or 4-bit codebook is used for the feedback of the selected precoder index. We can see that the multimode curve is close to the quantized linear precoding curves. Furthermore, we can check that all the antenna selection schemes achieve full diversity. As studied in [31], the probability of mode selection in the multimode strategy is varying with N_t, N_r and R. In this case, due to the limited number of possible modes, the two-stream mode is the most selected mode.

4.6   Codebook Design for Spatial Correlated Channel

In the previous section, we have mostly assumed that the entries of $\mathbf{H}$ are i.i.d. Gaussian random variables. While the performance analysis is tractable, the i.i.d. channel is often unrealistic in practice. In Chap. 2, we have introduced different channel models taking into account the second-order statistics of the MIMO channel.

Let's first assume that the channel is spatially correlated Rayleigh distributed with a single antenna at the receiver and that the channel experiences transmit correlation defined by the transmit correlation matrix $\mathbf{R}_t$. This simplified model corresponds to the case where the transmit antennas are separated by less than the coherence distance. If we assume a MISO system, the correlated channel vector can be modeled by

$$\mathbf{h} = \mathbf{A}\mathbf{h}_w \tag{4.137}$$

where $\mathbf{h}_w$ is spatially white of dimension $Nt \times 1$ channel and $\mathbf{A}$ is a $N_t \times N_t$ matrix such that $\mathbf{R}_t = \mathbf{A}\mathbf{A}^H$.

If the transmit correlation matrix is known to the transmitter, the optimal beamforming vector is given by the maximum ratio transmission as follows:

$$\mathbf{w}_{full} = \frac{\mathbf{A}\mathbf{g}_w}{||\mathbf{A}\mathbf{g}_w||} \tag{4.138}$$

where $\mathbf{g}_w = \frac{\mathbf{h}_w}{||\mathbf{h}_w||}$

Assuming that $\mathbf{A}$ is fixed and known by the transmitter, Love and Heath [59] have shown that the modified codebook $\mathcal{W}$ exploiting the transmit correlation matrix can be obtained from the codebook for i.i.d channel by just multiplying each precoding vectors by the matrix $\mathbf{A}$. Since this approach doesn't maintain the unit vector power constraint requirement, a normalization must be done. The modified codebook is then given by

$$\mathcal{W} = \left\{ \frac{\mathbf{A}\mathbf{w}_1}{||\mathbf{A}\mathbf{w}_1||}, \ldots, \frac{\mathbf{A}\mathbf{w}_N}{||\mathbf{A}\mathbf{w}_N||} \right\} \tag{4.139}$$

The extension to channel with both transmit and receive correlation using the Kronecker channel model has been addressed in [72]. Instead of adjusting the

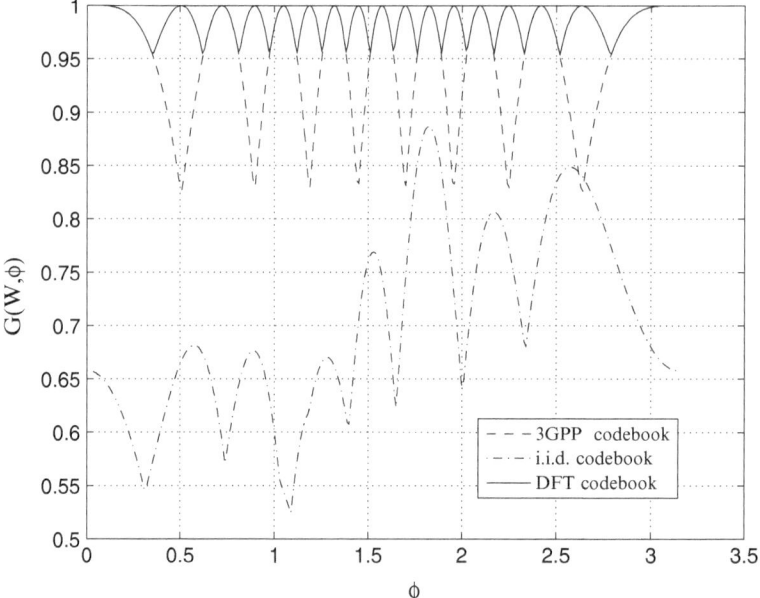

Fig. 4.18 Comparison of the robustness of different codebooks in correlated channels

codebook directly according to the covariance matrix, Raghavan and Heath [72] have built a local packing from a root codebook by performing both scaling and rotation. We will describe those operations in the next section dedicated to time correlated channels.

When the AoD at the transmitter has a small angle spread $\sigma_t \approx 0$, the spatial correlation coefficient for an uniform linear array can be simplified using relation (2.63). The transmit antenna array is given by

$$\mathbf{a}(\phi) = \frac{1}{\sqrt{N_t}} [1 \quad e^{-j2\pi\Delta_t \cos\phi} \quad e^{-j4\pi\Delta_t \cos\phi} \ldots e^{-j2(N_t-1)\pi\Delta_t \cos\phi}]^T \qquad (4.140)$$

where $\Delta_t = d/\lambda$ and ϕ is the AoD.

From Eq. (4.16), we have seen that the design of the codebook for correlated channels must minimize $1 - |\mathbf{u}_1^H \mathbf{w}_{opt}|^2$ for all the AoD. We can evaluate the robustness of the codebook in correlated channels using the following distance metric [15]:

$$G(\mathcal{W}, \phi) = \max_{\mathbf{w}_i \in \mathcal{W}} |\mathbf{a}(\phi)^T \mathbf{w}_i|^2 \qquad (4.141)$$

In Fig. 4.18, for uniform linear array, we compare the robustness versus AoD of i.i.d , DFT-based and 3GPP codebooks for $N_t = 4$, $N_r = 1$ and $N = 16$. As expected, the i.i.d. codebook achieves poor performance in correlated channel. On the contrary, the DFT-based codebook is well adapted [15].

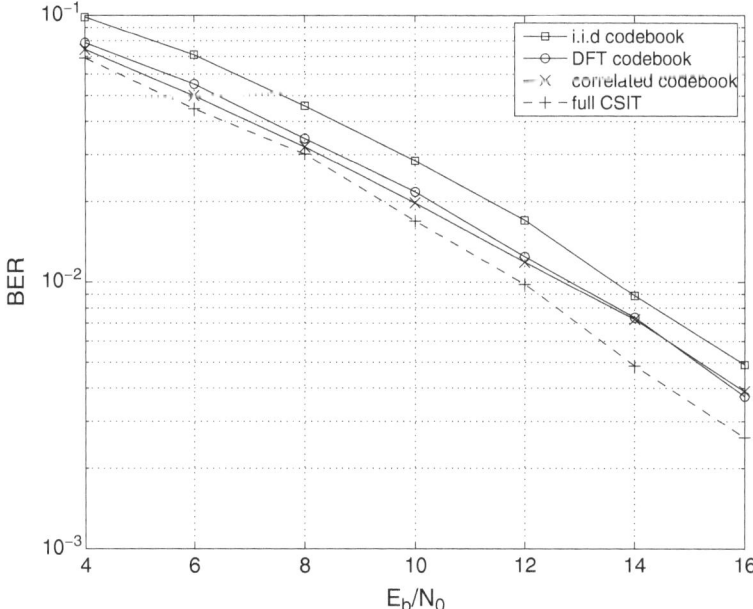

Fig. 4.19 Performance of different codebooks in correlated channels

In Fig. 4.19, we show the BER probability for a four transmit antenna beamforming system using QPSK modulation. The correlation was computed from [19] using one cluster and a 5° angular spread. The limited feedback codebooks were restricted to four bits of feedback. We have considered the i.i.d Grassmannian codebook, the DFT-based codebook and the modified codebook obtained by rotating the i.i.d codebook as introduced in Eq. (4.139). The modified codebook and DFT-based codebook outperform the i.i.d codebook of more than 1 dB. The performance obtained exploiting full channel state information is also given for comparison.

4.7 Codebook Design for Time-Correlated Channels

Up to now, we have assumed in this chapter block fading and consequently, we have ignored the temporal correlation. However, in practical systems, the block fading assumption for the downlink channel is inaccurate and the channel vectors are time correlated. Consequently, the base station can exploit this correlation in order to reduce the distortion or the amount of feedback bits by performing compression. In a temporally correlated channel, the channel variation of the user can be modeled as a first-order Gauss–Markov process as follows:

$$\mathbf{h}[t] = \kappa \mathbf{h}[t-1] + \sqrt{1-\kappa^2}\mathbf{b}[t] \qquad (4.142)$$

where $\mathbf{b}[t] \in \mathbb{C}^{N_t \times 1}$ is the innovation process at time index t with i.i.d. entries distributed according to $CN(0, 1)$. κ is the time correlation between elements $\mathbf{h}[t-1]$ and $\mathbf{h}[t]$.

As we will see in this section, a common framework for most of the proposed codebook designs for time correlated channel is the tracking of the channel using perturbation and projection. However, different approaches have been proposed in the literature. Channel-statistic-based codebook adaptation methods for a temporally correlated channel to reduce feedback requirements have been proposed with several studies appearing in the literature [3, 32, 35, 45, 77, 95, 96]. Banister and Zeidler [3,4] have proposed a transmission subspace tracking method using a single bit of feedback by employing a stochastic perturbation approach. This approach is extended in [95] where the trajectory of the channel subspace variation is modeled by a geodesic on the Grassmannian manifold.

Another approach consist in tracking the channel itself or the CDI. The two main classes of adaptive codebook are the rotation-based differential codebook and the polar-cap codebook. The rotation-based differential codebook has been investigated in [45] and proposed for the standard 802.16 m [103]. The polar-cap codebook initially introduced for a spatially correlated channel [72] has been extended to temporally correlated channels in [13, 32, 104] with a numerically optimized polar-cap size. Assuming that the channel variation is random within a cap of a given radius, it is better to have uniformly distributed codewords situated in the circumference of the cap since they can cover all the directions from the basis vector (the previous channel vector in ideal case) in order to track this channel variation efficiently with a small number of codewords.

In the next sections, we will give a more in-depth description of the differential feedback, the differential polar-cap differential and the rotation-based differential codebooks. We will then compare the performance of these codebooks and will show that in terms of distance the polar-cap differential codebook is preferable.

4.7.1   Differential Feedback Codebook

Assuming that the transmitter knows the previous precoding vector $\mathbf{w}[t-1]$ and that the channel is time correlated, the precoding vector at time instant t can be obtained from the directional variation from $\mathbf{w}[t-1]$ to $\mathbf{g}[t]$ the normalized channel vector. This variation between these vectors at adjacent time instants is along geodesics in the Grassmann manifold. Instead of quantizing the normalized vector itself like we have done previously, it is possible to quantize the geodesic trajectory connecting two subspaces. More specifically we can quantize the angular velocity matrix [3, 95].

The most well-known optimization technique is the gradient descent where at each iteration the estimate of the optimal vector is updated to move in the direction of the gradient vector. Since it is not possible to feedback the exact gradient vector, in [3], the authors have proposed an algorithm similar to the sign least mean square algorithm where an approximation of the gradient vector is performed using two perturbation vectors. We define the differential feedback function g such as

$$\mathbf{w}[t] = g(\mathbf{w}[t-1], \beta \Delta \mathbf{w}) \tag{4.143}$$

where $\Delta \mathbf{w}$ is a perturbation vector containing i.i.d. entries used to perturb the tangent space of $\mathbf{w}[t-1]$ and β is the length of the arc or the step size used to adjust the expected velocity.

The adaptive precoding vector is computed as follows:

$$\tilde{\mathbf{w}}[t] = \mathbf{w}[t-1] + \text{sign}(s[t]) \beta \Delta \mathbf{w} \tag{4.144}$$

$$\mathbf{w}[t] = \frac{\tilde{\mathbf{w}}[t]}{||\tilde{\mathbf{w}}[t]||} \tag{4.145}$$

Equations (4.144) and (4.145) correspond to the perturbation and projection phases, respectively, and

$$s[t] = d^2(\mathbf{g}[t], \mathbf{w}[t-1] + \Delta \mathbf{w}) - d^2(\mathbf{g}[t], \mathbf{w}[t-1] - \Delta \mathbf{w}) \tag{4.146}$$

The extension to precoding matrix is straightforward by performing the projection phase using a Procrustes or Gram–Schmidt orthonormalization [24].

Before introducing the polar-cap differential and rotation-based differential codebooks, we will first define the concept of the local codebook and scaling and rotation operations [32,72].

Spherical Cap: A spherical cap on $G(N_t, 1)$ or on $\mathbb{O}^{N_t}$ via equivalence partitioning with center $\mathbf{o}$ and square radius ϵ where $0 < \epsilon < 1$ is an open set:

$$\mathscr{B}_\epsilon(\mathbf{o}) = \{\mathbf{f} \in G(N_t, 1) : d^2(\mathbf{f}, \mathbf{o}) \leq \epsilon\} \tag{4.147}$$

Local Codebook: a local codebook $\mathscr{W}_{loc}$ is a codebook composed of the root or basis vector $\mathbf{o} = [1, 0, \ldots, 0]^T$ and $N-1$ other codewords $\mathbf{w}_i$ where $2 \leq i \leq N$.

$$\mathscr{W}_{loc} = \{\mathbf{o}, \mathbf{w}_2, \ldots, \mathbf{w}_N\} \tag{4.148}$$

The $N-1$ non-root vectors should satisfy the following radius constraint:

$$d^2(\mathbf{w}_i, \mathbf{o}) \leq \epsilon \tag{4.149}$$

where ϵ is the square radius of the local codebook.

Scaling a Local Codebook

While a local codebook with a given radius can be designed using for example
vector quantization, a practical approach consists in using a scaling function to
scale the vectors in the local codebook to obtain a modified local codebook with
a different radius.

Let α be the scaling factor and the vector $\mathbf{w} = [r_1 e^{j\theta_1}, r_2 e^{j\theta_2}, \ldots, r_{N_t} e^{j\theta_{N_t}}]^T$
with $||\mathbf{w}|| = 1$. The scaling operation $\mathbf{s}$ is given by

$$\mathbf{s} : \mathbb{C}^{N_t \times 1} \times \mathbb{R}[0, 1] \mapsto \mathbb{C}^{N_t \times 1}$$

$$\mathbf{s}(\mathbf{w}, \alpha) = \left[\sqrt{1 - \alpha^2 (1 - r_1^2)} e^{j\theta_1}, \alpha r_2 e^{j\theta_2}, \ldots, \alpha r_{N_t} e^{j\theta_{N_t}} \right]^T \tag{4.150}$$

We can check that the scaling operation preserves the unit norm property
$||\mathbf{s}(\mathbf{w}, \alpha)|| = 1$. We can also check that $d^2(\mathbf{s}(\mathbf{w}, \alpha), \mathbf{s}(\mathbf{o}, \alpha)) = \alpha^2 d^2(\mathbf{w}, \mathbf{o})$ [72].

The scaled codebook function for a given scaling factor α is defined as follows:

$$S(\mathcal{W}_{loc}, \alpha) := \{\mathbf{o}, \mathbf{s}(\mathbf{w}_2, \alpha), \ldots, \mathbf{s}(\mathbf{w}_{N_t}, \alpha)\} \tag{4.151}$$

Rotating a Local Codebook

Let's define the rotation function $\mathbf{r}$ that rotates a vector $\mathbf{w}$ to a vector $\mathbf{w}_{rot}$:

$$\mathbf{r} : \mathcal{B}_\epsilon(\mathbf{o}) \mapsto \mathcal{B}_\epsilon(\mathbf{o}_{rot})$$

$$\mathbf{w}_{rot} = \mathbf{r}(\mathbf{o}_{rot})\mathbf{w}$$

$$= \mathbf{U}_{rot}\mathbf{w} \tag{4.152}$$

where $\mathbf{U}_{rot}$ is a unitary matrix that rotates $\mathbf{o}$ to $\mathbf{o}_{rot}$ thus $\mathbf{o}_{rot} = \mathbf{U}_{rot}\mathbf{o}$.

The rotated codebook function for a given rotation matrix $\mathbf{U}_{rot}$ is defined as
follows:

$$R(\mathcal{W}_{loc}, \mathbf{o}_{rot}) := \{\mathbf{U}_{rot}\mathbf{o}, \mathbf{U}_{rot}\mathbf{w}_2, \ldots, \mathbf{U}_{rot}\mathbf{w}_{N_t}) \tag{4.153}$$

There are different ways to compute a rotation matrix such as using the SVD
decomposition or the Householder matrix.

The rotation matrix consists of two unitary matrices as

$$\mathbf{U}_{rot} = [\mathbf{o}_{rot} \quad \mathbf{o}_{rot}^\perp][\mathbf{o} \quad \mathbf{o}^\perp]^H$$

$$= [\mathbf{o}_{rot} \quad \mathbf{o}_{rot}^\perp] \tag{4.154}$$

since we have $[\mathbf{o} \quad \mathbf{o}^\perp] = \mathbf{I}_{N_t}$ for $\mathbf{o} = [1, 0, \ldots, 0]^T$. $\mathbf{o}_{rot}^\perp$ refers to the $N_t \times (N_t - 1)$
space orthogonal to the vector $\mathbf{o}_{rot}$.

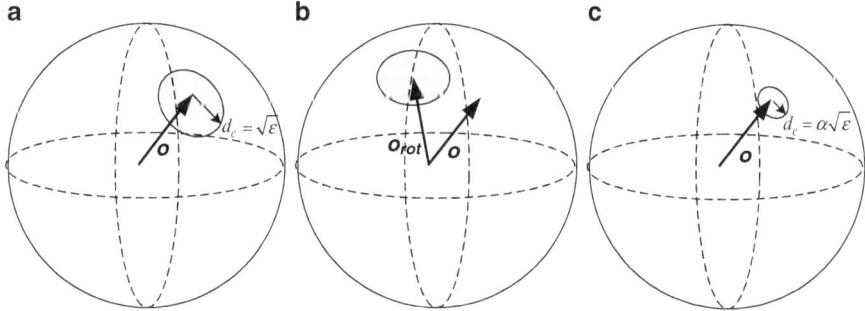

Fig. 4.20 Illustration of the rotation and scaling operations

Let's define $\mathbf{u} = \mathbf{o} - \mathbf{o}_{rot}$. The complex Householder matrix is given by [14]:

$$\mathbf{U}_{rot} = \mathbf{I}_{N_t} - \frac{1}{\mathbf{u}^H \mathbf{o}_{rot}} \mathbf{u}\mathbf{u}^H \qquad (4.155)$$

As expected, the first column of the Householder matrix is $\mathbf{o}_{rot}$ and the remaining $N_t - 1$ columns are orthogonal to $\mathbf{o}_{rot}$.

In Fig. 4.20 we illustrate the operation of rotation and scaling.

Generation of a Ring Codebook

A ring codebook with square radius ϵ is composed of the centroid vector $\mathbf{o}$ and $N-1$ vectors that are equidistant from $\mathbf{o}$ [32].

$$d^2(\mathbf{w}_i, \mathbf{o}) = \epsilon \qquad 2 \leq i \leq N \qquad (4.156)$$

Since the first vector of the ring codebook is $\mathbf{o}$, any vector $\mathbf{w}_i = [\sqrt{1-\epsilon}, \tilde{\mathbf{w}}_i^T]^T$ where $\tilde{\mathbf{w}}_i$ is a $(N_t - 1) \times 1$ unit norm vector has a square chordal distance ϵ from $\mathbf{o}$ and can consequently be chosen to construct the ring codebook. So we can fix the first element of each vector $\mathbf{w}_i$ as $\sqrt{1-\epsilon}$ for $2 \leq i \leq N$. From this observation, a ring codebook can be built as follows:

$$\mathscr{W}_{loc} := \left\{ \mathbf{o}, \begin{bmatrix} \sqrt{1-\epsilon} \\ \sqrt{\epsilon}\tilde{\mathbf{w}}_2 \end{bmatrix}, \dots, \begin{bmatrix} \sqrt{1-\epsilon} \\ \sqrt{\epsilon}\tilde{\mathbf{w}}_N \end{bmatrix} \right\} \qquad (4.157)$$

where $\{\tilde{\mathbf{w}}_2, \dots, \tilde{\mathbf{w}}_N\}$ is a Grassmannian codebook with vectors of size $(N_t - 1) \times 1$ and size $N - 1$.

Another way to build a ring codebook is to use a Kerdock codebook [32] as described in Sect. 4.2.2.4.

4.7.2 Polar-Cap Differential Codebook

We now describe the polar-cap differential codebook design [13, 32]. We define the polar-cap differential function g such as

$$\mathbf{w}[t] = g(\mathbf{w}[t-1], \alpha_t, \mathcal{W}_{loc}) \tag{4.158}$$

where $\mathcal{W}_{loc}$ is the base codebook and α_t is the scale factor at time index t where $1 \leq t \leq t_{max}$

The algorithm is described as follows:

- At time instant $t = 0$, the codeword is selected from the base codebook $\mathcal{W}$. The base codebook $\mathcal{W}$ is built for example using VQ to maximize the minimum chordal distance like in Sect. 4.2.2.5:

$$\mathbf{w}[0] = \arg \min_{\mathbf{w}_i \in \mathcal{W}} d(\mathbf{g}[0], \mathbf{w}_i) \tag{4.159}$$

where $\mathbf{g}[t] = \mathbf{h}[t]/||\mathbf{h}[t]||$ is the normalized channel.

$$\mathbf{c}[1] = \mathbf{w}[0] \tag{4.160}$$

- For time $t = 1, \ldots, t_{max}$:

  Form the scaled codebook $\mathcal{W}^S = S(\mathcal{W}, \alpha_t)$.
  Form the rotated codebook $\mathcal{W}^R = R(\mathcal{W}^S, \mathbf{c}[t])$.
  Compute the estimated vector.

$$\mathbf{w}[t] = \arg \min_{\mathbf{w}_i^R \in \mathcal{W}^R} (d(\mathbf{g}[t], \mathbf{w}_i^R)) \tag{4.161}$$

  Update the centroid $\mathbf{c}[t+1] = \mathbf{w}[t]$.

- At time index $t_{max} + 1$ the algorithm is reset and we set $t = 0$ to clear possible accumulated errors.

The sequence of scale factors $\{\alpha_1, \ldots, \alpha_{t_{max}}\}$ can be adjusted depending on the terminal speed. The higher the terminal speed is, the larger should be the scale factor since it is necessary to track the channel variation.

The complexity of this algorithm is high since for each time index $N-1$ rotation and scaling operations have to be performed. In order to reduce this complexity, Heath et al. [32] have proposed to avoid the rotation of the entire local codebook at each time index t by rotating the observation to match the local codebook with root $\mathbf{o}$. Another reduction of complexity can be obtained by avoiding the rescaling of the codebook using a ring codebook instead of local codebook.

4.7.3 Rotation-Based Differential Codebook

We now describe a differential feedback codebook design approach based on rotation [45]. We define the rotation-based evolution function g such as

$$\mathbf{w}_j[t] = g(\mathbf{w}[t-1], \alpha_t \Theta_j) \tag{4.162}$$

where $\Theta_j \in U(N_t, N_t)$ is a rotation matrix in the rotation codebook composed of a set of $N = 2^B$ unitary matrices $\mathscr{Q} = [\Theta_1, \ldots, \Theta_N]$ and α_t is a scaling factor.

Design of the Rotation Codebook

In [45], the authors have shown that in order to minimize the average distortion, the rotation codebook $\mathscr{Q}$ should be designed to maximize the minimum distance $d_T(\Theta_i, \Theta_j)$ defined in the Riemannian manifold:

$$\mathscr{Q} = \arg\max \min_{1 \leq l < k \leq N} d_T(\Theta_i, \Theta_j) \tag{4.163}$$

where

$$d_T(\Theta_i, \Theta_j) = \sqrt{1 - \frac{1}{N_t}|\text{tr}(\Theta_i^H \Theta_j)|} \tag{4.164}$$

Spherical Cap Codebook Adaptation Strategies

By adjusting the scalar coefficient α_t and using this rotation codebook we can create a spherical cap codebook. Two different methods for the generation of the perturbations have been considered in [45].

Method 1: perturbation in Euclidean space $\mathbb{C}^{N_t \times 1}$

Given the previous precoder $\mathbf{w}[t-1]$, $\mathbf{w}_j[t]$ is obtained by perturbation as follows:

$$\mathbf{w}_j[t] = proj(\Psi_j[t]) \tag{4.165}$$

where

$$\Psi_j[t] = \sqrt{1 - \alpha_t^2}\mathbf{w}[t-1] + \alpha_t \Theta_j \mathbf{w}[t-1] \tag{4.166}$$

The projection is required since $\Psi_j[t]$ is not a unitary matrix. The projection can be performed using the Procrustes orthonormalization or the Gram–Schmidt column orthogonalization [24]. Denote the singular value decomposition of $\Psi_j[t]$ as $\mathbf{U}_j \Sigma_j \mathbf{V}_j^H$. The solution to the Procrustes orthonormalization is given by

$\mathbf{w}_j[t] = \mathbf{U}_j \mathbf{V}_j^H$. In order to avoid the computation of Eqs. (4.165) and (4.166) at each time instant since it depends on $\mathbf{w}[t-1]$, the second method consists in designing the rotation codebook offline.

Method 2: perturbation in Euclidean space $\mathbb{C}^{N_t \times N_t}$

$$\mathbf{R}_j[t] = \sqrt{1 - \alpha_t^2}\mathbf{I}_{N_t} + \alpha_t \Theta_j \tag{4.167}$$

$$\Theta_j[t] = proj(\mathbf{R}_j[t]) \tag{4.168}$$

As previously, the Procrustes orthonormalization or the Gram–Schmidt column orthogonalization can be used for the projection.

The basic rotation-based limited feedback algorithm using method 2 is described as follows:

- Design the rotation codebook $\mathscr{Q} = \{\Theta_1, \ldots, \Theta_N\}$ according to Eq. (4.163).
- At time instant $t = 0$, the codeword is selected from the base codebook $\mathscr{W}[0]$. Like for polar-cap differential codebook, the base codebook $\mathscr{W}[0]$ is built for example using VQ to maximize the minimum chordal distance.

$$\mathbf{w}[0] = \arg \min_{\mathbf{w}_i \in \mathscr{W}[0]} d(\mathbf{g}[0], \mathbf{w}_i) \tag{4.169}$$

$$\mathbf{c}[0] = \mathbf{w}[0] \tag{4.170}$$

- For time $t = 1, \ldots, t_{max}$.
- Design the set of unitary matrix according to Eqs. (4.167) and (4.168) $\mathscr{Q}[t] = \{[\Theta_1[t], \ldots, \Theta_N[t]]\}$.

  Form the rotated codebook $\mathscr{W}^R[t] = [\Theta_1[t]\mathbf{c}[t-1], \ldots, \Theta_N[t]\mathbf{c}[t-1]]$.
  Compute the estimated vector

$$\mathbf{w}[t] = \arg \min_{\mathbf{w}_i^R \in \mathscr{W}^R[t]} (d(\mathbf{g}[t], \mathbf{w}_i^R)) \tag{4.171}$$

  Update the centroid $\mathbf{c}[t+1] = \mathbf{w}$.

- At time index $t_{max} + 1$ the algorithm is reset and we set $t = 0$ to clear possible accumulated errors.

We have performed Monte Carlo simulations to illustrate the performance gain of 4-bit polar-cap and rotation-based differential codebooks in time-varying channel for a MISO system with $N_t = 4$ for different values of α_t. The time correlation in the channel model in Eq. (4.142) is fixed to $\kappa = 0.9987$ corresponding to a speed $v = 1$ km/h assuming $f_c = 2.5$ GHz and a time period of 5 ms. As shown in Fig. 4.21 the rate increases with time instant and is depending on the scaling factor. We have also plot the performance obtained using the optimized time-varying sequence of α_t given in [13]. In that case, α_t is relatively large to track the

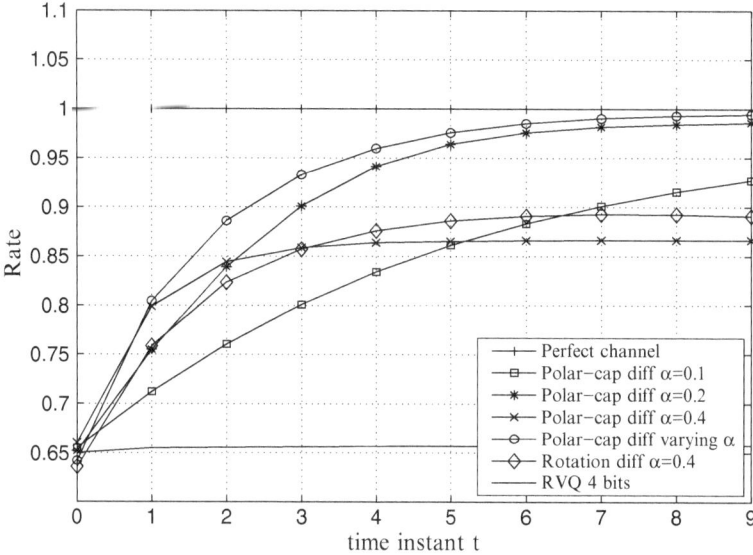

Fig. 4.21 Rate versus time instant in temporally correlated channel for $v = 1\,\text{km/h}$ and SNR=10 dB

channel for small t and become smaller when t increases. As expected, polar-cap differential codebooks slightly outperform rotation-based differential codebooks. We also plot the performance using a 4-bit RVQ codebook and assuming perfect channel knowledge (Fig. 4.21).

It has been shown in [13] that the distance between the update codewords cannot be maintained when using the rotation-based codebook while the polar-cap differential codebook preserves the same distance between codewords. The main drawback of the differential codebook is the error propagation. If an error occurs during the feedback, this error will propagate during the remaining part of the transmission until the algorithm is reset.

Extending the work on prediction for the SISO case presented in Chap. 3 to the MISO or MIMO case is not straightforward. When analog feedback of the channel vector **h** or **g** is performed, a Kalman filter can be implemented to predict the channel vector at the transmitter. Using the Jakes model, the elements of the channel vector **h** can be modelled using an AR model since the Doppler spectrum is frequency limited. On the other hand for the normalized channel **g** due to the norm constraint, it is possible to exploit the dependance between the elements. When quantized feedback is performed, it is more difficult to track the channel since the state space model is strongly nonlinear and information is only available at each transition between the codewords. This research field is still open.

4.8 Extension to Wideband Communication Systems

In this section we will extend the study of limited feedback link for MIMO system to the case of frequency selective channels. We have seen in Chap. 2 that OFDM enables low complexity implementation of MIMO scheme over frequency selective MIMO channels and consequently we will restrict our work to MIMO-OFDM. In that case, optimal beamforming or linear precoding requires the knowledge of channel state information for all the OFDM subcarriers. Different techniques have been proposed to reduce the amount of feedback information, but a classical approach consists in sending feedback information only for the pilot subcarriers. In that case, a challenge is to determine the precoders for the non-pilot subcarriers using interpolation. The precoders for the non-pilot subcarriers can be determined using the modified spherical interpolation [11], the geodesic interpolation [68] or the QR decomposition [10].

Another classical approach is to determine the mean of all the precoders in a cluster and feed back the index of the nearest precoder to the transmitter [68, 69].

The system model given in Eq. (4.84) can be extended to the MIMO-OFDM system at the nth subcarrier where $0 \leq n \leq N - 1$ and N is the number of subcarriers.

$$\mathbf{y}_n = \mathbf{H}_n \mathbf{x}_n + \mathbf{n}_n \tag{4.172}$$

where

$$\mathbf{x}_n = \mathbf{W}_n \mathbf{s}_n \tag{4.173}$$

Using uniformly distributed pilots, the receiver can estimate the channel response $\mathbf{H}_n$ at the pilot positions. Then the receiver feeds back a set of precoding matrices corresponding to subcarriers that are possibly pilot positions. Given this information, the challenge is to determine the precoders for the nontransmitted precoding subcarriers. The classical technique is to feed back only one precoding matrix per cluster corresponding to the subcarrier situated in the center of the cluster. At the transmitter side, the simplest approach called clustering consists for the transmitter in using directly these precoding matrices for all the subcarriers of the clusters. As a consequence, a performance degradation is observed at the cluster boundaries.

A more efficient solution consists in performing an interpolation at the transmitter given the fraction of received precoding vectors.

In the case of beamforming, given the set of beamforming vectors $\mathbf{w}_{(q-1)N_Q}$ where $1 \leq q \leq Q$, $N_Q = N/Q$ is the number of subcarriers per cluster and Q is the number of clusters, inspired by the spherical interpolation [9], Choi and Heath [11] have proposed to add a phase rotation parameter in order to compute $\mathbf{w}$ for all subcarriers. The computation of $\mathbf{w}$ for subcarrier $(q - 1)N_Q + p$ is given by

$$\mathbf{w}_{(q-1)N_Q+p} = \frac{(1 - c_p)\mathbf{w}_{(q-1)N_Q} + c_p e^{j\phi_q} \mathbf{w}_{qN_Q}}{||(1 - c_p)\mathbf{w}_{(q-1)N_Q} + c_p e^{j\phi_q} \mathbf{w}_{qN_Q}||} \tag{4.174}$$

where $0 \leq p \leq N_Q - 1$, ϕ_q is the parameter for phase rotation and $c_p = p/N_Q$ is the linear weight value corresponding to the normalized distance in number of subcarriers from the left nearest known position. The main idea of this solution is to weight and sum together the fed back beamforming vectors from the two nearest known positions. When $\phi_q = 0$, the interpolation is a simple first-order interpolator. To maximize the performance of this interpolator the phase rotation ϕ_q must be optimized according to a performance criterion. In order to maximize the minimum effective channel gain, the phase can be computed as

$$\phi_q = \arg\max_{\Theta} \min_{0 \leq p \leq N_Q - 1} ||\mathbf{h}_{(q-1)N_Q+p} \mathbf{w}_{(q-1)N_Q+p}||^2 \qquad (4.175)$$

The set $\{\phi_q, 1 \leq q \leq Q\}$ is then fed back with the set of beamforming vectors $\{\mathbf{w}_{(q-1)N_Q}, 1 \leq q \leq Q\}$. In practice, the phase is uniformly quantized using $\log_2 P$ bits $\Theta = \{0, \frac{2\pi}{P}, \frac{4\pi}{P}, \ldots, \frac{2(P-1)\pi}{p}\}$ to limit the complexity of the search and the feedback load. The phase can also be determined to maximize the sum rate of all OFDM subcarrier, considering waterfilling across subcarriers as proposed in [11].

When considering linear precoding in MIMO-OFDM, different methods have also been proposed including clustering and interpolation as in beamforming. Compared to beamforming, for each subcarrier n, the precoding matrix must satisfy the unitary property $\mathbf{W}_n^H \mathbf{W}_n = \frac{1}{N_s} \mathbf{I}_{N_s}$, i.e., the columns of $\mathbf{W}_n$ should be orthogonal. Due to the invariance to right multiplication by a unitary square matrix $\mathbf{Q}$, the precoders $\mathbf{W}_n$ and $\mathbf{W}_n \mathbf{Q}$ give the same performance. Consequently, like in the beamforming case, the optimal solution is not unique. Choi et al. [12] have proposed the following spherical interpolation where the phase rotation optimization has been replaced by the optimization of a square unitary matrix $\mathbf{Q}_q$ of size $N_s \times N_s$ as follows:

$$\mathbf{V}_{(q-1)N_Q+p} = (1 - c_p)\mathbf{W}_{(q-1)N_Q} + c_p \mathbf{Q}_q \mathbf{W}_{qN_Q} \qquad (4.176)$$

followed by the orthonormalization:

$$\mathbf{W}_{(q-1)N_Q+p} = \mathbf{V}_{(q-1)N_Q+p} \times \{\mathbf{V}_{(q-1)N_Q+p}^H \mathbf{V}_{(q-1)N_Q+p}\}^{-1/2} \qquad (4.177)$$

where $c_p = p/N_Q$ is the linear weight value like in the beamforming case. In Eq. (4.176), if $\mathbf{Q}_q = \mathbf{I}_{N_s}$, then the interpolation is again a simple first-order interpolator.

In practice, the unitary matrix $\mathbf{Q}_q$ is selected from a codebook $\mathscr{Q}$ composed of unitary matrices. Given this codebook, the selection of the matrix $\mathbf{Q}_q$ can be done to maximize the minimum effective channel gain of the subcarrier halfway between the two known positions as follows:

$$\mathbf{Q}_q = \arg\max_{\mathbf{Q} \in \mathscr{Q}} ||\mathbf{H}_{(q-1/2)N_Q} \mathbf{W}_{(q-1/2)N_Q}||^2 \qquad (4.178)$$

Another interpolation consists in performing a geodesic interpolation, i.e., linear interpolation on the Grassmann manifold $G(N_t, N_s)$. Geodesic interpolation has been first proposed for reduced feedback MIMO-OFDM precoding by Pande et al. [69].

Let's describe this approach. Given two points on a manifold (a Grassmannian manifold in our case), a geodesic is the curve on the manifold of shortest length between these two points. Please refer to the seminal paper of Edelman et al. [18] for a comprehensive introduction to differential geometry and manifolds. Given a point $\mathbf{W}(0) \in G(N_t, N_s)$ we can reach the point $\mathbf{W}(f)$ of the geodesic using the following parametric equation:

$$\mathbf{W}(f) = \mathbf{Q} \exp(f\mathbf{B}) \begin{bmatrix} \mathbf{I}_{N_s} \\ \mathbf{0}_{(N_t - N_s) \times N_s} \end{bmatrix} \tag{4.179}$$

where the matrix $\mathbf{Q}$ is an orthogonal matrix of size $N_t \times N_t$ given by

$$\mathbf{Q} = [\mathbf{W}(0) \; \mathbf{W}^{\perp}(0)] \tag{4.180}$$

where $\mathbf{W}^{\perp}(0) \in G(N_t, N_t - N_s)$ spans the column null space of $\mathbf{W}^T(0)$. As seen in Sect. 4.8, the computation of $\mathbf{Q}$ can be done via Householder reflectors. In the case of geodesics in the Grassmann manifold, the matrix $\mathbf{B}$ is a skew-symmetric matrix of the form

$$\mathbf{B} = \begin{bmatrix} 0 & \mathbf{A}^H \\ -\mathbf{A} & 0 \end{bmatrix} \tag{4.181}$$

where the matrix $\mathbf{A}$ of size $(N_t - N_s) \times N_s$ specifies the direction and the speed of the geodesic curve with parameter f.

Let θ be the vector of principal subspace angles between the two points $\mathbf{W}(0)$ and $\mathbf{W}(1)$ $\theta = [\theta_1 \ldots \theta_{N_s}]$ where θ_i be the ith principal angle. The singular decomposition of $\mathbf{W}(1)^H \mathbf{W}(0)$ is then given by

$$\mathbf{W}(1)^H \mathbf{W}(0) = \mathbf{U}_1 (\cos \Theta) \mathbf{V}_1^H \tag{4.182}$$

where $\mathbf{U}_1$ and $\mathbf{V}_1 \in U(N_t, N_t)$ and Θ is the diagonal matrix of principal angles θ_i.

The velocity matrix $\mathbf{A}$ relates how the principal subspace angles between $\mathbf{W}(f)$ and $\mathbf{W}(0)$ evolve with f and can also be decomposed using singular value decomposition as follows:

$$\mathbf{A} = \mathbf{U}_2 \hat{\Sigma} \mathbf{U}_1^H \tag{4.183}$$

where $\mathbf{U}_2 \in U(N_t - N_s, N_s)$, $\hat{\Sigma} \in \mathbb{R}^{N_s \times N_s}$ is a real diagonal matrix and $\mathbf{U}_1$ is the above matrix.

Given a starting point $\mathbf{W}(0)$ and an ending point $\mathbf{W}(1)$ depending on the application and the number of samples of the geodesic that we want to compute, different solutions can be derived. A first method consists in explicitly calculating Eq. (4.179) after performing the SVD decomposition of the direction matrix $\mathbf{A}$. Another method proposed in [20] does not require to calculate $\mathbf{A}$. We will now describe this last method.

Exploiting the invariance to right unitary multiplication, a convenient way to connect the two points $\mathbf{W}(0)$ and $\mathbf{W}(1)$ is by defining canonical bases $\bar{\mathbf{W}}(0)=\mathbf{W}(0)\mathbf{U}_1$ and $\bar{\mathbf{W}}(1)=\mathbf{W}(1)\mathbf{V}_1$. The geodesic can be rewritten in terms of the canonical bases by right multiplication of Eq. (4.179) given by

$$\bar{\mathbf{W}}(f) = \mathbf{W}(f)\mathbf{U}_1$$

$$= \mathbf{Q}\exp(f\mathbf{B})\begin{bmatrix} \mathbf{I}_{N_s} \\ 0_{(N_t-N_s)\times N_s} \end{bmatrix}\mathbf{U}_1$$

$$= \mathbf{Q}\exp(f\mathbf{B})\mathbf{Q}^H\bar{\mathbf{W}}(0) \tag{4.184}$$

The matrix $\exp(f\mathbf{B})$ can be decomposed as follows:

$$\exp(f\mathbf{B}) = \mathbf{U}\mathbf{R}(f)\mathbf{U}^H \tag{4.185}$$

where

$$\mathbf{U} = \begin{bmatrix} \mathbf{U}_1 & 0 \\ 0 & \hat{\mathbf{U}}_2 \end{bmatrix} \quad \text{and} \quad \mathbf{R}(f) = \begin{bmatrix} \cos(f\Theta) & \sin(f\Theta) & 0 \\ -\sin(f\Theta) & \cos(f\Theta) & 0 \\ 0 & 0 & \mathbf{I}_{N_t-2N_s} \end{bmatrix} \tag{4.186}$$

where $\hat{\mathbf{U}}_2 \in U(N_t - N_s, N_t - N_s)$ and $\mathbf{U}_1$ is the above matrix.

By substituting for $\mathbf{R}(f)$ in Eq. (4.184), we obtained

$$\bar{\mathbf{W}}(f) = \mathbf{Q}\mathbf{U}\mathbf{R}(f)\mathbf{U}^H\mathbf{Q}^H\bar{\mathbf{W}}(0)$$

$$= \mathbf{Q}\mathbf{U}\mathbf{R}(f)\mathbf{U}^H\begin{bmatrix} \mathbf{I}_{N_s} \\ 0_{(N_t-N_s)\times N_s} \end{bmatrix}\mathbf{U}_1$$

$$= \mathbf{Q}\begin{bmatrix} \mathbf{U}_1\cos(f\Theta) \\ -\mathbf{U}_2\sin(f\Theta) \end{bmatrix}$$

$$= \mathbf{W}(0)\mathbf{U}_1\cos(f\Theta) - \mathbf{W}^\perp(0)\mathbf{U}_2\sin(f\Theta) \tag{4.187}$$

where $\mathbf{U}_2$ is a $N_t - N_s \times N_t$ matrix made up of the first N_t columns of $\hat{\mathbf{U}}_2$.

Without loss of generality we can link $\bar{\mathbf{W}}(0)$ to the precoder matrix $\bar{\mathbf{W}}_0$ and $\bar{\mathbf{W}}(1)$ to the precoder matrix $\bar{\mathbf{W}}_{N_Q}$ assuming that the pilots are spaced by N_Q subcarriers. The aim of the geodesic interpolation is then to determine $\mathbf{W}_n = \bar{\mathbf{W}}_n$ with $0 < n < N_Q$.

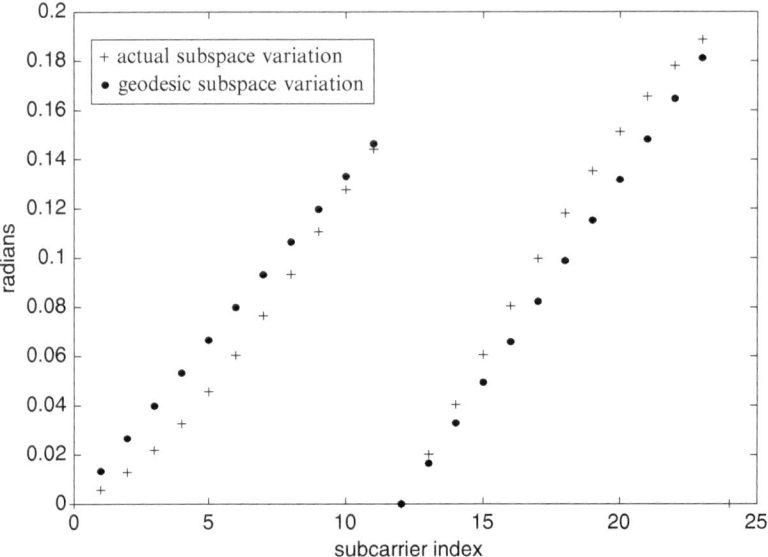

Fig. 4.22 Subspace angular variation of the data precoder using geodesic interpolation with pilot spacing of 12 subcarriers

Using Eq. (4.187), $\bar{\mathbf{W}}_{N_Q}$ is given by

$$\bar{\mathbf{W}}_{N_Q} = \mathbf{W}_0 \mathbf{U}_1 \cos(\Theta) - \mathbf{W}_0^{\perp} \mathbf{U}_2 \sin(\Theta) \tag{4.188}$$

Then we can derive the unknown matrix $\mathbf{W}_0^{\perp} \mathbf{U}_2$ from $\bar{\mathbf{W}}_0$ and $\bar{\mathbf{W}}_{N_Q}$ as follows:

$$\mathbf{W}^{\perp}(0)\mathbf{U}_2 = \left(\bar{\mathbf{W}}_0 \cos(\Theta) - \bar{\mathbf{W}}_{N_Q}\right)\left(\sin(\Theta)\right)^{-1} \tag{4.189}$$

Finally, the precoder matrices can be determined for $0 < n < N_Q$ from

$$\mathbf{W}_n = \bar{\mathbf{W}}_n = \bar{\mathbf{W}}_0 \cos\left(\frac{n}{N_Q}\Theta\right) - \mathbf{W}_0^{\perp} \mathbf{U}_2 \sin\left(\frac{n}{N_Q}\Theta\right) \tag{4.190}$$

Another interpolation based on the QR decomposition has been proposed in [10].

Following [69], we show in Fig. 4.22, for a single realization of frequency selective channels, the subspace angular variation for the right singular vector corresponding to the dominant singular value in a MIMO-OFDM system with $N_t = 4$, $N_r = 4$ and 1,024 subcarriers corresponding to a 10 MHz bandwidth and a pilot spacing of 12 subcarriers. The channel model used is the ITU Vehicular A with a maximum delay spread of 2.5 μs as defined in Table 2.2. For clarity, we only plot the subspace angular variation associated to the first two clusters.

We can observe a slight difference between the actual subspace variation and the straight line approximation of connecting the end points obtained using the geodesic interpolation.

While in current clustering-based methods for each cluster, the precoding matrix of the center subcarrier is reused for the other subcarriers of the cluster, and an alternative approach proposed in [69] computes the mean of all the precoding matrices in a cluster and determines the associated quantized precoding matrix taken from the codebook.

An interpolation framework for the difficult multimode precoding problem where both the precoding matrix and the rank of the matrix evolve over the subcarriers has been proposed in [42].

4.9 Conclusion

In this chapter, we have presented in detail the different feedback strategies for the case of a single user MIMO wireless communication system.

In the first part of this chapter, we have considered the simplest case where only one stream is sent from the transmitter to the receiver corresponding to the MISO case or MIMO with receiver combining. In order to limit the amount of feedback, and since the information is contained in the channel direction, we have mainly considered the case where the normalized MISO channel is quantized and feedback and beamforming is applied at the transmitter.

We have seen that the codebook composed of vectors can also be seen as a finite set of subspaces in the Grassmannian manifold. We have derived the codebook criterion and different bounds on performance. Then we have reviewed the different methods of construction of codebooks including vector quantization.

For the MIMO systems where multiple streams are transmitted to the receiver, linear precoding extends the beamforming precoding by premultiplying the different transmitted data streams using a precoding matrix. In that case, the CDI is a unitary matrix and depending on the selected criterion the codebook design should be adapted.

While the first part of this chapter was dedicated to single- or multiple-stream transmission using spatial multiplexing, we have shown that space-time codes can also benefit from CSIT when they are associated with a precoder. We have also studied the different possible antenna selection strategies in MIMO systems. We have shown that antenna selection can also be seen as a special case of linear precoding.

In the last part of this chapter, we have studied the different techniques to exploit the spatial and time and frequency correlation of the channel. We have shown that the codebook design should be adapted as spatial correlation by performing a local packing. For time correlation, the same idea can be exploited and similarly to correlated source coding, the class of differential feedback strategies are the most

promising. Finally in the case of selective MIMO channels we have shown that it
is possible to reduce the quantity of feedback information or improve the wideband
channel estimation quality using interpolation techniques.

References

1. Akhtar J, Gesbert D (2004) Extending orthogonal block codes with partial feedback. IEEE
   Trans. Wireless Commun. 3 : 1959–1962
2. Au-Yeung C K, Love D J (2005) Performance Analysis of Random Vector Quantization
   Limited Feedback Beamforming. Proc. of IEEE Asilomar Conf. on Signals, Systems, and
   Computers. Pacific Grove, CA
3. Banister B C, Zeidler J R (2003) Feedback assisted transmission subspace tracking for MIMO
   systems. IEEE Jour. Select. Areas in Commun. 21: 452–463
4. Banister B C, Zeidler J R (2003) A simple gradient sign algorithm for transmit antenna weight
   adaptation with feedback. IEEE Trans. Sig. Proc. 51 : 1156–1171
5. Barg A, Nogin D Y (2002) Bounds on packings of spheres in the Grassmann manifold. IEEE
   Trans. Inform. Theory. 48 : 2450–2454
6. Blum R S, Winters J H (2002) On optimum MIMO with antenna selection. Proc. International
   Conference Communication. 386–390
7. Boothby W M (1986) An Introduction to Differential Manifolds and Riemannian Geometry.
   Academic Press. 2nd ed. San Diego, CA.
8. Brierley S, Weigert S, Bengtsson I (2010) All mutually unbiased bases in dimensions two to
   five Journal Quantum Information and Computation. 10 : 803–820
9. Buss S R, Fillmore J P (2001) Spherical averages and applications to spherical splines and
   interpolation. ACM Trans. Graphics. 20 : 95–126
10. Cescato D, Borgmann M, Bolcskei H, Hansen J, Burg A (2005) Interpolation-based QR
    decomposition in MIMO-OFDM systems. Proc. IEEE Int. Workshop on Signal Proc. Adv.
    Wireless Commun. 945–949
11. Choi J, Heath Jr R W (2005) Interpolation based transmit beamforming for MIMO-OFDM
    with limited feedback. IEEE Trans. Sig. Proc. 53 : 4125–4135
12. Choi J, Mondal B, Heath Jr R W (2006) Interpolation based unitary precoding for spatial
    multiplexing MIMO-OFDM with limited feedback. IEEE Trans. Signal Processing. 54 :
    4730–4740
13. Choi J, Clerck B, Lee N, Kim G (2012) A new design of polar-cap differential codebook for
    temporally/spatially correlated MISO channels. IEEE Trans. on Wireless Communications.
    11 : 703–711
14. Chung K L, Yan W M (1997) The complex householder transform. IEEE Transactions on
    Signal Processing. 45 : 2374–2376
15. Clerckx B, Zhou Y, Kim Y (2008) Practical codebook design for limited feedback spatial
    multiplexing. IEEE Int. Conf. Communication Proc. 1 :3982–3987
16. Conway J H, Hardin R H, Sloane N J A (1996) Packing lines, planes, etc.: Packings in
    Grassmannian spaces. Experimental Math. 5:139–159
17. Conway J H, Sloane N J A (1998) Sphere Packing, Lattices and Groups. Number 290 in
    Grundlehren der mathematischen Wissenschaften. Springer Verlag, 3rd edition
18. Edelman A, Arias T A, Smith S T (1998) The geometry of algorithms with orthogonality
    constraints. SIAM J. Matrix Anal. Appl. 20 : 303–353
19. Erceg V et al, Indoor MIMO WLAN Channel Models (2003) IEEE 802.11-03/161r4.
20. Gallivan K, Srivastava A, Liu X, Dooren P (2003) Efficient algorithms for inferences on
    Grassmann manifolds. Proc. IEEE Workshop Statistical Signal Process. 315–318

21. Gaspar M, Rimoldi R, Vetterli M (2003) To code, or not to code: lossy source channel communication revisited . IEEE Trans. on Information Theory. 19 : 1147–1158
22. Gerlach D, Paulraj A (1994) Adaptive transmitting antenna arrays with feedback. IEEE Sig. Proc. Lett. 1 . 150–152
23. Gersho A, Gray R M (1992) Vector Quantization and Signal Compression. Kluwer Academic Publishers.
24. Golub G H, Loan C F V (1996) Matrix Computation, 3rd edition. Johns Hopkins University Press
25. Gore D, Nabar R, Paulraj A (2000) Selecting an optimal set of transmit antennas for a low rank matrix channel, in Proc. Int. Conf. Acoust., Speech and Sig. Proc. (ICASSP). 5 : 2785–2788
26. Gore D, Heath Jr R W, Paulraj A J (2002) Transmit selection in spatial multiplexing systems. IEEE Commun. Lett. 6 : 491–493
27. Gore D, Paulraj A (2002) MIMO antenna sub-set selection with space-time coding. IEEE Trans. Signal Processing. 50 : 2580–2588
28. Gorokhov A, Collados M, Gore D, Paulraj A (2004) Transmit/receive MIMO antenna subset selection. Proc. IEEE Int. Conf. Acoust. Speech and Sig. Proc (ICASSP). 2 : 13–16
29. Gutierrez I, Bader F, Mourad A (2009) Joint Transmit Antenna and Space-Time Coding Selection for WiMAX MIMO System. Proc. of 20th IEEE Personal, Indoor and Mobile Radio Communications (PIMRC) Symposium. Tokyo, Japan. 3138–3143
30. Heath Jr R W, Paulraj A (1998) A simple scheme for transmit diversity using partial channel feedback. Proc. IEEE Asilomar Conf. on Signals, Systems, Comp. 2 : 1073–1078
31. Heath Jr R W, Love D J (2005) Multimode antenna selection for spatial multiplexing systems with linear receivers. IEEE Trans. Signal Processing. 53 : 3042–3056
32. Heath Jr R W, Wu T, Soong A C K (2009) Progressive refinement of beamforming vectors for high-resolution limited feedback. EURASIP J. Advances Signal Process. article ID 463823, 13 pages.
33. Hochwald B M, Marzetta T L, Richardson T L, Sweldens W, Urbanke R (2000) Systematic design of unitary space-time constellations. IEEE Trans. on Information Theory. 46 : 1962–1973
34. Hottinen A, Tirkkonen O, Wichman R (2000) Closed-loop transmit diversity techniques for multi-element transceivers. Proc. IEEE Veh. Technol. Conf. (VTC). 1 : 70–73
35. Huang K, Mondal B, Heath R W, Andrews J G (2006) Multiantenna limited feedback for temporally correlated channels: feedback compression. Proc. IEEE Global Telecommunications Conference (GLOBECOM). 1–5
36. Inoue T, Heath Jr R W (2008) Kerdock codes for limited feedback MIMO systems. IEEE Trans. Signal Proc. 57 : 3711–3716
37. Isukapalli Y, Zheng J, Rao B D (2007) Average SEP loss analysis of transmit beamforming for finite rate feedback MISO systems with QAM constellation. Proc. IEEE Int. Conf. Acoust., Speech and Sig. Proc. (ICASSP). 3 : 425–428
38. Jafar S A, Goldsmith A J (2004) Transmit optimization and optimality of beamforming for multiple antenna systems with imperfect feedback. IEEE Trans. Wireless Commun. 3 : 1165–1175
39. Jindal N (2006) MIMO broadcast channels with finite-rate feedback. IEEE Transactions on Information Theory. 52 : 5045–5060
40. Jongren G, Skoglund M, Ottersten B (2002) Combining beamforming and orthogonal space-time block coding. IEEE Trans. Inform. Theory. 48 : 611–627
41. Jorswieck E A, Boche H (2004) Channel Capacity and Capacity-range of Beamforming in MIMO Wireless Systems under Correlated Fading with Covariance Feedback. IEEE Trans. Wireless Commun. 3 : 1543–1553
42. Khaled N, Mondal B, Leus G, Heath Jr R W, Petre F (2007) Interpolation-based multi-mode precoding for MIMO-OFDM systems with limited feedback. IEEE Trans. Wireless Comm. 6 : 1003–1013
43. Kaviani S, Tellambura C (2006) Closed-form BER analysis for antenna selection using orthogonal space-time block codes. IEEE Commun. Lett. 10 : 704–706

44. Kerdock A (1972) Studies of low-rate binary codes. IEEE Trans. Inf. Theory. 18 : 316–316
45. Kim T, Love D J, Clerckx B (2010) MIMO system with limited rate differential feedback in slow varying channel. IEEE Trans. Commun. 59 : 1175–1189
46. Klappenecker A, Roetteler M (2004) Constructions of mutually unbiased bases, Finite Fields and Applications. Lecture Notes in Computer Science. 2948 : 137–144
47. Ko Y, Tepedelenlioglu C (2008) Threshold-based substream selection for closed-loop spatial multiplexing. IEEE Trans. Veh. Technol. 57 : 215–226
48. Larsson E G, Ganesan G, Stoica P, Wong W H (2002) On the performance of orthogonal space-time block coding with quantized feedback. IEEE Commun. Lett. 6 : 487–489
49. Le Ruyet D, Ozbek B (2007) Partial and Analog Feedback for MISO Precoding Systems. Proceedings of IEEE Intern. Conf. On Communications (ICC), Glasgow, UK. 2767–2772
50. Li Y, Mehta N B, Molisch A F, Zhang J (2007) Optimal signaling and selection verification for single transmit-antenna selection. IEEE Trans. Commun. 55 : 778–789
51. Liu L, Jafarkhani H (2005) Application of quasi-orthogonal space time block codes in beamforming. IEEE Trans. Signal Process. 53 : 54–63
52. https://engineering.purdue.edu/~djlove/grass.html
53. Love D J, Heath Jr R W (2003) Equal gain transmission in multiple-input multiple-output wireless systems. IEEE Trans. Commun. 51 : 1102–1110
54. Love D J, Heath Jr R W, Strohmer T (2003) Grassmanian beamforming for multiple input multi output wireless systems. IEEE Trans. on Information Theory. 49 : 2735–2747
55. Love D J, Heath Jr R W (2005) Limited feedback unitary precoding for orthogonal space-time block codes. IEEE Trans. Signal Process. 53 : 64–73
56. Love D J, Heath Jr R W (2005) Limited feedback unitary precoding for spatial multiplexing systems. IEEE Trans. Info. Theory. 51 : 2967–2976
57. Love D J, Heath Jr R W (2005) Multimode Precoding for MIMO Wireless Systems. IEEE Trans. Signal Process. 53 : 3664–3678
58. Love D J, (2005) On the probability of error of antenna-subset selection with space-time block codes. IEEE Trans. Commun. 53 :1799–1803
59. Love D J, Heath Jr R W (2006) Limited feedback diversity techniques for correlated channels. IEEE Trans. on Veh. Technol. 55 : 718–722
60. Marzetta T L, Hochwald B M (2006) Fast transfer of channel state information in wireless systems. IEEE Trans. on Information Theory. 54 :1268–1278
61. Mehta N B, Molisch A F, Kashyap S (2012) Antenna selection in LTE from motivation to specification. IEEE Communications Magazine. 50 : 144–150
62. Molisch A F, Win M Z (2004) MIMO systems with antenna selection. IEEE Microwave Mag., 5 : 46–56
63. Mondal B, Thomas T A, Harrison M (2007) Rank independent codebook design from a quaternary alphabet. Proceedings of the Asilomar Conference on Signals, Systems and Computers, Pacific Grove, Calif, USA. 297–301
64. Moustakas A L, Simon S H (2003) Optimizing Multiple-Input Single-Output (MISO) Communication Systems with General Gaussian Channels : Nontrivial Covariance and Nonzero Mean. IEEE Trans. Inf. Theory. 49 : 2770–2780
65. Mukkavilli K K, Sabarwal A, Erkip E, Aazhang B (2003) On beamforming with finite rate feedback in multiple antenna systems. IEEE Trans. on Information Theory . 49 : 2562–2579
66. Nagaraj S, Huang Y F (2000) Downlink transmit beamforming with selective feedback. Proc. IEEE Asilomar Conf. on Signals, Systems, Comp. 2 : 1608–1612
67. Narula A, Lopez M J, Trott M D, Wornell G W (1998) Efficient use of side information in multiple antenna data transmission over fading channels. IEEE Journal on selected areas in communications. 1423–1436
68. Pande T, Love D J, Krogmeier J V (2006) A weighted least squares approach to precoding with pilots for MIMO-OFDM. IEEE Trans. Sig. Proc. 54 : 4067–4073
69. Pande T, Love D J, Krogmeier J V (2007) Reduced feedback MIMO-OFDM precoding and antenna selection. IEEE Trans. Sig. Proc. 55 : 2284–2293

70. Phan K T, Tellambura C (2007) Capacity analysis for transmit antenna selection using orthogonal space-time block codes. IEEE Commun. Lett. 11 : 423–425

71. Pitaval R A, Tirkkonen O, Blostein S D (2011) Density and Bounds for Grassmannian Codes with Chordal Distance. IEEE International Symposium on Information Theory (ISIT), Saint Petersburg, Russia. 2298–2302

72. Raghavan V, Heath Jr R W, Sayeed A M (2007) Systematic codebook designs for quantized beamforming in correlated MIMO channels. IEEE Jour. Select. Areas in Commun. 25 : 1298–1310

73. Rankin R A (1947) On the closest packing of spheres in n dimensions. Ann. Math. 48 : 1062–1081

74. Roh J C, Rao B D (2006) Transmit beamforming in multiple-antenna systems with finite rate feedback: A VQ-based approach. IEEE Trans. Info. Theory. 52 : 1101–1112

75. Rouquette S, Merigeault S, Gosse K (2002) Orthogonal full diversity space-time block coding based on transmit channel state information for 4 Tx antennas. IEEE ICC. 1 : 558–562

76. Ryan D J, Clarkson I V L, Collings I B, Guo D, Honig M L (2009) QAM and PSK codebooks for limited feedback MIMO beamforming. IEEE Trans. Commun. 57 : 1184–1196

77. Sacristan D, Kaltenberger F, Pascual-Iserte A, Perez A (2009) Differential feedback in MIMO communications: performance with delay and real channel measurements. Proc. Int. ITG Workshop on Smart Antennas.

78. Sampath H, Stoica P, Paulraj A (2001) Generalized linear precoder and decoder design for MIMO channels using the weighted MMSE criterion. IEEE Trans. Commun. 49 : 2198–2206

79. Sampath H, Paulraj A (2002) Linear precoding for space-time coded systems with known fading correlations. IEEE Commun. Lett. 6 : 239–241

80. Sanayei S, Nosratinia A (2004) Antenna selection in MIMO systems. IEEE Comm. Mag. 42 : 68–73

81. Santipach W, Honig M L (2003) Asymptotic performance of MIMO wireless channels with limited feedback. Proc. IEEE Mil. Comm. Conf. 1 : 141–146.

82. Scaglione A, Stoica P, Barbarossa S, Giannakis G B, Sampath H (2002) Optimal designs for space-time linear precoders and decoders. IEEE Trans. Sig. Proc. 50 : 1051–1064

83. Simon M K, Alouini M S (1998) A unified approach to the performance analysis of digital communication over generalized fading channels. Proceedings of the IEEE. 86 : 1860–1877

84. http://www.research.att.com/~njas/grass/index.html

85. Sollenberger N R (1993) Diversity and automatic link transfer for a TDMA wireless access link. Proc. IEEE Glob. Telecom. Conf. 1 : 532–536

86. Strohmer T, Heath Jr R W (2003) Grassmannian frames with applications to coding and communications. Applied and Computational Harmonic Analysis. 14 : 257–275

87. Tammes P M L (1930) On the origin of number and arrangement of the places of exit on the surface of pollen grains. Rec. Trav. bot. neerl. 27:1–84

88. Toker C, Lambotharan S, Chambers J (2004) Closed-loop quasi-orthogonal STBCs and their performance in multipath fading environments and when combined with turbo codes. IEEE Trans. Wireless Commun. 3 : 1890–1896

89. Tropp J, Dhillon I, Heath Jr R W, Strohmer T (2008) Constructing packings in grassmannian manifolds via alternating projections. Experimental Mathematics. 17 : 9–35

90. Visotsky E, Madhow U (2001) Space-time transmit precoding with imperfect feedback. IEEE Trans. Inf. Theory. 47 : 2632–2639

91. Welch L (1974) Lower bounds on the maximum cross correlation of signals. IEEE Trans. Inf. Theory. 20 : 397–399

92. Winters J H (1983) Switched diversity with feedback for dpsk mobile radio systems. IEEE Trans. on Veh. Tech. 32 : 134–150

93. Xia P, Zhou S, Giannakis G B (2005) Achieving the Welch bound with difference sets. IEEE Trans. Info. Theory. 51 : 1900–1907

94. Xia P, Giannakis G B (2006) Design and analysis of transmit beamforming based on limited-rate feedback. IEEE Trans. Sig. Proc. 54 : 1853–1863

95. Yang J, Williams D (2007) Transmission subspace tracking for MIMO systems with low-rate feedback. IEEE Trans. Commun. 55 : 1629–1639
96. Zhang L, Song L, Ma M, Jiao B (2012) On the minimum differential feedback for time-correlated MIMO Rayleigh block-fading channels. IEEE Trans. on Communications. 60 : 411–420
97. Zheng L, Tse D (2002) Communication on the Grassmann manifold: a geometric approach to the noncoherent multiple-antenna channel. IEEE Trans. Inform. Theory. 48 : 359–383 .
98. Zheng J, Rao B D (2007) Analysis of multiple antenna systems with finite-rate channel information feedback over spatially correlated fading channels. IEEE Trans. Sig. Proc. 55 : 4612–4626
99. Zhou S, Giannakis G B (2002) Optimal transmitter eigen-beamforming and space-time block coding based on channel mean feedback. IEEE Trans. Signal Process. 50 : 2599–2613
100. Zhou S, Giannakis G B (2003) Optimal transmitter eigen-beamforming and space-time block coding based on channel correlations. IEEE Trans. Inf. Theory. 49 : 1673–1690
101. Zhou S, Wang Z, Giannakis G B (2004) Performance analysis for transmit-beamforming with finite rate feedback. Conf. on Information Sciences and Systems (CISS), Princeton, USA. 17–19
102. Zhou S, Li B (2006) BER criterion and codebook construction for finite rate precoded spatial multiplexing with linear receivers. IEEE Trans. Sig. Proc. 54 : 1653–1665
103. IEEE C802.16m-08/1187 (2008) Evaluation of CL SU and MU MIMO Codebooks.
104. IEEE C802.16m-09/0058r3 (2009) Differential Feedback for IEEE 802.16m MIMO Schemes.

Chapter 5
Feedback Strategies for Multiuser Systems

5.1 Introduction

Multiple access techniques divide up the total signaling dimensions into channels and then assign these channels to different users. The most common methods to divide up the signal space are along the time, frequency or code axes. The different user channels are then created by an orthogonal division along these axes: Time-division multiple access (TDMA) and frequency-division multiple access (FDMA) are orthogonal channelization methods whereas code-division multiple access (CDMA) can be orthogonal or non-orthogonal, depending on the code design. Multiuser systems refer to transmission system where the resources are shared among multiple users. In multiuser systems, the channel is allocated to the users adaptively by employing different scheduling techniques to achieve multiuser diversity.

This chapter gives background on different transmission techniques according to the knowledge of the users' channels at the transmitter side and reduced feedback information strategies for multiuser systems. Firstly, an overview of the previous works that are specially derived in information-theoretic view when the users' channel state information (CSI) are fully known at the transmitter. In Sect. 5.3, user scheduling algorithms are introduced by taking into account different criteria. Then, the reduced and limited feedback algorithms are examined and the performance evaluations are provided for single-carrier and multicarrier-based multiuser systems.

5.2 Multiuser Systems

In point-to-point systems, the channel capacity is a single number that defines the maximum theoretical achievable data rate between the transmitter and the receiver. In a multiuser system such as broadcast or multiple access channel, where different wireless channels fade independently among the users, the transmitter or receiver

B. Özbek and D. Le Ruyet, *Feedback Strategies for Wireless Communication*, 165
DOI 10.1007/978-1-4614-7741-9__5, © Springer Science+Business Media New York 2014

Fig. 5.1 K users uplink
transmission

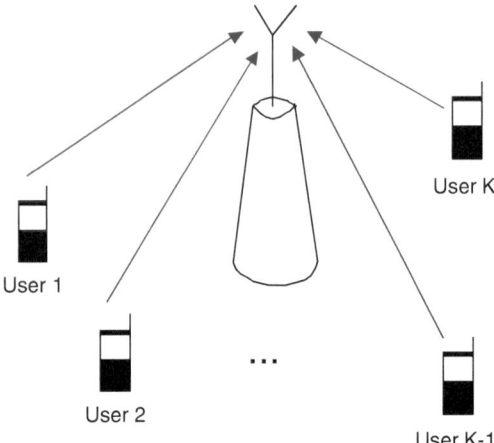

can simultaneously transmit or receive more than one user. Thus, the channel
capacity is a set of all simultaneously achievable rate vectors called capacity region.
When the base station (BS) transmits data to the mobile users, the channel is referred
as a downlink or broadcast channel. Conversely, when the mobile users transmit
data to the BS, the channel is referred as an uplink or multiple access channel. The
broadcast channel (BC) was introduced by Cover in the early 1970s [4], whereas the
multiple-access channel (MAC) dates back to Shannon [19].

5.2.1   Gaussian MAC Communication System

The MAC communication system where multiple users are transmitting their
individual messages to one BS is shown in Fig. 5.1 for K users.

The baseband discrete-time model for the uplink Gaussian channel with $K = 2$
users is given by

$$y = h_1 x_1 + h_2 x_2 + n \tag{5.1}$$

where n is i.i.d. complex Gaussian noise and average power constraint is
$\mathbb{E}\{|x_k|^2\} \le P_k$ with $k = 1, 2$.

The capacity region where user 1 at rate R_1 and user 2 at rate R_2 can reliably
communicate simultaneously is obtained for the set of all pairs (R_1, R_2). Since the
two users share the same bandwidth, there is naturally a trade-off between reliable
transmission rates of users. If user 1 requires to communicate at higher data rates,
user 2 should reduce its data rate.

As the capacity of the single user case, there is a very simple characterization
of the capacity region of the uplink channel. These rates, (R_1, R_2), satisfy the
following three constraints:

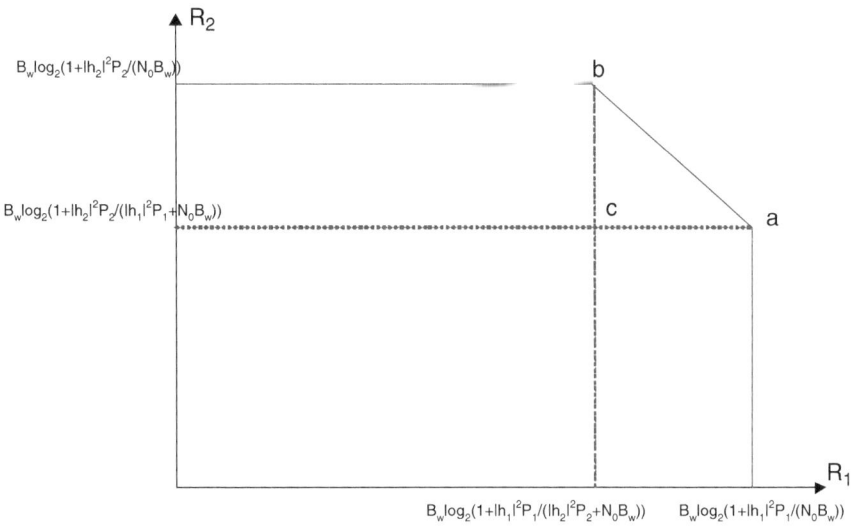

Fig. 5.2 Capacity region of the $K = 2$ users Gaussian MAC transmission

$$R_1 < B_W \log_2 \left(1 + \frac{|h_1|^2 P_1}{N_0 B_W} \right) \tag{5.2}$$

$$R_2 < B_W \log_2 \left(1 + \frac{|h_2|^2 P_2}{N_0 B_W} \right) \tag{5.3}$$

$$R_1 + R_2 < B_W \log_2 \left(1 + \frac{|h_1|^2 P_1 + h_2|^2 P_2}{N_0 B_W} \right) \tag{5.4}$$

The first two constraints say that the rate of the individual user cannot exceed the capacity of point-to-point link when the other user is not in the system. The third constraint says that the total throughput cannot exceed the capacity of the point-to-point Gaussian channel with the sum of the received powers of the two users. Without the third constraint, the capacity region would have been a rectangular and both users could simultaneously transmit at the rate of point-to-point system as if the other user did not exist. However, it is not possible as indicated in the third constraint. There is a trade-off when the two users start to communicate at the same time. As shown in Fig. 5.2, the capacity region is a pentagon. When user 1 can achieve its single user bound as in point a while at the same time user 2 can have a rate higher than zero which is shown in point b.

In the Gaussian MAC with K users, each user transmits its signal that is subject to an average power constraint, $\mathbb{E}[|x_k|^2] \le P_k$ where $k = 1, \dots, K$, and the received signal is equal to the sum of the transmitted signals and the additive white Gaussian noise. Each user is assumed to suffer from a constant channel coefficient, h_k. Mathematically, the received signal in the MAC is equal to

$$y = \sum_{k=1}^{K} h_k x_k + n \tag{5.5}$$

From [6], the capacity region of a Gaussian MAC with channel gains $\mathbf{h} = [h_1, \ldots, h_K]$ and power constraints $\mathbf{P} = [P_1, \ldots, P_K]$ is denoted by $C_{MAC}(\mathbf{h}, \mathbf{P})$:

$$C_{MAC}(\mathbf{h}, \mathbf{P}) = \{\mathbf{R} : \sum_{k=1}^{K} R_k \leq B_W \log_2 \left(1 + \frac{\sum_k h_k^2 P_k}{N_0 B_w}\right) \tag{5.6}$$

The capacity region of the MAC is a K-dimensional polyhedron, and successive decoding with interference cancelation can be used to achieve all $K!$ corner points for a given decoding order which corresponds to a different corner point of the capacity region. Given a decoding order $[\pi_1, \pi_2, \ldots, \pi_K]$ in which user π_1 is decoded first, user π_2 is decoded second, etc., the rates of the corresponding corner point become

$$R_{\pi_k} = B_w \log_2 \left(1 + \frac{h_{\pi_k}^2 P_{\pi_k}}{\sum\limits_{j=k+1}^{K} h_{\pi_j}^2 P_{\pi_j} + N_0 B_w}\right) \tag{5.7}$$

The ordering is stated from the stronger user, π_1, and then move towards the weakest user, π_K. Then, all corner points achieve the same optimal sum rate with successive decoding.

5.2.2 Gaussian BC Communication System

The BC communication system where a single transmitter is sending separate information to K receivers is shown in Fig. 5.3.

The baseband discrete-time model for Gaussian BC with $K = 2$ users is given by

$$y_k = h_k x + n_k; \quad k = 1, 2 \tag{5.8}$$

where $x = \sum_{k=1}^{K} x_k$, n_k is i.i.d. complex Gaussian noise, h_k is a constant channel coefficient and the transmitter is subject to an average transmitted power constraint P as $\mathbb{E}[|x|^2] \leq P$.

The region where two users can simultaneously establish reliable communications is the rates (R_1, R_2). If user 2 has better channel gain than user 1, user 2 can decode any data that user 1 can successfully decode. Thus, the superposition coding scheme is applied: First the transmit signal is the linear superposition of the signals of the two users. User 1 treats the signal of user 2 as noise and decodes its data from

Fig. 5.3 Gaussian BC transmission with K users

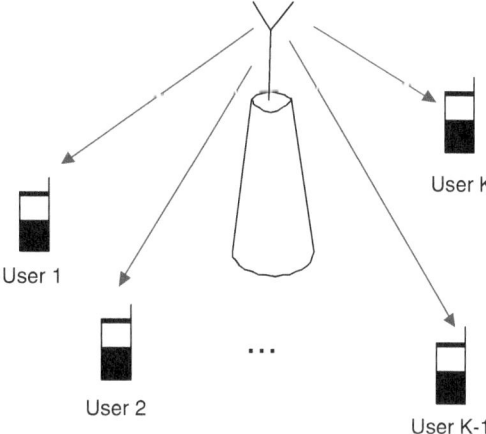

the received signal y_1. User 2 decodes its data by subtracting the signal of user 1 from its received signal y_2. With each possible power split of $P = P_1 + P_2$, the following rate pair can be achieved:

$$R_1 = B_w \log_2 \left(1 + \frac{P_1 |h_1|^2}{P_2 |h_2|^2 + N_0 B_w} \right) \tag{5.9}$$

$$R_2 = B_w \log_2 \left(1 + \frac{P_2 |h_2|^2}{N_0 B_w} \right) \tag{5.10}$$

For Gaussian BC transmission with $K = 2$, the capacity region consists of a set of all rate pairs (R_1, R_2) that can be achievable without errors. Similarly, the rate region for a K-user BC is a K-dimensional region.

In the Gaussian BC with K users, the transmitter sends independent information to each receiver by broadcasting a signal $\mathbf{x}$ to K different receivers simultaneously, where $\mathbf{x}$ contains information for all K receivers such as some part of it can be for user 1, another part can be for user 2, etc. Mathematically, the kth received signal in the BC can be described by

$$y_k = h_k x_k + \sum_{j \neq k} h_k x_j + n_k; \quad k = 1, 2, \dots, K \tag{5.11}$$

where each receiver is assumed to suffer from a fixed gain represented by channel coefficients $h_k; k = 1, 2, \dots, K$.

In [2], Bergmans showed that the capacity region of the Gaussian BC with channel $\mathbf{h} = [h_1, \dots, h_K]$ and power constraint $\sum_{k=1}^{K} P_k \leq P$ is denoted by

$$R_k \leq \log_2 \left(1 + \frac{|h_k|^2 P_k}{N_0 B_w + |h_k|^2 \sum_{j=1}^{K} P_j \mathbf{1}[|h_j| > |h_k|]} \right) ; \forall k \tag{5.12}$$

where if $|h_j| < |h_k|$, then $\mathbf{1}$ operator returns to 0, otherwise it returns as 1.

Any set of rates in the capacity region is achievable using successive decoding, in which users decode and subtract out signals intended for other users before decoding their own signal. To achieve the boundary points of the BC capacity region, the signals are encoded in a way that the strongest user can decode all users' signals, then the second strongest user can decode all users' signals except for the strongest user's signal, etc. The strongest user refers to the user with the largest channel gain $|h_{k*}|$.

As seen in [29], the capacity region of the BC is also achievable via dirty paper coding [5], in which the transmitter pre-subtracts (similar to precoding for inter-symbol interference mitigation) certain users' codewords instead of receivers decoding and subtracting out other users' signals. When users are encoded in that way, this technique achieves capacity and is equivalent to successive decoding. Assuming encoding order $[\pi_1, \pi_2, \ldots, \pi_K]$ in which the codeword of user π_1 is encoded first, the rates achieved in the BC are

$$R_{\pi_k} = \frac{1}{2} \log_2 \left(1 + \frac{|h_{\pi_k}|^2 P_{\pi_k}}{N_0 B_w + |h_{\pi_k}|^2 \sum_{j=k+1}^{K} P_{\pi_j}} \right) \tag{5.13}$$

5.2.3 Fading MAC Transmission

The received signal of the uplink flat fading channel with K users is given in (5.5). In the case of fading MAC transmission, h_k is complex channel coefficient for user k and modeled as i.i.d. fading process with $\mathbb{E}[|h_k|^2] = 1$. When the full CSI is available at the transmitter, it is possible to dynamically allocate powers to the users as a function of the channel states. The optimal capacity-achieving power allocation strategy which allocates powers to the users as a function of the joint channel state $\mathbf{h} = [h_1, \ldots, h_K]$:

$$P_{k*}(\mathbf{h}) = \begin{cases} \left(\frac{1}{\lambda} - \frac{N_0 B_w}{|h_{k*}|^2} \right)^+, & \text{if } |h_{k*}|^2 = \max_k |h_k|^2 \\ 0, & \text{else.} \end{cases} \tag{5.14}$$

with λ chosen to satisfy the power constraint as

$$\sum_{k=1}^{K} \mathbb{E}[P_{k*}(\mathbf{h})] = KP \qquad (5.15)$$

This formula is valid only when there is exactly one user with the strongest channel. Then, the resulting sum capacity is given by

$$C_{SMAC} = \mathbb{E}\left[\log_2 \left(1 + \frac{P_{k*}(\mathbf{h})|h_{k*}|^2}{N_0 B_w} \right) \right] \qquad (5.16)$$

where $k*$ is the index of the user with the strongest channel.

5.2.4   Fading BC Transmission

The received signal of the downlink flat fading channel with K users is given in (5.8). In the case of fading BC transmission with K users, the h_k is the complex channel coefficient of user k and is modeled as i.i.d processes with $\mathbb{E}[|h_k|^2] = 1$.

For the case of full CSI available at the transmitter, the transmitter tracks all the channels of the users and the powers can be allocated to the users as a function of the channel fade level.

The sum capacity is achieved by transmitting only to the best user for each time. In addition to that, an appropriate power subject to a constraint on the average power is further allocated by

$$C_{SBC} = \mathbb{E}\left[\log_2 \left(1 + \frac{P_{k*}(\mathbf{h}) \max_{k=1,2,\dots,K} |h_k|^2}{N_0 B_W} \right) \right] \qquad (5.17)$$

where

$$P_{k*}(\mathbf{h}) = \frac{1}{\lambda} - \frac{N_0 B_W}{\max_{k=1,2,\dots,K} |h_k|^2} \qquad (5.18)$$

5.3   User Scheduling

The usual forms of diversity in single user channels include time, frequency and space diversity. In a multiuser environment with multiple independent wireless links, it is highly probable that at any given point in time, at least one of those links has high quality. This advantage is called multiuser diversity. In multiuser systems, users compete for resources to ensure larger rates and/or better reliability.

In a cellular framework with a single antenna, the maximum throughput is achieved by transmitting to the users with the largest receive SNR at each channel use with the multiuser diversity gain for the uplink transmission raised in [13] and for the downlink transmission provided in [24].

At any point in time, it is likely that at least one user will have a very good channel realization. If the BS is aware of the user channels, it can schedule data transmission to the user with the best instantaneous channel at a high rate, thereby achieving better performance. The quality of the selected channel increases with the number of users, and the spectral efficiency increases by $\log\log(K)$ for a large number of users K [27, 28]. Multiuser diversity is also a key component of contemporary cellular systems such as UMTS and LTE. The scheduler decides which user to transmit information to at each time slot. The simplest scheduler transmits data to each user in a Round-Robin (RR) fashion regardless of the channel conditions of the users. The scheduling algorithm in channel-dependent manner is used to exploit multiuser diversity. Obviously, multiuser diversity requires that the BS knows the channel coefficients for all users, which is usually estimated at the mobiles and fed back to the BS.

In this section, we introduce scheduling algorithms assuming full CSI is available at both the transmitter and receiver. In order to achieve multiuser diversity, we will examine a system consisting of one BS with single antenna and K active users with single antenna.

For downlink fading channel, the baseband representation of the received signal at each user is given by

$$y_k = h_k \sqrt{d_k} x_k + \sum_{j \neq k}^{K} h_k \sqrt{d_k} x_j + n_k \tag{5.19}$$

where d_k is the attenuation in the power caused by path loss.

At each user, the received instantaneous SNR for flat fading channel is

$$\gamma_k = \frac{|h_k|^2 d_k P_k}{N_0 B_W} \tag{5.20}$$

where P_k is the transmitted power for each user and h_k is channel coefficient and modeled as Rayleigh distribution with $\mathbb{E}[|h_k|^2] = 1$.

The average SNR at each user, $\bar{\gamma}_k$, can be calculated as

$$\bar{\gamma}_k = \mathbb{E}_{h_k}\left[\frac{|h_k|^2 d_k P_k}{N_0 B_W}\right] \tag{5.21}$$

For the case of i.i.d. Rayleigh fading channels assumption, the attenuation factor of all users are assumed to be the same and normalized to 1, such that $d_k = 1$; $k = 1, 2, \ldots, K$.

5.3.1   Opportunistic Scheduling

Wireless spectrum efficiency is becoming increasingly important with the growing demand for wideband wireless services. Since the wireless resource is scarce and mobile users perceive time-varying channel conditions, resource allocation and scheduling policies are critical in these wireless networks. Since users experience time-varying and location-dependent channel conditions in wireless environments, we can schedule users opportunistically so that a user can exploit more of its good channel conditions and avoid (as far as possible) bad times, at least for applications (e.g., data service) that are not time-critical. Opportunistic scheduling [13] increases the system spectral efficiency by giving priority to mobile users when they have good channel quality. The fact that in a system with many users whose channels vary independently, there is likely to be a user whose channel is near its peak at any one time. Therefore, a gain called multiuser diversity can be achieved by this system. In the literature, multiuser diversity can be mainly achieved by using opportunistic scheduling.

Opportunistic scheduling exploits the variation of channel conditions and thus provides an additional degree of freedom in the time domain. Moreover, it can be coupled with other resource management mechanisms to further increase network performance. Opportunistic scheduling gives users chance to transmit when the users' CSI are in good conditions. Therefore, a natural question is how long is a user willing to wait for having good conditions. Hence, there exists trade-off between scheduling and short-term performance.

In summary, opportunistic scheduling exploits the channel fluctuations of users. Hence, the larger the channel fluctuation, the higher the scheduling gain. In the previous section, it is clearly shown that sum rate of downlink transmission is achieved when the strongest user is selected at each time as follows:

$$\sum_{k=1}^{K} R_k \le C_{OS} = \mathbb{E}\left[B_W \log_2 \left(1 + \frac{P}{N_0 B_W} \max_{k=1,\ldots,K} |h_k|^2 \right) \right] \qquad (5.22)$$

where equal power is allocated to each user as $P_k = P$.

In the case of the user with i.i.d Rayleigh fading channel, the $|h_k|^2$ are all i.i.d exponential random variables with zero mean. The extreme value theory shows that for large K the maximum of this expression behaves as $\log K$ with high probability. Thus, for large K, we have

$$C_{OS} = \log \left(\frac{P}{N_0 B_W} \right) \log K + o(1) \qquad (5.23)$$

where $o(1)$ represents the terms that vanish as K grows.

The above expression explains the $\log \log K$ nature of the multiuser gain. The effect of transmitting to the strongest user results in a $\log K$-fold increase in the SNR and therefore a $\log \log K$-fold increase in the sum rate.

In wireless systems, opportunistic scheduling can be implemented to schedule the best user at every time slot in the downlink. Based on SNR selection, it can be expressed as

$$k^* = \arg \max_{k=1,\dots,K} \gamma_k \qquad (5.24)$$

If the users experience i.i.d. Rayleigh fading channels where each user experiences the same average received SNR, the cumulative density function (cdf) of received SNR over the time slot is given by

$$F_{\gamma_{k*}}(x) = \prod_{k=1}^{K} F_{\gamma_k}(x) \qquad (5.25)$$

After performing opportunistic scheduling, the pdf of the received SNR becomes

$$p_{\gamma_{k*}}(x) = K p_{\gamma_k}(x)[F_{\gamma_k}(x)]^{K-1} \qquad (5.26)$$

Mathematically, the ergodic capacity can be calculated by averaging the instantaneous capacity over the distribution of the received SNR as

$$C = \int_0^\infty \log_2(1 + \gamma) p_{\gamma_{k*}}(\gamma) d\gamma \qquad (5.27)$$

As a result, the capacity with multiuser diversity for i.i.d. Rayleigh fading channels is given as

$$C_{ASNR} = K \log_2(e) \sum_{k=0}^{K-1} (-1)^k \frac{(K-1)!}{(k+1)!(K-k-2)!} \exp\left(\frac{k+1}{\bar{\gamma}}\right) E_1\left(\frac{k+1}{\bar{\gamma}}\right) \qquad (5.28)$$

where $E_1(.)$ is the exponential integral function of the first order which is defined by $E_1(x) = \int_1^\infty \frac{e^{-xt}}{t} dt; x \geq 0$.

In order to provide fairness when the users have different average received values, the scheduling is based on their normalized SNR rather than absolute SNR as in (5.24). The normalized SNR scheduling algorithm [11] is given by

$$k^* = \arg \max_{k=1,\dots,K} \tilde{\gamma}_k \qquad (5.29)$$

where the normalized SNR per user is

$$\tilde{\gamma}_k = \frac{\gamma_k}{\bar{\gamma}_k} \qquad (5.30)$$

The pdf of the normalized SNR under i.i.d Rayleigh fading channel model is given by

$$p_{\tilde{\gamma}_k}(\tilde{\gamma}_k) = \exp(-\tilde{\gamma}_k) \qquad (5.31)$$

Then, the pdf of the received SNR at the scheduled user is

$$p_{\gamma_{k*}}(x) = \sum_{i=1}^{K} \frac{1}{\bar{\gamma}_i} p_{\tilde{\gamma}_i}(\tilde{\gamma}_k) \prod_{k=1, k\neq i}^{K} F_{\tilde{\gamma}_k}(\tilde{\gamma}_k) \qquad (5.32)$$

The sum capacity with normalized SNR based scheduling on this criterion is given by

$$C_{NSNR} = \log_2(e) \sum_{i=1}^{K} \sum_{k=0}^{K-1} \frac{(-1)^k}{1+k} \binom{K-1}{k} \exp\left(\frac{k+1}{\bar{\gamma}_i}\right) E_1\left(\frac{k+1}{\bar{\gamma}_i}\right) \qquad (5.33)$$

5.3.2   *Proportional Fair Scheduling*

The opportunistic scheduling denotes the ability to schedule users based on good channel conditions. However, the potential to transmit at higher data rates opportunistically (i.e., when channel conditions permit) also introduces an important trade-off between wireless resource efficiency and level of satisfaction among different users. For example, allowing only users close to the BS to transmit at high transmission power may result in very high throughput, but sacrifices the transmissions of other users. Such a scheme cannot satisfy the increasing demand for quality of service (QoS) provisioning in the emerging high-rate data wireless networks.

In order to maintain fairness over a given finite time slot or in the case of unequal average SNR conditions among users (such as due to distance to the BS), the scheduler must exploit a metric that takes into account the accumulated throughput up to time slot i.

At time slot i, we schedule user $k^*(i)$ with maximum normalized capacity, such that

$$k^*(i) = \arg\max_k \frac{C_k(i)}{T_k(i)} \qquad (5.34)$$

where $T_k(i)$ is the actual transmission throughput of user k over the link up to time slot i and $C_k(i)$ is the capacity of user k at time slot i. The throughput is updated on a per time slot according to

$$T_k(i+1) = T_k(i)\left(1 - \frac{1}{t_c}\right); \quad k \neq k^* \tag{5.35}$$

$$T_{k^*}(i+1) = \frac{T_{k^*}(i)}{t_c} + C_{k^*}(i) \tag{5.36}$$

where t_c is a time constant adjusted to maintain fairness over a predetermined time duration. The larger t_c is, longer time is needed to provide fairness. However, there can be a risk that the fairness among the users may be destroyed. Besides, the proportional fairness scheduling (PFS) and max SNR algorithms are equivalent for large t_c. When all users experience the same SNR distribution, max SNR or opportunistic scheduler gives the same performance.

5.4   Feedback Strategies

Although multiuser diversity can provide significant benefits, there is also a non-negligible cost associated with obtaining instantaneous CSI at the BS. Such CSI is obtained through explicit feedback of the instantaneous SNR from each of the users or through utilization of uplink pilots when the channel is reciprocal. Thus, in terms of the system resources, the cost is power and bandwidth which is used to provide acquired CSI and the required bandwidth increases with the number of users. Hence there is a cost-benefit trade-off associated with multiuser diversity.

The feedback cost associated with obtaining CSI at the BS depends on the specific CSI feedback method used and the number of users which feeds back. The two main directions have previously been pursued to reduce the degradation due to feedback, namely,

• Feedback load reduction (reduced feedback strategies)
• Feedback quantization (limited feedback strategies).

The studies investigating the first approach have shown that heavy quantization of the CSI being fed back will not lead to a significant reduction of the system gain [7,15]. In the second approach, the algorithms are trying to reduce the feedback load, i.e., the number of users feeding back CSI. The CSI can be either an instantaneous SNR value, or it can be the MCS corresponding to the instantaneous SNR as shown in Chap. 3. In the literature, these feedback load reduction algorithms are based mainly on SNR thresholds [11, 14] and opportunistic approaches. In the following sections, we will examine these strategies in detail.

5.4.1   Reduced Feedback Strategies

The availability of users' instantaneous CSI is essential to the implementation of multiuser diversity transmission. For the downlink scenario, the BS needs to collect the channel SNRs corresponding to all users in order to select the best user for transmission at each frame depending on the channel variations. This leads to a huge amount of channel quality feedback, which increases signaling overhead instead of useful data traffic. Therefore, the feedback load can be reduced while maintaining nearly the same multiuser gain. In this context, the so-called selective multiuser diversity scheme is an effective solution [10, 30]. The basic idea of selective multiuser diversity is to allow only those users whose channel qualities are good enough to feed back their SNR. These users are called as qualified users. Note that with opportunistic scheduling, the BS will select a single user for transmission and only users with good channel will have the chance to be selected. Therefore, it is expected that selective multiuser diversity approach can achieve the same diversity gain as conventional multiuser diversity if at least one user feeds back its SNR. In the following, we will examine the strategies for selection of the users at the received side and analyze the feedback load [16] and threshold determination for different scheduling algorithms.

5.4.1.1 Threshold-Based Schemes

In selective multiuser diversity scheduling, each user compares its channel gain to a threshold. Then, only the users whose channel gain are above this threshold are allowed to feed back their SNR, channel gain or achievable downlink transmission rate [11]. The other users that fail to satisfy this condition remain silent and do not feed back any information. Because the user to be scheduled for transmission is the one with the best relative SNR, it is unlikely that a user with bad relative SNR will be selected by the scheduler under the condition of a reasonable total number of users K. Therefore, the feedback resource provisioned for this user is wasted bandwidth and should be avoided to feed back at the user side.

In the particular case where all users have the same average SNR, $\bar{\gamma}_k = \bar{\gamma}$ for all k, we may define the threshold in terms of the instantaneous SNR, γ_{th}. At a time slot, user k will feed back its channel gain or SNR to the BS if it satisfies the following:

$$\gamma_k \geq \gamma_{th} \tag{5.37}$$

Note that in this case, the threshold is chosen according to the SNR. Clearly it could also be applied to other quality metrics such as the capacity or normalized quality metrics when not all users have the same average SNR, such as the normalized SNR. A feedback reduction scheme in which each user compares its PFS metric to a threshold instead of channel gain has been investigated in [12] for a PFS.

Feedback Load

The reduction in the feedback load obtained by adopting the multiuser diversity scheme instead of the classical full feedback multiuser diversity algorithm will be determined. While the load is fixed and is equal to the number of users K for full feedback, it can range anywhere from zero to K at each time slot with the threshold-based feedback scheme. The average feedback load $\bar{D}$ is defined as the average number of users that feed back their SNR per time slot and represents the number of users whose channel gain is equal or higher than a given threshold. Mathematically this can be simply written as

$$\bar{D} = \sum_{k=1}^{K} k Pr(\text{k users feedback}) \tag{5.38}$$

The conditional probability that k out of K users are preselected during a particular time slot is equal to the conditional probability that the SNRs of these k users equal or exceed the threshold γ_{th} is $(1 - F_\gamma(\gamma_{th}))^k (F_\gamma(\gamma_{th}))^{K-k}$.

For i.i.d. fading channel among users, the probability that $D = k$ is equal to

$$Pr(D = k) = \binom{K}{k} (1 - F_\gamma(\gamma_{th}))^k (F_\gamma(\gamma_{th}))^{K-k} \tag{5.39}$$

Therefore, the average feedback load is given as

$$\bar{D} = \sum_{k=1}^{K} \frac{K!}{(K-k)!(k-1)!} (1 - F_\gamma(\gamma_{th}))^k (F_\gamma(\gamma_{th}))^{K-k} \tag{5.40}$$

For Rayleigh fading, this simplifies to

$$\bar{D} = K \exp\left(-\frac{\gamma_{th}}{\bar{\gamma}}\right) \tag{5.41}$$

When the users' channel are independent but not identically distributed, it is possible to use normalized SNR scheduling as it has been shown in (5.29) instead of absolute SNR. In this case, the average feedback load can be determined by replacing γ_{th} by $\tilde{\gamma}$ in (5.40).

Threshold Determination

The selection of optimal γ_{th} is very important for this scheduling. Intuitively, with a too small γ_{th}, the qualified set will be contaminated by users experiencing unfavorable channel conditions, while with a too large γ_{th}, we risk to have an empty set. Different criteria can be used to determine the threshold such as sum capacity, outage capacity, loss in capacity, for a fixed feedback load, etc.

In order to find the optimal threshold to maximize the sum capacity, it is not possible to obtain a numerical solution since a closed-form solution is in general not tractable.

The other strategy is to chose the threshold γ_{th} to reach a predetermined scheduling outage probability P_{out}. It occurs when all users' channel gain is lower than a given threshold. Under the assumption of i.i.d. faded user channels, it is calculated as,

$$P_{out} = \prod_{k=1}^{K} F_{\gamma_k}(\gamma_{th}) \tag{5.42}$$

where $F_{\gamma_k}(\gamma_{th})$ denotes the cdf of received SNR at the kth user.

For the case of i.i.d Rayleigh fading channel, it becomes

$$P_{out} = \left(1 - \exp\left(-\frac{\gamma_{th}}{\bar{\gamma}}\right)\right)^K \tag{5.43}$$

For a given P_{out}, the γ_{th} can be calculated as

$$\gamma_{th} = -\bar{\gamma} \ln[(1 - P_{out})^{1/K}] \tag{5.44}$$

For non-i.i.d. fading channels under normalized SNR scheduling, the scheduling outage probability is calculated as

$$P_{out} = [F_{\tilde{\gamma}}(\tilde{\gamma}_{th})]^K = [1 - \exp(-\tilde{\gamma}_{th})]^K \tag{5.45}$$

Then, the threshold value can be obtained as

$$\gamma_{th} = -\ln((1 - P_{out})^{1/K}) \tag{5.46}$$

If the size of the feedback channel is limited, it may be also of interest to choose the threshold in order to meet a certain average feedback load specification. In i.i.d Rayleigh fading this can be obtained as

$$\gamma_{th} = -\bar{\gamma} \ln(\bar{D}) \tag{5.47}$$

5.4.1.2 1-Bit Feedback Scheme

The simplest method is scheduling by using a 1-bit feedback [20, 26]. The BS sets a threshold γ_{th} for all users and each user compares the absolute value of their channel gain to this threshold. Whenever the channel gain exceeds the threshold a 1 will be transmitted to the BS to construct qualified set; otherwise no feedback bit is transmitted.

$$b_k = 1, \text{ if } \gamma_k \geq \gamma_{th} \tag{5.48}$$

The BS receives feedback from all users and then randomly picks a user whose feedback bit was set to 1 to start a data transmission at the data rate calculated based on threshold SNR. If there is no feedback received by the BS, then no data is transmitted in that interval or any user is selected randomly.

In [20], the ergodic sum rate of a downlink single antenna channel with K users is analyzed in the presence of Rayleigh flat fading, where only 1-bit feedback per fading block per user is available at the BS. It is shown that in reducing the CSI feedback to 1 bit, and even when subject to feedback delay, the scaling law of the ergodic sum rate is the same as that of a system with full CSI at the transmitter. The sum-rate capacity of a wireless network with 1-bit feedback and optimal choice of threshold behaves as $O(\log(\log K) + \log \gamma)$, exactly the same as the sum-rate capacity of a network with having full CSI. The impact of feedback delay on the 1-bit feedback system has been analyzed in [17] by expressing ergodic capacity as a function of the fading temporal correlation coefficient.

The average feedback load in terms of the number of users whose channel gain is higher as the given threshold is the same than given in the previous section and consequently the same derivations can be used to calculate feedback load as well as threshold value determination.

Multiple Thresholds

The algorithm in [11] does not always schedule the user with the highest SNR since it will always be a possibility that all users are below the threshold value and a random user has to be chosen. By employing multiple feedback thresholds, the BS can conduct the feedback collection process by polling the users sequentially from the highest threshold value down to the lowest threshold value until feedback from one or more users is received [8, 9]. This algorithm also adapts the feedback threshold value according to the scheduling metric of the scheduling algorithm and multiple feedback thresholds are used to collect feedback from the preferred user, i.e., the user that the scheduling algorithm prefers to schedule.

The multiple thresholds are denoted as $\gamma_{th,L} > \gamma_{th,L-1}, \ldots, \gamma_{th,0}$. The highest threshold can be chosen as infinity and the lowest one can be set at zero to cover all ranges. To initiate the feedback process, the BS sends out a query containing the number of thresholds employed and a list of the relevant users. From the number of feedback thresholds (L) and the number of users (K) in the query, each user can look up the threshold values in tables. After the query is sent, each threshold value is assigned the duration of a predefined minislot. In the first minislot the BS requests feedback from those users whose SNR is above $\gamma_{th,L-1}$. If there are none, the threshold is successively lowered to $\gamma_{th,L-2}, \gamma_{th,L-3}$ down to $\gamma_{th,0}$. Since $\gamma_{th,0}$ equals zero, the best user is always found.

The feedback load for the threshold-based user selection using L different threshold level is calculated as

$$\bar{D} = \sum_{\ell=1}^{L} (F_\gamma(\gamma_{th,\ell+1}) - F_\gamma(\gamma_{th,\ell}))(F_\gamma(\gamma_{th,\ell+1}))^{K-1} \qquad (5.49)$$

Depending on the number of users as well as the number of feedback thresholds, it is possible to determine the threshold levels in a recursive manner.

$$\gamma_{th,\ell} = F_\gamma^{-1}(S_\ell F_\gamma(\gamma_{th,\ell})); \ell = 1, 2, \ldots, L - 1 \qquad (5.50)$$

where $F^{-1}(x)$ is the inverse cdf of x for a single user. S_ℓ is calculated for $K \geq 2$ as

$$S_\ell = \begin{cases} K^{\frac{1}{1-K}}, & \ell = 1 \\ (K - (K-1)S_{\ell-1})^{\frac{1}{1-K}}, & \ell = 2, 3, \ldots, L - 1 \end{cases} \qquad (5.51)$$

In order to calculate the threshold levels, first $\gamma_{th,L-1}$ is easily calculated since $\gamma_{th,L}$ is defined to be infinity and $F_\gamma(\gamma_{th,L}) = 1$. Then, with knowledge of $\gamma_{th,L-1}$, all threshold values down to $\gamma_{th,1}$ are calculated.

5.4.1.3 Performance Evaluations

We show the sum rate and the feedback load for the full feedback- and threshold-based selection algorithms in i.i.d and non-i.i.d. Rayleigh wireless channels. As shown in Figs. 5.4 and 5.5, the performances of sum capacity and feedback load for different average SNR and threshold values have been illustrated by employing absolute SNR scheduling algorithm. When the average SNR is increased, the sum capacity increases while the feedback load also increases for a same threshold for SNR-based selection and the sum capacity does not change for 1-bit feedback-based reduced schemes. The performance of sum capacity as shown in Fig. 5.6 is degraded depending on the chosen normalized threshold value for non-i.i.d. Rayleigh fading channels. In addition to that the gain in terms of feedback load is significantly high as shown in Fig. 5.7.

5.4.1.4 Opportunistic Feedback Schemes

In this section, we will consider opportunistic feedback schemes where the feedback load is not directly proportional to the total number of users. In [25], a random access-based feedback protocol that uses threshold-based feedback is presented using fixed number of minislots. Users contend for minislots using a common threshold that is chosen to maximize the sum throughput of the system. The threshold and the probability that a user makes a feedback attempt is adjusted based on the number of users and minislots in the system. In order to achieve

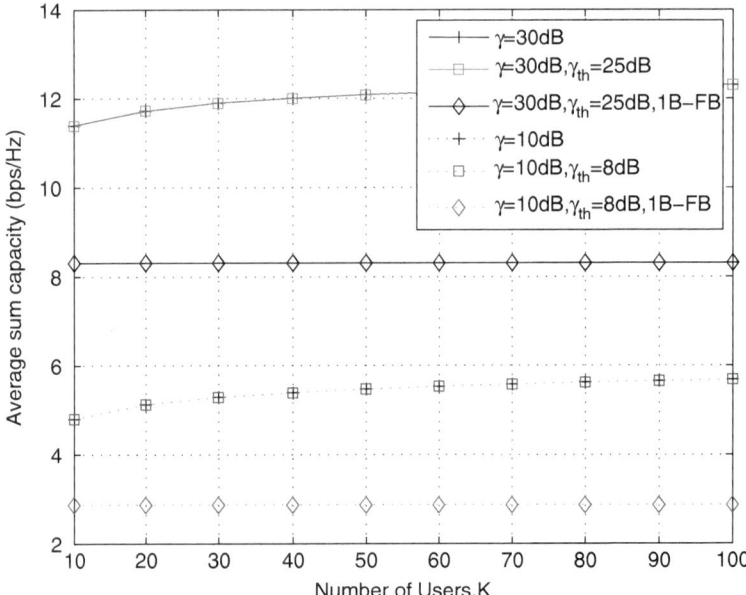

Fig. 5.4 The sum capacity versus number of users for i.i.d. Rayleigh wireless channels

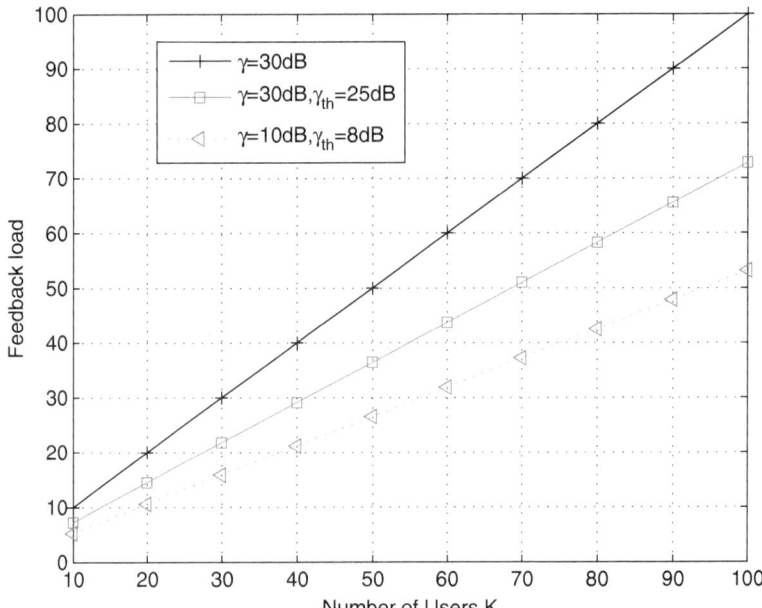

Fig. 5.5 The total feedback load versus number of users for i.i.d. Rayleigh wireless channels

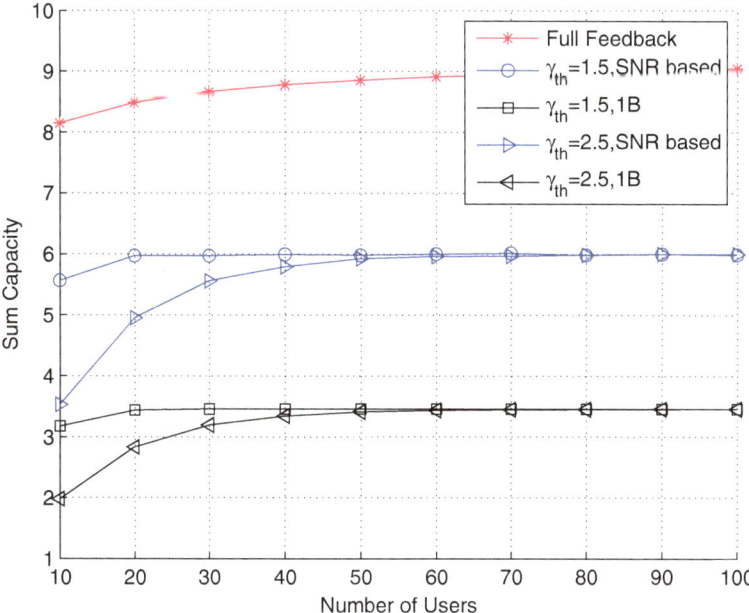

Fig. 5.6 The sum capacity versus number of users for non-i.i.d. Rayleigh wireless channels

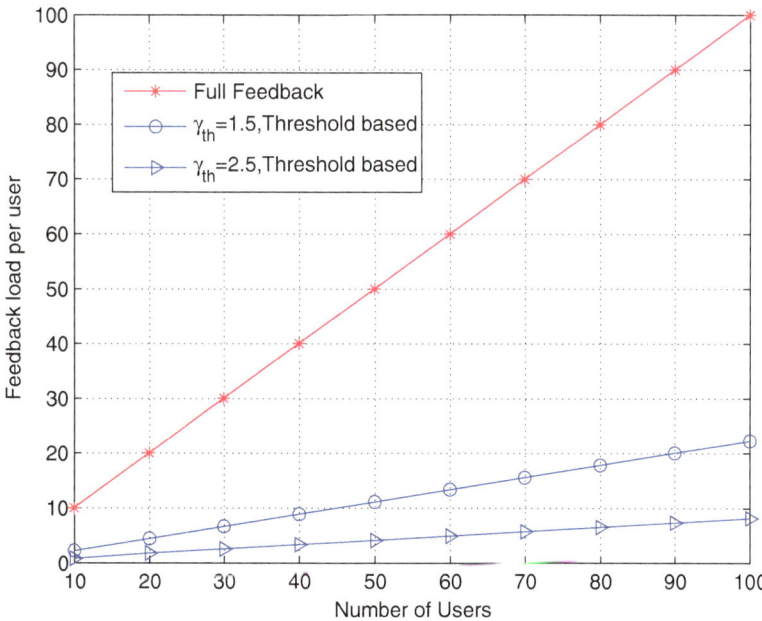

Fig. 5.7 The total feedback load versus number of users for non-i.i.d. Rayleigh wireless channels

multiuser diversity with PFS using opportunistic feedback schemes, the uplink feedback period is divided into two subperiods and users randomly access the feedback minislots according to their scheduling metric in [21]. The generalization of [25] by using multiple access probabilities and multiple thresholds is presented in [22]. The idea is to assign a different access probability to each user depending on their traffic class to improve the sum-rate capacity with a minimal amount of feedback. In [3], opportunistic feedback scheme is examined to achieve low packet delays in the context of a heterogeneous wireless system based on queue/weight-based opportunistic scheduling techniques. In this systems, the feedback thresholds are assigned to each user with determination of the weights which are designed to capture their queue length.

In every minislot, $i = 1, 2, .., I$, when a user finds that its channel power is higher than a given threshold, it randomly feeds back a data containing its user identification information with probability $p_k(i)$. Otherwise, it does not feed back during that frame, i. The random access probability is obtained as

$$p_k(i) = \begin{cases} p, & |h_k[i]|^2 > \alpha_{th}; \\ 0, & \text{otherwise.} \end{cases} \tag{5.52}$$

where α_{th} is a threshold in channel power and p is the probability that indicates channel access attempt for each user. It can be chosen as the same for all users, or unequal depending on different throughput requirements or channel statistics.

When performing the algorithm, it is assumed that the random access attempts are independent among users and also independent among random access minislots. If there are multiple users who successfully feed back, the BS randomly selects one of the successful users. A more optimistic approach is to allow the BS to select the user with the best channel gain among the users who successfully feed back.

A group random access-based feedback scheme [18] is presented for multiuser diversity systems, where users are divided into several feedback groups according to their measured channel states. Different random access probabilities and number of feedback minislots are assigned to each feedback group. Each user competes to send a feedback message through random access if its scheduling metric exceeds a threshold. The BS jointly controls the random access probabilities, the thresholds and the number of feedback minislots to adjust the feedback load.

In the opportunistic feedback scheme, which regulates feedback signaling by adopting feedback threshold, determining the optimum feedback threshold and allocating the optimum resource are important problems. When the feedback threshold is too low, too many users send feedback information. Consequently, many feedback signals will collide with each other, resulting in the decrease in the multiuser diversity gain. On the contrary, when the feedback threshold is too high, very few users send feedback information, and it also results in the decrease in the multiuser diversity gain. Therefore, in the opportunistic feedback scheme, both the feedback threshold and the amount of resources allocated in feedback signaling should be optimized to maximize the sum capacity.

5.4.2 Limited Feedback Strategies

In order to manage priorities among all users, each user measures its instantaneous SNR and feeds it back to the scheduler. In the previous section, an unquantized feedback case was assumed; however, in order to apply these algorithms in practical systems, quantized SNR values or quantized channel gain should be fed back instead of holding the availability of unquantized values.

An adaptive transmission is used when information about specific transmission modes is given. Otherwise we will simply use the term affordable-rate transmission to describe any systems that adapt the transmission rate according to channel condition, so that the transmission reaches the maximum data rate that can be accommodated by a given channel realization. Note that the switching thresholds of the adaptive-transmission modes are functions of the modulation and coding schemes and target error performance.

The impact of the quantization of SNR measurements on the throughput of a multiuser diversity scheme for affordable-rate transmission is investigated under a block-Rayleigh fading assumption in [7]. In the downlink, each user measures its SNR, maps it into a quantization level and feeds it back to the transmitter, which transmits a packet to the user with the highest quantized SNR. In the case of several users having the same quantized SNR, one of them is selected at randomly. It is concluded that using only a few quantization levels can yield a throughput that is only slightly less than the throughput obtained by using unquantized feedback.

An optimal discrete rate switch based multiuser diversity (DSMUDiv) scheduling scheme that reduces the feedback load, while preserving most of the performance of opportunistic scheduling [1].

The absolute SNR-based scheduling algorithm with quantized feedback is described as follows:

- We assume that there are L quantized modulation levels.
- Each user knows the L quantized modulation levels.
- The BS sends a pilot randomly to user k which measures its instantaneous received SNR and feeds back a quantized value indicating its modulation level.
- If the quantized SNR value of that user indicates that he can transmit with the possible highest level, then the BS will grant its access to the channel.
- If not, then modulation level will be stored for later comparison and another pilot is sent to another random user.
- The scenario goes on and the channel access is given to the first user who can transmit with the highest level L or to the user with the highest level compared to others.
- In case a tie occurs, a random pick is performed.

In order to perform the algorithm, the target average bit error rate (BER) is determined by BER_o and the M-QAM thresholds or switching thresholds are given by

$$\gamma_{th}^{(1)} = [\text{erfc}^{-1}(2BER_o)]^2 \tag{5.53}$$

$$\gamma_{th}^{(m)} = -\frac{2}{3}(2^m - 1)\ln(5BER_o); m = 2, 2, ..., L \tag{5.54}$$

$$\gamma_{th}^{(L+1)} = +\infty \tag{5.55}$$

where erfc^{-1} denotes the inverse complementary error function.

Let D be the number of information sent until the channel is accessed (feedback load) and let $Pr[D = k]$ be the probability that the feedback load equals k. The average feedback load $\mathbb{E}[D]$ consists of two terms: the first term, $\bar{D}_1$, captures the possibility that at least one user has an instantaneous SNR, γ, lying in the L modulation level. In this case, the BS picks that user and stops searching through the other users. The probability that the kth user will be granted the channel access is

$$Pr[D = k] = [F(\gamma < \gamma_{th}^{(L)})]^{k-1}[F(\gamma \geq \gamma_{th}^{(L)})] \tag{5.56}$$

$$= [F_\gamma(\gamma_{th}^{(L)})]^{k-1}[1 - F(\gamma_{th}^{(L)})] \tag{5.57}$$

where $F_\gamma(\gamma) = (1 - \exp(\gamma/\bar{\gamma}))$ is the cdf. Therefore, the average feedback load is this case is,

$$\bar{D}_1 = \sum_{k=0}^{K-1} k[F_\gamma(\gamma_{th}^{(L)})]^{k-1}[1 - F(\gamma_{th}^{(L)})] \tag{5.58}$$

The second term, $\bar{D}_2$, captures the possibility that no user has an instantaneous SNR, γ, lying in L modulation level. The user will search through all users and picks the user with the highest modulation level. The average feedback load in this case is

$$\bar{D}_2 = K[F_\gamma(\gamma_{th}^{(L)})]^{K-1} \tag{5.59}$$

Then, the average feedback load becomes

$$E[D] = \bar{D}_1 + \bar{D}_2 \tag{5.60}$$

5.5 Feedback in OFDMA-Based Wireless Systems

Feedback reduction in multiuser OFDM systems has become an important issue due to the excessive amount of feedback required to use opportunistic scheduling, particularly when the number of users and subcarriers is large.

In order to reduce the amount of overhead without sacrificing too much in performance, several information can be taken into account for OFDMA-based systems. Since the correlation between adjacent subcarriers is generally high, the subcarriers can be divided into clusters of adjacent subcarriers, which can be used as feedback units. The feedback information is a measure of the channel quality in the cluster, for instance, the minimum subcarrier supportable rate within the cluster. In a well-designed system, the cluster size is chosen so that the subcarriers within one cluster are highly correlated. Having small clusters offers better feedback accuracy for the subcarriers in the cluster but does not reduce feedback much. Having large clusters, on the other hand, reduces feedback more. However, increases the risk of users feeding back supportable rates lower than necessary for some subcarriers. Hence, finding a suitable cluster size is a critical problem.

An OFDMA system with Q clusters and K users is considered. A cluster structure where the correlation is high among the subcarriers is employed so that the feedback of only one value is sufficient, e.g., the CSI value belonging to the subcarrier of minimum channel gain. For this model, the channel coefficient between the BS and the kth user for the qth cluster is defined by $H_{k,q}$ and the channel coefficient of a cluster is determined by

$$H_{k,q} = \bar{H}_{k,n^*} \tag{5.61}$$

where $\bar{H}_{k,*}$ is channel coefficient associated to the kth user and the n^*th subcarrier and n^* is

$$n^* = (q-1)N_Q + \arg \min_{0 \leq i \leq N_Q-1} \{|\bar{H}_{k,(q-1)N_Q+i}|^2\}, \quad q = 1, 2, \ldots, Q \tag{5.62}$$

where N_Q is the subcarrier in one cluster and calculated as $N_Q = N/Q$ with N which is the total number of subcarriers in an OFDM symbol.

In order to maximize sum data rate for OFDM-based systems, it is possible to use the threshold-based algorithms described in the previous section by applying threshold per clusters at each user. As a result of this reduced feedback scheme, the users having good clusters will feed back more information than the users that have weak clusters. However, an opportunistic scheduler usually does not schedule users on their weakest clusters. Hence, the amount of feedback information can be further reduced by letting each user feed back information only about its strongest clusters. In the S-best feedback scheme, each user selects S best subchannels having the highest channel gains and sends the CQIs of these subchannels to the BS. Under this scheme, it can be conjectured that the fairness-oriented scheduling policies such as the PFS will behave as designed since all users are given an equal amount of feedback opportunity. However, the performance of the maximum rate scheduler (MRS) (which serves the user with the highest achievable rate in each time slot) can be degraded, because some of the subchannels can be scheduled to a user in deep fading, thereby yielding throughput loss, unless S is close to the total number of subchannels.

For the clustered S-best criterion, each user selects independently a set $\mathbb{S}_k$ composed of the S clusters with the highest channel gain $|H_{k,q}|^2$. Then, each user feds back only its CSI associated to the selected clusters to the BS [23].

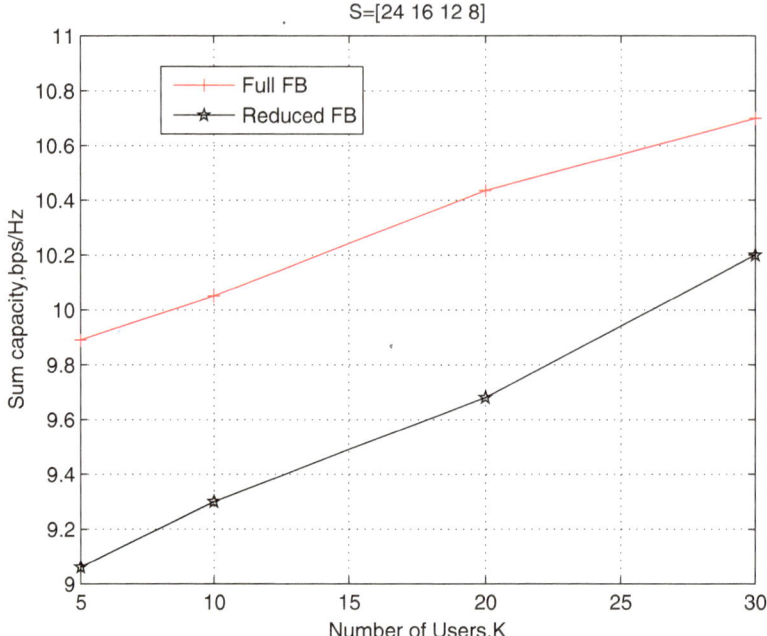

Fig. 5.8 The sum capacity versus number of users for non-i.i.d. Rayleigh wireless channels

Let $\mathbb{T}_q$ be the set of users that feed back their CSI associated to the cluster q as,

$$\mathbb{T}_q = \{k \in \{1, 2, \ldots, K\} : q \in \mathbb{S}_k\} \tag{5.63}$$

Then, for each cluster q, the BS selects the set $\mathbb{S}_q$ from the set $\mathbb{T}_q$ to maximize the sum capacity. Since the total feedback rate is proportional to KS, it is reasonable to adjust S in function of K according to a desired function $S = f(K)$.

In order to obtain the comparison results for full and reduced feedback schemes, a single antenna OFDMA system is used with $Q = 48$ clusters using proportional fair scheduling in non-i.i.d. Rayleigh wireless channel. The total feedback load is fixed to approximately 200 clusters which corresponds to feedback [20 10 7 5 4] clusters for the number of users [10 20 30 40 50] respectively. As shown in Fig. 5.8, the sum capacity with full feedback is achieved with the degradation of 1bps/Hz while reducing the total feedback load of 60 to 90 % depending on the number of users in the systems.

5.6 Conclusion

In this chapter, we have examined the multiuser systems including user scheduling and reduced and limited feedback strategies to achieve multiuser diversity for wireless systems. The user scheduling has been explained by considering different criteria such as the maximization of sum rate, proportionally fair scheduling and opportunistic scheduling. In order to achieve multiuser diversity, the users' CSI is required at the BS and the amount of feedback information increases with the number of users. Therefore, it is important to perform a selection at the user side considering the feedback load. The user selection strategies based on threshold criterion have been examined for single-carrier and multicarrier systems. In addition to that, in order to feed back the users' CSI, the quantization issues and the effect of the sum capacity performance has been shown for both full and reduced feedback schemes.

References

1. Al-Harthi Y S, Tewfik A H, Alouini M S (2007) Multiuser diversity with quantized feedback. IEEE Transactions on Wireless Communications. 6 :330–337.
2. Bergmans P (1974) A simple converse for broadcast channels with additive white Gaussian noise. IEEE Trans. Inform. Theory, 20: 279–280.
3. Baek S J, Veciana G D (2012) Opportunistic Feedback and Scheduling to Reduce Packet Delays in Heterogeneous Wireless Systems. IEEE Trans. on Vehicular Technology. 61:3282–3289.
4. Cover T (1972) Broadcast channels. IEEE Trans. Inform. Theory, 18: 2–14.
5. Costa M (1983) Writing on dirty paper. IEEE Trans. Inform. Theory, 29: 439–441.
6. Cover T M, Thomas J A (1991) Elements of Information Theory. Wiley.
7. Floren F, Edfors O, Molin B A (2003). The effect of feedback quantization on the throughput of a multiuser diversity scheme. Proc. IEEE Global Telecommunication Conference, 1: 497–501.
8. Hassel V, Alouini M-S, Gesbert D, Oien G E (2005)Exploiting multiuser diversity using multiple feedback thresholds. Proc. IEEE Vehicular Tech.COnf. 2: 1302–1306.
9. Hassel V, Gesbert D, Alouini M-S, Oien, G E (2007) A Threshold-Based Channel State Feedback Algorithm for Modern Cellular Systems. IEEE Trans. Wireless. Commun.6:2422–2426.
10. Gesbert D, Alouini M-S (2003) Selective multiuser diversity. Proceedings of the 3rd IEEE International Symposium on Signal Processing and Information Technology, 1: 162–165.
11. Gesbert D, Alouini M-S (2004) How much feedback is multi-user diversity really worth? Proc. IEEE Int. Conf. Commun., 1: 234–238.
12. Kim H, Han Y (2007) An opportunistic channel quality feedback scheme for proportional fair scheduling. IEEE Communications Letters. 11: 501–503.
13. Knopp R, Humblet P (1995) Information capacity and power control in single cell multiuser communications, in Proc.IEEE International Conference on Communications (ICC), USA, 1:331–335.
14. Qin X, Berry R (2004) Opportunistic splitting algorithms for wireless networks. Proc. IEEE International Conf. Computer Commun. 1: 1662–1672.
15. Johansson M (2003) Benefits of multiuser diversity with limited feedback. Proc. IEEE Workshop Signal Processing Advances Wireless Commun., 1:155–159.
16. Nam S S, Alouini M-S, Yang H-C, Qaraqe K A (2009)Threshold-Based Parallel Multiuser Scheduling. IEEE Trans. on Wireless Comm. 8: 2150–2159.
17. Niu B, Simeone O, Somekh O, Haimovich A M. (2010).Ergodic and Outage Performance of Fading Broadcast Channels With 1-Bit Feedback. IEEE Trans. on Vehicular Technology. 59:1282–1293
18. So J, Cioffi J M (2009) Feedback reduction scheme for downlink multiuser diversity. IEEE Trans. Wireless. Commun., 8: 668–672.

19. Shannon C (1974) Two-way communication channels. Reprinted in Key Papers in the Development of Information Theory, IEEE Press: 339–372.
20. Sanayei S, Nosratinia A (2005) Exploiting Multiuser Diversity with Only 1-Bit Feedback. Proc. IEEE Wireless Communications and Networking Conference (WCNC). 2:978–983
21. So J-W, Cioffi John M. (2008) Capacity and Fairness in Multiuser Diversity Systems with Opportunistic Feedback.IEEE Communications Letters.12:648:650.
22. So J (2009) Opportunistic Feedback with Multiple Classes in Wireless Systems. IEEE Communications Letters. 13:384–386
23. Svedman P, Wilson S K, Jr. Cimini L J, Ottersten B E (2007) Opportunistic Beamforming and Scheduling for OFDMA Systems. IEEE Transactions on Communications. 55: 941–952.
24. Tse D, Viswanath P (2005) Fundamentals of Wireless Communication, Cambridge University Press.
25. Tang T, Heath R (2005). Opportunistic feedback for downlink multiuser diversity. IEEE Communications Letters. 9: 948–950.
26. Xue Y, Kaiser T (2007) Exploiting Multiuser Diversity With Imperfect One-Bit Channel State Feedback. IEEE Trans. on Vehicular Technology. 56:183–193
27. S. Vishwanath, N. Jindal, A. Goldsmith, "Duality, achievable rates and sum rate capacity of Gaussian MIMO broadcast channel". IEEE Trans.on Information Theory, vol.49, no.10, pp. 2658–2668, October 2003.
28. Viswanath P, Tse D (2003). Sum capacity of the vector gaussian broadcast channel and uplink downlink duality. IEEE Trans. on Info. Theory. 49: 1912–1921.
29. Yu W, Cioffi J (2001) Trellis precoding for the broadcast channel. In Proceedings of Global Commun. Conf., 1: 1344–1348.
30. Yang L, Alouini, M-S (2006)Performance Analysis of Multiuser Selection Diversity. IEEE Trans. on Vehicular Tech. 55:1848–1861.

Chapter 6
Feedback Strategies for Multiantenna Multiuser Systems

6.1 Introduction

Over the last years, the interest in high data rate transmission has significantly increased as stated in Chap. 4. It is now well known that the capacity grows linearly with $\min(N_t, N_r)$ for point-to-point communication system using N_t transmit and N_r receive antennas. The broadcast channel where N_t transmit antennas are used to transmit to K users with N_r receive antennas is illustrated in Fig. 6.1. The total transmission rate to the K users is called the sum capacity which grows also linearly with $\min(N_t, K)$ if the transmitter and the receivers have full knowledge about the channel state information (CSI) [1, 55, 59].

The achievability proof is based on Costa's dirty coding [2] which states that the capacity of the channel is the same as the capacity of the channel without interference in an AWGN channel with interference if the transmitter noncausally knows the interference. Caire and Shamai first applied the concept of dirty paper coding (DPC) [2] to the multiple antenna broadcast channel [1]. The optimization problem of the multiantenna broadcast channels under power constraint is a non-convex optimization problem. In [18], this problem is transformed into a convex optimization by establishing a duality between the sum capacity of the broadcast channel (BC) and the sum capacity of multiple-access channels (MAC). This duality can greatly simplify the power allocation computation for the BC using a simple linear transformation from the MAC solution. Nevertheless, this optimal coding scheme is difficult to implement especially when the number of users is large. Therefore, suboptimal techniques are presented for practical implementation [3, 12, 52].

Different transmission strategies can be applied depending on the knowledge about CSI at the transmitter side: (1) full and perfect CSI at the transmitter, (2) without CSI at the transmitter, and (3) partial CSI at the transmitter.

This chapter will describe precoding, scheduling and feedback strategies for MIMO multiuser systems with one and multiple receive antennas in flat fading and frequency selective wireless channels. Firstly, we will analyze the capacity of

B. Özbek and D. Le Ruyet, *Feedback Strategies for Wireless Communication*, DOI 10.1007/978-1-4614-7741-9__6, © Springer Science+Business Media New York 2014

Fig. 6.1 Downlink multiuser
MIMO transmission

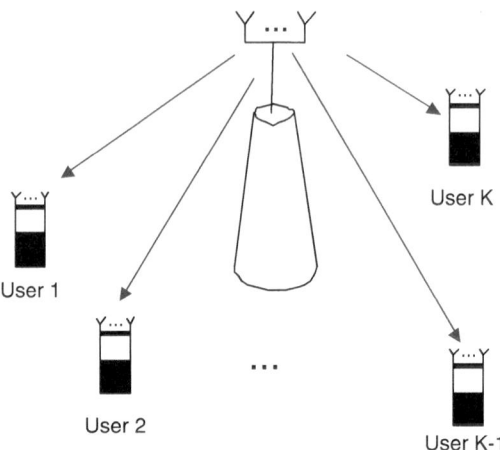

the multiuser MIMO system with single receive antenna for uplink and downlink. Secondly, we will examine the precoding and user selection algorithms assuming perfect CSI at both transmitter and receiver sides. Lastly, we will show the effect of reduced and limited feedback information including user selection at the receiver side and quantization for both single-carrier and multicarrier transmissions.

6.1.1 Capacity for Uplink Transmission

Firstly, we consider an uplink transmission system composed of a base station (BS) with N_r antennas and K users with one antenna for flat fading channels. For this system, the received vector of dimension of $N_r \times 1$ is given by

$$\mathbf{y} = \sum_{k=1}^{K} \mathbf{h}_k x_k + \mathbf{n} \tag{6.1}$$

where $\mathbf{h}_k$ is an uplink channel response of user k with dimension of $N_r \times 1$, x_k is the transmitted signal of user k and $\mathbf{n}$ is i.i.d complex Gaussian process.

The use of multiple receive antennas in the uplink is often called space division multiple access (SDMA) which can fully exploit the total number of degrees of freedom (DoF) $\min(N_r, K)$ of the uplink channel if the overall channel matrix $\mathbf{H} = [\mathbf{h}_1 \ \mathbf{h}_2 \ \dots \ \mathbf{h}_K]$ of size $N_r \times K$ is well conditioned.

The two users SDMA capacity region, for the multiple receive antenna case, is given by

$$R_1 < B_W \log_2 \left(1 + \frac{P_1 \|\mathbf{h}_1\|^2}{N_0 B_W} \right) \tag{6.2}$$

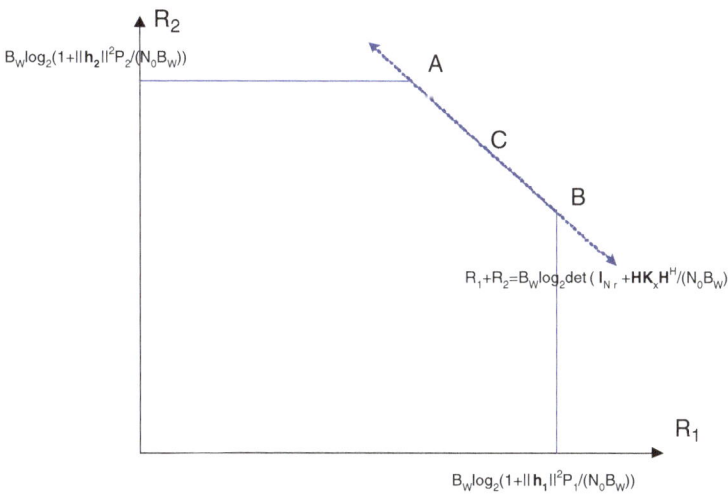

Fig. 6.2 Capacity region of two users in SDMA link

$$R_2 < B_W \log_2 \left(1 + \frac{P_2 \|\mathbf{h}_2\|^2}{N_0 B_W} \right) \tag{6.3}$$

$$R_1 + R_2 < B_W \log_2 \det \left(\mathbf{I}_{N_r} + \frac{\mathbf{H K}_x \mathbf{H}^H}{N_0 B_W} \right) \tag{6.4}$$

where $\mathbf{K}_x = \text{diag}(P_1, P_2)$ with P_1 and P_2 are the transmitted powers of user 1 and user 2, respectively. This capacity region is drawn in Fig. 6.2.

This two users capacity region can be achieved by using minimum mean-square-error (MMSE)-successive interference cancellation (SIC) receivers since the total rate of the point-to-point channel with two users acts as one user with two transmit antennas. If user 1 is canceled firstly, user 2 has only the Gaussian noise and its performance meets corner point A in Fig. 6.2. If user 2 is canceled as a first, corner point B is achieved. Thus, MMSE-SIC receivers are theoretically optimal for SDMA in the sense of achieving rate pairs corresponding to the two corner points of A and B.

The extension of this capacity to K users is given by

$$\sum_{k=1}^K R_k < B_W \log_2 \det \left(\mathbf{I}_{N_r} + \frac{1}{N_0 B_W} \sum_{k=1}^K P_k \mathbf{h}_k \mathbf{h}_k^H \right) \tag{6.5}$$

The capacity region is now a K dimensional polyhedron: the set of rates $(R_1, R_2, \ldots, R_K)$ is calculated for any combination of $1, 2, \ldots, K$. Each corner on the boundary of the capacity region is specified by ordering of the K users. Then, the corresponding rates are achieved by implementing an MMSE-SIC receiver with the ordering process.

For the case of fading channels where $\mathbf{h}_k$ with dimension of $N_r \times 1$ is a fading process for each user k, the uplink sum capacity for $K = 2$ becomes

$$C_{sum} = \max_{P_k(\mathbf{h}_1,\mathbf{h}_2); k=1,2} \mathbb{E}\left[B_W \log_2 \det\left(\mathbf{I}_{N_r} + \frac{1}{N_0 B_W} \mathbf{H}\mathbf{K}_x\mathbf{H}^H \right) \right] \qquad (6.6)$$

where power allocations are subject to the average constraint as $\mathbb{E}\left\{ P_k(\mathbf{h}_1, \mathbf{h}_2) \right\} \leq P$ with $k = 1, 2$.

The capacity region for K users has $K!$ vertices, each one corresponding to a specific ordering of the successive cancelation of the users. Each user adjusts its transmit power as a function of the channel states of all the users.

We consider how the sum capacity benefits from multiuser diversity, i.e., the optimal power allocation policy in terms of the sum of the user rates. Previously, for point-to-point single antenna channel, a simple form of power allocation is found as waterfilling. For a single antenna uplink, the policy is to allow only the best user to transmit and its power is allocated by applying waterfilling over its channel quality. In the uplink with multiple receive antennas, there is no such a simple expression in general. However, with both N_r and K large, the following simple policy is very close to the optimal one. Every user transmits with the power allocated according to waterfilling over its own channel state as

$$P_k(\mathbf{H}) = \left(\frac{1}{\Gamma_0} - \frac{I_0}{\|\mathbf{h}_k\|^2} \right)^2 ; k = 1, 2, \ldots, K$$

where Γ_0 is chosen such that the average power constraint meets. In this optimal scheme, since all the users are simultaneously transmitting, the waterfilling is done according to the background interference and noise which is denoted I_0 unlike in the uplink case with single antenna where the waterfilling is done over the channel quality with respect to only the noise.

As already discussed in Chap. 5, in the uplink transmission employing a single receive antenna at the BS, the power allocation that maximizes the sum capacity allows only the best user to transmit. In order to achieve multiuser diversity in uplink channel employing multiple receive antennas at the BS, a suboptimal strategy which is to transmit from only one user at each time can be considered. In this case, the multiple antennas at the BS translate into receive beamforming gain for the users. Thus, by analogy to the strongest user in the single antenna situation, the user which has the largest receive beamforming gain is chosen such as the user with the largest $\|\mathbf{h}_k\|^2$.

Assuming i.i.d. user channel statistics, the sum rate with this strategy and power allocation is calculated by

$$R_{SR} = \mathbb{E}\left[B_W \log_2\left(1 + \frac{P_{k*}^*(\|\mathbf{h}_{k*}\|^2)}{N_0 B_W} \right) \right] \qquad (6.7)$$

where $k^* = \arg\max_{k=1,2,\ldots,K} \|\mathbf{h}_k\|^2$ and $P_{k^*}^*$ is the power obtained by performing waterfilling as described in Chap. 5. The only difference compared to the single antenna case is that the scalar channel gain $|\mathbf{h}_k|^2$ is replaced by the receive beamforming gain $\|\mathbf{h}_k\|^2$.

In the case of an uplink MIMO channel, it is possible to achieve the gain of SDMA with multiple transmit antennas. This is an extension of SDMA with one transmit antenna to multiple antennas.

$$y = \sum_{k=1}^{K} \mathbf{H}_k \mathbf{x}_k + \mathbf{n} \tag{6.8}$$

where $\mathbf{H}_k$ is the channel matrix with the dimension of $N_r \times N_t$.

For the simple case which includes two users, each user splits its data and encodes it into independent streams of information by employing $N_s = \min(N_t, N_r)$ number of streams at the transmitter side. Powers $P_{k_1}, P_{k_2}, \ldots, P_{k_{N_s}}$ are allocated to the N_s data streams, passed through a rotation matrix $\mathbf{U}_k$ and sent over the transmit antenna array at user k. The rotation matrix $\mathbf{U}_k$ is chosen to correspond to the right unitary matrix in the singular value decomposition (SVD) of the channel and the powers allocated to the data streams correspond to the waterfilling allocations over the squared singular values of the channel matrix. The BS uses the MMSE-SIC receiver to decode the data streams of the users.

The rates R_1 and R_2 achieved by this architecture must satisfy the following conditions:

$$R_k \leq B_W \log_2 \det \left(\mathbf{I}_{N_r} + \frac{1}{N_0 B_W} \mathbf{H}_k \mathbf{K}_k \mathbf{H}_k^H \right) ; k = 1, 2 \tag{6.9}$$

$$R_1 + R_2 < B_W \log_2 \det \left(\mathbf{I}_{N_r} + \frac{1}{N_0 B_W} \sum_{k=1}^{2} \mathbf{H}_k \mathbf{K}_k \mathbf{H}_k^H \right) \tag{6.10}$$

where $\mathbf{K}_k = \mathbf{U}_k \Lambda_k \mathbf{U}_k^H$ and Λ_k to be a diagonal matrix with the N_k entries that equals to the allocated powers to the data streams as $P_{k_1}, P_{k_2}, \ldots, P_{k_{N_s}}$ (if $N_r < N_t$ then the remaining diagonal entries are equal to zero). The rate region is a pentagon.

The extension to K users is straightforward. The capacity region is now K dimensional and there are $K!$ corner points on the boundary region of the achievable rate region; each corner point is specified by an ordering of the K users and the corresponding rate is achieved by linear MMSE filter followed by successive cancelation of users and streams within a user's data.

6.1.2 Capacity for Downlink Transmission

Based on the principle of duality between BC and MAC, this case is analogous to the uplink with multiple receive antennas and there are up to N_t spatial DoF.

We consider a broadcast channel with N_t transmit antennas and K users equipped with N_r receive antennas. The received signal y_k at each user can be expressed as follows:

$$\mathbf{y}_k = \mathbf{H}_k \mathbf{x} + \mathbf{n}_k \tag{6.11}$$

where $\mathbf{y}$ is the transmitted vector of size $N_t \times 1$, $\mathbf{y}_k$ is the received vector with dimension of $N_r \times 1$, $\mathbf{H}_k$ is a channel matrix of user k with dimension of $N_r \times N_t$ and each element of $\mathbf{n}_k$ is i.i.d complex Gaussian noise.

From a multiuser information-theoretic perspective, the capacity region boundary is achieved by serving all K active users simultaneously. The allocated resource to each user, such as P_k is dependent on the instantaneous channel conditions and may vary greatly from user to user. If the transmitter knows all the channels, the interference at each receiver is also known at the transmitter. Therefore, the transmitter can potentially pre-subtract some interference.

There has been a lot of work carried out for calculation of the capacity region of the K users broadcast channel. In [55, 56, 58], the achievable rate region has been found for any number of users by applying directly coding for known interference. It is possible to apply DPC at the transmitter while choosing different codewords for each user. Let $(\pi_1, \pi_2, \ldots, \pi_K)$ denotes a permutation of the user indices. The transmitter first picks a codeword for user π_1. The transmitter then chooses a codeword for user π_2 with full (non causal) knowledge of the codeword intended for user π_1. Therefore, the codeword of user π_1 can be pre-subtracted such that user π_2 does not see the codeword intended for user π_1 as interference. Similarly, the codeword for user π_3 is chosen such that user π_3 does not see the signals intended for users π_1 and π_2 as interference. This process continues for all K users.

For a given permutation, the following rate vector is achievable:

$$RDPC_{\pi_k} = B_W \log_2 \frac{\det\left(\mathbf{I}_{N_r} + \mathbf{H}_{\pi_k}\left(\sum_{j \geq k} \mathbf{Q}_{\pi_j}\right) \mathbf{H}_{\pi_k}^H\right)}{\det\left(\mathbf{I}_{N_r} + \mathbf{H}_{\pi_k}\left(\sum_{j > k} \mathbf{Q}_{\pi_j}\right) \mathbf{H}_{\pi_k}^H\right)}, \qquad k = 1, \cdots, K \tag{6.12}$$

where $\mathbf{Q}_1, \ldots, \mathbf{Q}_K$ are the input covariance matrices of size $N_t \times N_t$ satisfying the input power constraint $\mathrm{tr}\left(\sum_k \mathbf{Q}_k\right) \leq P$.

The so-called dirty paper region is defined as the union of all such vector rates over all covariance matrices satisfying the input power constraint:

$$C_{DPC}(P, \mathbf{H}) = \bigcup_{\pi_k, \mathbf{Q}_k, \mathrm{tr}(\sum_k \mathbf{Q}_k) \leq P} RDPC_{\pi_k}(\mathbf{Q}_k) \tag{6.13}$$

And the dirty paper sum rate is given by

$$C_{DPC} = \max_{\pi_k, \mathbf{Q}_k, \text{tr}(\sum_k \mathbf{Q}_k) \le P} \sum_{k=1}^{K} RDPC_{\pi_k} \tag{6.14}$$

In [55, 56, 58], it has been proved that the dirty paper region of the broadcast system with power constraint P is equal to the capacity region of the dual multiple-access system with sum power constraint P.

$$C_{DPC}(P, \mathbf{H}) = C_{MAC}(P, \mathbf{H^H}) \tag{6.15}$$

Thanks to this BC-MAC duality, the sum rate capacity is given by

$$C_{SUM} = \max_{\mathbf{S}_k \ge 0, \sum_{k=1}^{K} \text{tr}(\mathbf{S}_k) \le P} B_W \log_2 \det \left(\mathbf{I}_{N_t} + \sum_{k=1}^{K} \mathbf{H}_k^H \mathbf{S}_k \mathbf{H}_k \right) \tag{6.16}$$

where $\mathbf{S}_1, \dots, \mathbf{S}_K$ of size $N_r \times N_r$ are the uplink covariance matrices. In [56], a transformation is provided that maps uplink covariance matrices to downlink covariance matrices and achieves the same rates by using the same powers.

In [1, 55], it has been shown that the maximal achievable rate grows linearly with $\min(N_t, K)$ if the transmitter and the receivers know the channel by using DPC [2]. This technique is based on coding for known interference. DPC states that in an AWGN channel with interference, if the transmitter noncausally knows the interference, the capacity of the channel is the same as the capacity of the channel without interference. This category includes Costa precoding [2], Tomlinson–Harashima precoding [12, 52] and the vector perturbation technique [14]. With vector DPC [55, 56, 58], the multiuser interference can be pre-subtracted (at the transmitter) for a desired receiver to achieve the interference-free rate performance. The receiver can then decode message as if the interference is known (as if there is cooperation between receivers) and canceled especially without incurring a power penalty at the transmitter. In [58], generalized decision-feedback equalizer structure is used for precoding at the transmitter. Besides, the optimal sum capacity for general case under some power constraint is achieved using precoding techniques having a decision-feedback equalizer [58].

Jindal and Goldsmith [19] and Sharif and Hassibi [42] showed that in system having a very large number of users, the maximum throughput achieved by DPC, could be obtained by the beamforming strategy. In beamforming strategy, each user stream is coded independently and multiplied by a beamforming weight vector for transmission through multiple antennas. However, finding the optimal beamforming vectors is a non-convex optimization problem and the optimal solution for downlink channel with K users is given by exhaustive search over all possible combinations. Evidently, the complexity of the above problem becomes prohibitively high for large K. In [10], a less complex algorithm using a linear precoding zero forcing (ZF) with non-exhaustive selection of the N_t users among K users has been presented to reduce the system complexity.

6.2 Multiuser MIMO Systems with Single Receive Antenna

In this chapter, we will focus on precoding and user scheduling strategies for downlink multiuser transmission consists of multiple transmit antennas and only one receive antenna.

6.2.1 Precoding

DPC is a technique that reduces interference seen at each receiver by pre-subtracting interference at the transmitter. Using DPC, Caire and Shamai established an achievable rate region, which is a lower bound to the capacity region by definition, and showed that this rate region actually achieves the sum capacity of the multiple antenna broadcast channel for a two user transmission scheme in which only the transmitter has multiple antennas.

One suboptimal technique with much reduced complexity combines DPC with QR channel decomposition as examined in [1]. This scheme is named zero forcing dirty paper precoding (ZF-DP). Since the precoding matrix is chosen in order to force to zero the interference caused by other users, this solution is optimal when the number of transmit antennas N_t is higher than the number of users K equipped with a single receive antenna at high or low SNR and it is suboptimal in general for medium SNR. DPC can be implemented using lattice precoding [62] or trellis precoding [3]. A simpler one-dimensional implementation can also be performed using Tomlinson–Harashima precoders [12, 52]. These studies are based on the ideal assumption of an implemented system with very high complexity and perfect channel knowledge. However, for practical applications it is necessary to search for less complex schemes such as linear precoding techniques.

6.2.1.1 System Model

In this section, we consider a multiuser MIMO system with $K = N_t$ and in which each of the receivers is equipped with only one antenna. The case where $K > N_t$ will be discussed in the next sections.

The received signal at user k is given by the relation

$$y_k = \mathbf{h}_k^H \mathbf{x} + n_k \quad k = 1, \ldots K \tag{6.17}$$

where $\mathbf{h}_k$ is the channel vector of size $N_t \times 1$ and can be modeled as fading process, $\mathbf{x}$ is the transmitted vector. The entries of the noise n_k are i.i.d. Gaussian process. The transmit power is assumed to be constraint to P.

6.2.1.2 Nonlinear Precoding

Zero Forcing Dirty Paper Coding

Since the complexity of the optimal precoding strategy is too high for practical implementation, a reduced-complexity suboptimal solution to sum rate maximization has been proposed by Caire and Shamai in [1]. It suggests the use of the QR decomposition of the channel matrix combined with DPC at the transmitter.

The overall channel matrix, $\mathbf{H} = [\mathbf{h}_1 \ \mathbf{h}_2 \ \ldots \ \mathbf{h}_K]^H$ with dimension $K \times N_t$, is decomposed $\mathbf{H} = \mathbf{LQ}$ obtained by applying Gram–Schmidt orthogonalization (GSO) to the rows of $\mathbf{H}$. The matrix $\mathbf{L} = \{l_{i,j}\}$ is a lower triangular matrix (i.e., it has zeros above its main diagonal), and $\mathbf{Q}$ has orthonormal rows. With the assumption of $K = N_t$, the transmitted symbol $\mathbf{x}$ of size $N_t \times 1$ is given by

$$\mathbf{x} = \mathbf{FPs}$$

$$= \sum_{k=1}^{K} \sqrt{P_k} \mathbf{f}_k s_k \tag{6.18}$$

where $\mathbf{P} = \text{diag}\left(\sqrt{P_1}, \cdots, \sqrt{P_K}\right)$ is the matrix for power loading with the total transmit power constraint $\sum_{k=1}^{K} P_k \leq P$. $\mathbf{F} = [\mathbf{f}_1 \ \mathbf{f}_2 \ \ldots \ \mathbf{f}_K]$ is the precoding matrix where the columns of $\mathbf{F}$ are normalized to unit norm and $\mathbf{s} = [s_1 \ s_2 \ \ldots \ s_K]^T$ is the input symbol vector with $E\{\mathbf{ss}^H\} = \mathbf{I}_K$. The elements of $\mathbf{s}$ are generated by successive dirty paper encoding.

If we fix $\mathbf{F} = \mathbf{Q}^H$, the received signals are given by

$$y_k = l_{k,k} \sqrt{P_k} s_k + \sum_{j < k} l_{k,j} \sqrt{P_j} s_j + n_k, \qquad k = 1, \cdots, N_t \tag{6.19}$$

For each user k, the noncausally known interference signal is given by $\sum_{j < k} l_{k,j} \sqrt{P_j} s_j$. Since the precoding matrix is chosen in order to force to zero the interference caused by users $j > k$ on each user k, this scheme is called the ZF-DP coding.

From [1], the achievable sum rate of ZF-DPC is

$$R_{sum} = B_W \sum_{k=1}^{K} \log_2 \left(1 + \frac{P_k |l_{k,k}|^2}{N_0 B_W}\right) \tag{6.20}$$

where P_k are the solution of the optimal power loading via waterfilling as given in the following.

$$P_k = \begin{cases} \frac{P}{\gamma_0} - \frac{N_0 B_W}{P |l_{k,k}|^2} & \frac{P |l_{k,k}|^2}{N_0 B_W} \geq \gamma_0 \\ 0 & \frac{P |l_{k,k}|^2}{N_0 B_W} < \gamma_0 \end{cases} \tag{6.21}$$

The value of the water level γ_0 is computed from the power constraint. The maximum achievable sum rate is obtained by choosing the best user order.

If we accept a loss in performance, instead of performing a high dimension quantization for the precoding we can perform a scalar quantization with the Tomlinson–Harashima precoding technique [12, 52]. The Tomlinson–Harashima precoder originally developed for the intersymbol interference channel is a clever scheme to pre-subtract interference with minimal extra power.

We consider the scheme that transmitting uncoded symbols using an uncoded M-PAM where the constellation points are $\{(2m - 1 - M)a; m = 1, 2, \ldots, M\}$ and that s is the known interference signal at time instant t. In order to convey the uncoded symbol u, a naive precoder can transmit $x = u - s$ to compensate for the interference. However, s may be large, so x may exceed the power constraint. In order to avoid this problem, the Tomlinson–Harashima precoder forces the transmitted signal to lie within $[-aM, aM]$. The idea is to replicate the M-PAM constellation to get an infinite extended constellation. Each of the M symbols now corresponds to an equivalent class of points. Given the uncoded symbol u, the precoding scheme will first choose the symbol p in the extended constellation that is the closest to s. Then we transmit $x = p - s$. This last operation can be seen as a modulo operation. Likewise, the decoder also performs this modulo operation since all received symbols that differ by an integer multiple of $2aM$ are regarded as equivalent. The average transmit power is then $a^2 M^2/3$ if we assume that s is random so that the transmitted symbol is uniformly distributed over $[-Ma, +Ma]$. Since the M-PAM average transmit power is $a^2 M^2/3 - a^2/3$, there is a small power penalty of $M^2/(M^2-1)$ when using this precoding scheme. The Tomlinson–Harashima precoder can be seen as a one-dimensional implementation of the dirty paper precoding technique.

In [7], the authors have designed a Tomlinson–Harashima precoder based on minimizing the sum mean square error (MSE) of the individual users for imperfect CSI. The performance of this precoder has been investigated using random vector quantization (RVQ) in [49].

6.2.1.3 Linear Precoding

Although DPC provides an optimal solution, it has a prohibitively high computational complexity due to the associated encoding process. Therefore, it is a great practical interest to design MIMO multiuser systems with low complexity and a minimum CSI requirement at the transmitter side. One suboptimal approach is to apply linear precoding schemes, such as zero forcing beamforming (ZF-BF) [40, 60] or MMSE criterion [43].

With ZF-BF where the users' weight vectors are chosen to avoid interference among user streams, the precoding matrix is simply designed to be the pseudo inverse of the channel matrix of the selected users to create orthogonal channels between the transmitter and the receivers. ZF-BF is generally power inefficient since the beamforming weights are not matched to the users' channels. However, it has been shown that ZF-BF is asymptotically optimal when the number of users approaches infinity because of the multiuser diversity effect.

The transmitted signal is represented as

$$\mathbf{x} = \mathbf{FPs} \tag{6.22}$$

where the vector $\mathbf{F}$ can be determined using linear precoding schemes.

Zero Forcing Beamforming

In the ZF-BF, when $K \leq N_t$, assuming perfect CSI at the transmitter, the precoding matrix $\mathbf{F} = [\mathbf{f}_1, \ldots, \mathbf{f}_K]$ is determined as

$$\mathbf{F} = \eta \mathbf{H}^H (\mathbf{HH}^H)^{-1} \tag{6.23}$$

where $\mathbf{H} = [\mathbf{h}_1, \ldots, \mathbf{h}_K]^H$ is the channel matrix of size $K \times N_t$ with $\mathbf{h}_k \in \mathbb{C}^{N_t \times 1}$. The Moore-Penrose inverse of $\mathbf{H}$ is given by $\mathbf{H}^H (\mathbf{HH}^H)^{-1}$.

In order to keep the short-term power constant, the factor η is calculated as

$$\eta = \frac{1}{\sqrt{\mathrm{tr}((\mathbf{HH}^H)^{-1})}} \tag{6.24}$$

When $K = N_t$, the precoding matrix is proportional to the inverse of the channel matrix $\mathbf{H}$ as

$$\mathbf{F} = \eta \mathbf{H}^{-1} \tag{6.25}$$

Since the beamforming vectors are chosen perfectly orthogonal to all the channel vectors, the multi-user interference is completely suppressed. Assuming that the power is uniformly distributed $P_k = P/N_t$, the signal-to-interference-plus-noise ratio (SINR) is

$$\gamma_k = \frac{P}{N_t N_0 B_W} \left| \mathbf{h}_k^H \mathbf{f}_k \right|^2 \tag{6.26}$$

The resulting sum data rate R is calculated as

$$R = B_W \sum_{k=1}^{K} \log_2(1 + \gamma_k) \tag{6.27}$$

Another precoding matrix is constructed by choosing the beamforming vectors $f_k; k = 1, 2, \ldots, K$ as the normalized column of the matrix $\mathbf{H}^{-1}$.

Regularized Zero Forcing

In order to improve the performance of ZF linear precoding, we can regularize the inverse matrix. In this vector perturbation (VP) linear precoding, we add a multiple of identity matrix before inverting [36]. It has been shown that this regularization of the inverse matrix helps to reduce the effects of the largest eigenvalue. We have

$$\mathbf{F} = \eta \mathbf{H}^H (\mathbf{H}\mathbf{H}^H + \xi \mathbf{I}_K)^{-1} \tag{6.28}$$

where

$$\eta = \frac{1}{\sqrt{\mathrm{tr}((\mathbf{H}\mathbf{H}^H + \xi \mathbf{I}_K)^{-1})}} \tag{6.29}$$

A reasonable metric for choosing the regularization factor ξ is to maximize the SINR. Using this criterion, the ξ has been calculated in [36] as $K\frac{N_0 B_W}{P}$.

Simulation Results

We consider a broadcast multiuser system with $K = N_t = 4$ in block fading channels. The elements of channel vector are generated as i.i.d random complex Gaussian variables with unit variance.

In Fig. 6.3, we plot the sum rate performance versus SNR for using different precoding schemes with $K = N_t = 4$. As it is shown, there is no difference between the ZF-DP performance with and without waterfilling especially at high SNR. The reason for that is all users channels have good gains at high SNR, then power allocation values are very low levels. When the SNR is getting high, the ZF-DP approaches the sum rate capacity performance. The matrix inversion in the ZF precoding is performed by adding an identity matrix before inverting as given in vector perturbation precoding. As shown in the performance result, the VP significantly outperforms the ZF-BF.

Figure 6.4 shows the performance of bit error rate (BER) as a function of E_b/N_0 using QPSK modulation with perfect CSI at the transmitter with ZF-BF, VP and ZF-DP using Tomlinson Harashima precoding for $K = N_t = 4$. The ZF precoding gives always the worse performance compared to other scheme. ZF-DP with TH precoding gives better BER performance than vector perturbation precoding starting from 15 dB.

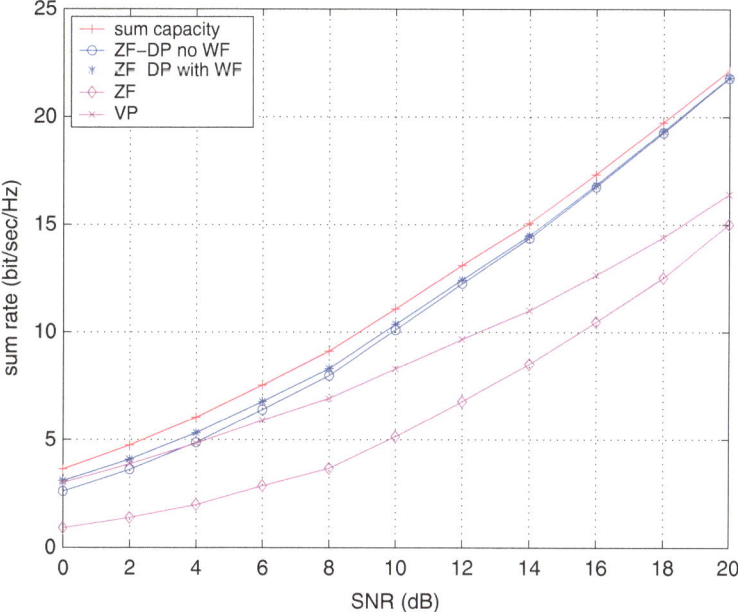

Fig. 6.3 Sum rate versus SNR for a multiuser system with $K = 4$ and $N_t = 4$ using ZF-BF, vector perturbation (VP) and ZF-DP with and without waterfilling (WF)

6.2.1.4 Precoding with Limited Feedback

In the finite rate feedback link case, each receiver quantizes the direction of the channel vector using B bits and if required channel gain information (CGI) to feed back to the BS. Compared to the single user case, the transmitter must then determine the precoding matrix using the quantized channel direction information (CDI). In this section, we will restrict our analysis to the ZF-BF precoding scheme.

Similar to the point-to-point single user multiple input single output (MISO) case, each user quantizes its CDI, $\mathbf{g}_k$ that is given in Chap. 4, with a vector $\mathbf{w}_k^{opt}$ that is selected from the codebook $\mathcal{W}$ of size $N = 2^B$ in order to maximize the instantaneous SNR or equivalently to minimize the chordal distance metric.

The optimum precoding vector is selected according to the following criterion:

$$
\begin{aligned}
\mathbf{w}_k^{opt} &= \arg \max_{\mathbf{w}_i \in \mathcal{W}} |\mathbf{h}_k^H \mathbf{w}_i|^2 \\
&= \arg \min_{\mathbf{w}_i \in \mathcal{W}} (1 - |\mathbf{g}_k^H \mathbf{w}_l|^2)
\end{aligned}
\tag{6.30}
$$

The selected vectors are fed back to the transmitter who builds a concatenated quantized CDI matrix $\tilde{\mathbf{W}} = [\mathbf{w}_1^{opt}, \dots \mathbf{w}_K^{opt}]$.

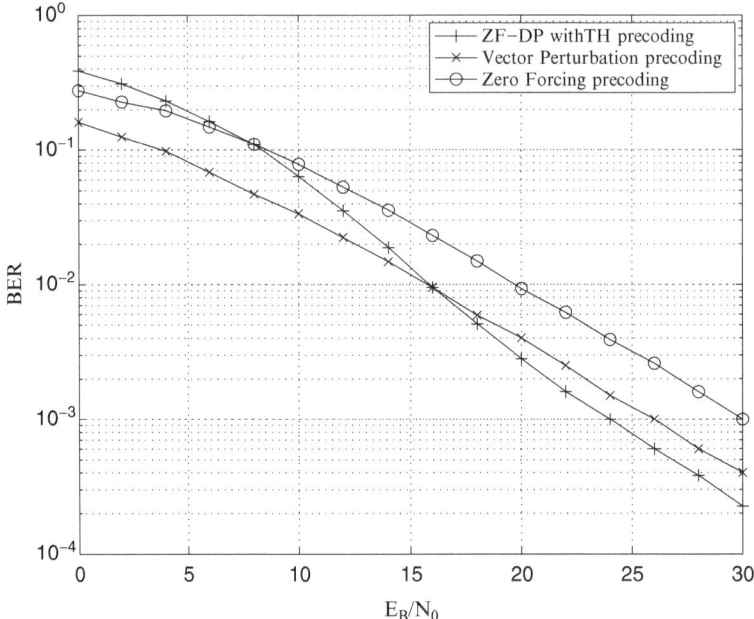

Fig. 6.4 BER versus E_b/N_0 for $K = N_t = 4$ using ZF, VP and ZF-DP with Tomlinson–Harashima with full CSI at the transmitter

The beamforming vectors $\mathbf{f}_1, \ldots, \mathbf{f}_K$ are chosen as the normalized columns of the matrix $\mathbf{F}$ by

$$\mathbf{F} = \tilde{\mathbf{W}}^H (\tilde{\mathbf{W}}\tilde{\mathbf{W}}^H)^{-1} \tag{6.31}$$

and are given by for each $k = 1, 2, \ldots, K$ as

$$\mathbf{f}_k = \frac{\mathbf{F}(:, k)}{||\mathbf{F}(:, k)||} \tag{6.32}$$

In the limited feedback case, since ZF-BF is performed from the quantized vectors, the multi-user interference is only partially removed. Then, the SINR becomes

$$\gamma_k = \frac{\frac{P}{N_t} \left|\mathbf{h}_k^H \mathbf{f}_k\right|^2}{N_0 B_W + \frac{P}{N_t} \sum_{j=1; j \neq k}^{N_t} \left|\mathbf{h}_k^H \mathbf{f}_j\right|^2} \tag{6.33}$$

and the instantaneous data rate R_k is calculated as

$$R_k = B_W \log_2(1 + \gamma_k) \tag{6.34}$$

We have shown in Chap. 4 that RVQ is a practical tool to estimate the performance of MISO point-to-point single user communication. For the multiuser MIMO broadcast channels, the throughput loss using finite rate feedback with RVQ compared to rate achieved with perfect CSI at the transmitter (CSIT) ZF-BF has been computed by Jindal in [20].

Theorem 1. *Finite-rate feedback with B feedback bits per user incurs a throughput loss compared to full CSIT ZF-BF that can be bounded by*

$$\Delta R < \log_2 \left(1 + \frac{P}{N_0 B_W} 2^{-\frac{B}{N_t - 1}} \right) \tag{6.35}$$

Proof. The throughput loss is bounded as

$$\Delta R = \mathbb{E}_{\mathbf{H}} \left[\log_2 \left(1 + \frac{P}{N_t N_0 B_W} \left| \mathbf{h}_k^H \mathbf{f}_k^{full} \right|^2 \right) \right]$$

$$- \mathbb{E}_{\mathbf{H,F}} \left[\log_2 \left(1 + \frac{\frac{P}{N_t} \left| \mathbf{h}_k^H \mathbf{f}_k \right|^2}{N_0 B_W + \frac{P}{N_t} \sum_{j=1; j \neq k}^{N_t} \left| \mathbf{h}_k^H \mathbf{f}_j \right|^2} \right) \right]$$

$$\leq \mathbb{E}_{\mathbf{H}} \left[\log_2 \left(1 + \frac{P}{N_t N_0 B_W} \left| \mathbf{h}_k^H \mathbf{f}_k^{full} \right|^2 \right) \right] - \mathbb{E}_{\mathbf{H,F}} \left[\log_2 \left(1 + \frac{P}{N_t N_0 B_W} \left| \mathbf{h}_k^H \mathbf{f}_k \right|^2 \right) \right]$$

$$+ \mathbb{E}_{\mathbf{H,F}} \left[\log_2 \left(1 + \frac{P}{N_t} \sum_{j=1; j \neq k}^{N_t} \left| \mathbf{h}_k^H \mathbf{f}_j \right|^2 \right) \right]$$

$$= \mathbb{E}_{\mathbf{H,F}} \left[\log_2 \left(1 + \frac{P}{N_t} \sum_{j=1; j \neq k}^{N_t} \left| \mathbf{h}_k^H \mathbf{f}_j \right|^2 \right) \right] \tag{6.36}$$

The inequality comes from the fact that $\sum_{j \neq k} \left| \mathbf{h}_k^H \mathbf{f}_j \right|^2 \geq 0$ and $\log(.)$ is a monotonically increasing function. The second equality is due to the isotropically distribution of $\mathbf{f}_k$ for both the full CSIT and the quantized case.

Then, applying the Jensen inequality and using Lemma 2 in [20] that states the random variable $\left| \mathbf{g}_k^H \mathbf{f}_j \right|^2$ for $k \neq j$ is equal to the product of the chordal distance $d^2(\mathbf{g}_k, \mathbf{f}_k)$ and is an independent $\beta(1, N_t - 2)$ random variable, we have

$$\Delta R \leq \log_2 \left(1 + \frac{P}{N_0 B_W} (N_t - 1) \mathbb{E}[||\mathbf{h}_k||^2] \mathbb{E}[|\mathbf{g}_k^H \mathbf{f}_j|^2] \right)$$

$$= \log_2 \left(1 + \frac{P}{N_0 B_W} (N_t - 1) \mathbb{E}[|\mathbf{g}_k^H \mathbf{f}_j|^2] \right)$$

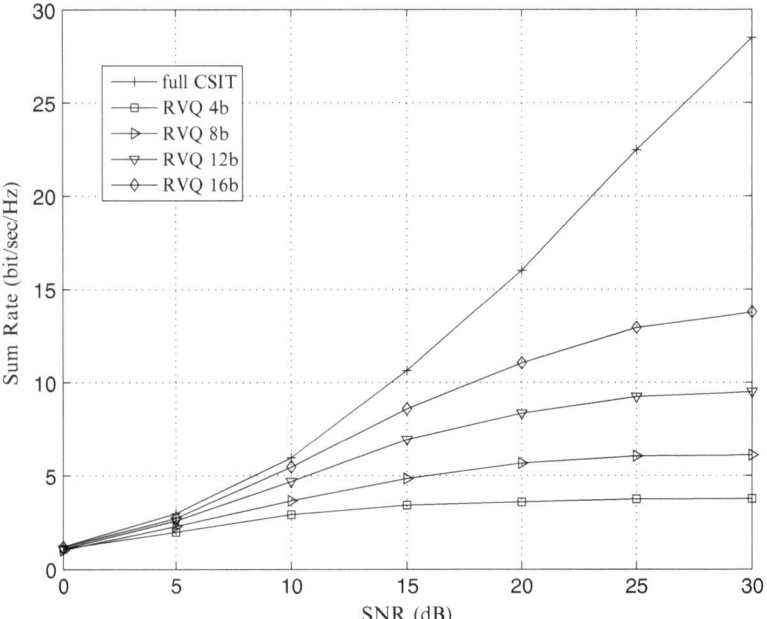

Fig. 6.5 Sum rate versus SNR for ZF-BF with full CSIT, ZF-BF with RVQ with $B = 4, 8, 12, 16$ bits for downlink multiuser MISO transmission with $K = N_t = 4$

$$\leq \log_2 \left(1 + \frac{P}{N_0 B_W} \mathbb{E}[d^2(\mathbf{g}_k, \mathbf{f}_k)] \right)$$

$$< \log_2 \left(1 + \frac{P}{N_0 B_W} 2^{-\frac{B}{N_t-1}} \right) \tag{6.37}$$

The last equation is obtained using the RVQ bound on the chordal distance between $\mathbf{g}_k$ and $\mathbf{f}_k$.

Another interesting theorem quantifies the scaling of feedback needed to keep the rate gap ΔR no larger than a constant $r > 0$ [20].

Theorem 2. *A rate gap ΔR no larger than a constant $r > 0$ is guaranteed at the transmit power P by fixing the number of bits B as*

$$B \approx \frac{N_t - 1}{3} P_{dB} - (N_t - 1) \log_2(2^r - 1) \tag{6.38}$$

Proof. Obtained by setting the upper bound ΔR equal to r to the SNR degradation in Theorem 1 and solving for B as a function of P.

In Fig. 6.5, we plot the sum rate versus SNR of ZF-BF with full CSIT and using RVQ with $B = 4, 8, 12$ and 16 bits for the case $K = N_t = 4$. When the SNR is small, limited feedback performs nearly as well as zero forcing with full CSI. However, when the SNR is increased, the limited feedback system becomes interference limited and the rates converge to an upper limit depending on the amount of the multi-user interference.

6.2.2 User Scheduling Algorithms

In the case of $K > N_t$ a multiuser diversity can be achieved by performing user scheduling before applying ZF-BF. Multiuser diversity is a form of selection diversity among users; when the number of users is large, the base station can schedule its transmission to those users with favorable channel fading conditions to improve the system throughput. When full CSI is available, a better choice of beamforming directions can be made and it provides fairly good performance under the zero-forcing strategy. Namely, the transmitter can choose a group of users with high channel magnitudes and for which their channel directions are matched to zero-forcing beam directions. In [1], it has been shown that random selection of N_t users incurs significant throughput loss for both ZF-DP and ZF-BF schemes. In [51] the authors emphasized the importance of user order in maximizing the sum rate using the ZF-DP scheme.

The multi-user multiantenna downlink with ZFBF requires a brute-force exhaustive search over all possible user sets and the complexity of an exhaustive search is prohibitive when the number of users is large. In order to decrease the complexity of this search, several suboptimal greedy user scheduling algorithms have been designed in the literature for ZF-BF. Generally, these algorithms fall into two categories: Capacity-based and Frobenius norm-based algorithm. The capacity-based algorithm, represented by the zero forcing with selection (ZFS) algorithm proposed by Dimic et al. [10], chooses users greedily based on the accurate sum rate variation. It chooses the first user with the highest capacity and then finds the next user that provides the maximum sum rate from the remaining unselected users. The Frobenius norm-based algorithm, represented by the semi-orthogonal user scheduling (SUS) algorithm proposed by Yoo et al. [60], chooses users greedily based on the approximate sum rate variations with respect to channel norm-related parameters. SUS adds the new user with the largest effective channel norm nearly orthogonal to the selected users in each iteration. The Frobenius norm-based algorithms have lower complexity by eliminating the calculation of sum rate, but pay a price in sum rate performance by not guaranteeing a positive sum rate increment in the user selection process.

Let $\mathcal{R} = \{1, 2, \ldots, K\}$ denotes the set of indices of all K active users in the cell and let $\mathcal{A} = \{a_b\}_{b=1}^{B} \subset \mathcal{R}$ denote the set of selected users where $|\mathcal{A}| = B \leq N_t$.

For downlink multiuser MIMO fading channel with one receive antenna for $K > N_t$, the baseband representation of the received signal at each selected user is given by

$$y_b(\mathcal{A}) = \mathbf{h}_b(\mathcal{A})^H \sqrt{P l_b(\mathcal{A})} \mathbf{F}(\mathcal{A}) s_b(\mathcal{A}) + n_b(\mathcal{A}) \qquad (6.39)$$

where $Pl_b(\mathscr{A})$ is the attenuation in the power caused by path loss, $P_b(\mathscr{A}) = Pl_b(\mathscr{A})P$ and $\mathbf{h}_b(\mathscr{A})$ is the channel vector with dimension of $N_t \times 1$ and each element of channel vector is modeled with Rayleigh distribution with $\mathbb{E}[|h_{k,b}|^2] = 1$ for $k = 1, 2, \ldots, K$ and $b = 1, 2, \ldots, N_t$. For the case of i.i.d. Rayleigh fading channel assumption, the attenuation factor of all users is assumed to be the same and normalized to 1, such that $Pl_k(\mathscr{A}) = 1$ for $k = 1, 2, \ldots, K$.

The ZF-BF vector is determined as

$$\mathbf{F}(\mathscr{A}) = \eta(\mathscr{A})\mathbf{H}(\mathscr{A})^H (\mathbf{H}(\mathscr{A})\mathbf{H}(\mathscr{A})^H)^{-1} \tag{6.40}$$

In order to keep the short-term power constant we have

$$\eta(\mathscr{A}) = \frac{1}{\sqrt{\mathrm{tr}((\mathbf{H}(\mathscr{A})\mathbf{H}(\mathscr{A})^H)^{-1})}} \tag{6.41}$$

The resulting instantaneous data rate R_b is calculated as

$$R_b = B_W \log_2 \left(1 + \frac{P}{N_t N_0 B_W} \left| \mathbf{h}_b^H(\mathscr{A})\mathbf{f}_b(\mathscr{A}) \right|^2 \right) \tag{6.42}$$

The sum rate achieved by the ZFBF scheme is

$$R_{ZFBF}(\mathscr{A}) = \sum_{a_b \in \mathscr{A}} R_b \tag{6.43}$$

In order to maximize the sum rate under average power constraint P, first of all it is necessary to choose the best combination of N_t users. The exhaustive search which consists in evaluating $\binom{K}{N_t}$ combinations quickly becomes prohibitory. In the following section, we will examine two main algorithms to select N_t out of K users, namely, capacity and semi-orthogonal-based user scheduling.

6.2.2.1 Capacity-Based User Scheduling Algorithm

Capacity-based user scheduling algorithm selects N_t out of K rows of $\mathbf{H} = [\mathbf{h}_1 \cdots \mathbf{h}_K]^H$ and orders the selected rows in the GSO, aiming to maximize the throughput.

Capacity-Based Greedy User Selection Algorithm

1. *Initialization phase:*

   - Set $b=1$.

- Let $r_{1,k} = \mathbf{h}_k^H \mathbf{h}_k, \forall k$. Find the best user a_1, according to its channel gain a_1 such that,

$$a_1 = \arg\max_{k \in \mathscr{R}} r_{1,k} \tag{6.44}$$

Then BS knows its first selected user, and the set of selected users will be $\mathscr{A}_1 = \{a_1\}$.

2. *Sum rate calculation phase:*

- For $b=2$ to N_t repeat

  - The objective is to find users who are orthogonal to the selected ones. Therefore, we project each remaining channel vector onto the orthogonal complement of the subspace spanned by the channels of the selected users. The projector matrix is obtained by

    $$\mathbf{P_b^\perp} = \mathbf{I}_{N_t} - \mathbf{H}(\mathscr{A}_{b-1})^H [\mathbf{H}(\mathscr{A}_{b-1})\mathbf{H}(\mathscr{A}_{b-1})^H]^{-1} \mathbf{H}(\mathscr{A}_{b-1}) \tag{6.45}$$

    where $\mathbf{I}_{N_t}$ is the $N_t \times N_t$ identity matrix, and $\mathbf{H}(\mathscr{A}_{b-1})$ denotes the matrix consisting of the channel vectors of the users selected in the first $b-1$ steps as $\mathbf{H}(\mathscr{A}_{b-1}) = [\mathbf{h}_{a_1} \cdots \mathbf{h}_{a_{b-1}}]^H$.
  - From all the orthogonal users, we select the user who maximizes the sum rate (according to the projection properties) as

    $$r_{b,k} = \mathbf{h}_k^H \mathbf{P_b^\perp} \mathbf{h}_k \tag{6.46}$$

  - Find user a_b such that

    $$a_b = \arg\max_{k \in \mathscr{R} \setminus \mathscr{A}_{b-1}} r_{b,k} \tag{6.47}$$

  - Set $\mathscr{A}_b = \mathscr{A}_{b-1} \cup \{a_b\}$.

3. *Beamforming:*
   Apply the ZF transmit beamforming as previously described by using the selected user set $\mathscr{A} = \mathscr{A}_b$.

It is also possible to simplify the preselection in [10] and then to reduce the complexity of the algorithm at the base station by reducing the user search space as described in the following.

Capacity-Based Greedy User Selection Algorithm with Reduced User Search Space

1. *Initialization phase:*

- Set $b=1$.
- Let $r_{1,k} = \mathbf{h}_k^H \mathbf{h}_k, \forall k$. Find the best user a_1, such that,

$$a_1 = \arg\max_{k \in \mathscr{R}} r_{1,k} \tag{6.48}$$

Then, set $\mathscr{A}_1 = \{a_1\}$ and compute $R_{ZF}(\mathscr{A}_1)$.

2. *Sum rate calculation phase:*

   • For $b=2$ to N_t repeat

     – Find a user s_n such that
     $$a_b = \arg\max_{k \in \mathscr{R} \setminus \mathscr{A}_{b-1}} R_{ZF}(\mathscr{A}_{b-1} \cup \{k\})$$
     – Set $\mathscr{A}_b = \mathscr{A}_{b-1} \cup \{a_b\}$ and denoted achievable rate as $R_{ZF}(\mathscr{A}_b)$.
     – If $R_{ZFBF}(\mathscr{A}_b) \leq R_{ZF}(\mathscr{A}_{b-1})$ break and set $\mathscr{A} = \mathscr{A}_{b-1}$.

3. *Beamforming:*
   The ZF transmit beamforming is performed as previously described by using the selected user set $\mathscr{A}$.

6.2.2.2 Semi-orthogonal User Group Scheduling

A suboptimal user group using a semiorthogonal user group selection (SUS) algorithm in [60] is performed as follows.

1. *Initialization phase:*

   • Set $b = 1$, $\mathscr{A} = \emptyset$, $\mathscr{R}_1 = \{1, 2, \ldots, K\}$.

2. *Semi-orthogonal user selection:*

   • Calculate $\mathbf{z}_k$, the component of $\mathbf{h}_k$ orthogonal to the subspace spanned by $\mathbf{o}_{(1)}, \mathbf{o}_{(2)}, \ldots, \mathbf{o}_{(b-1)}$:

   $$\mathbf{z}_k = \mathbf{h}_k \left(\mathbf{I} - \sum_{j=1}^{b-1} \frac{\mathbf{o}_{(j)}^H \mathbf{o}_{(j)}}{\|\mathbf{o}_{(j)}\|^2} \right), \forall k \in \mathscr{R}_b \tag{6.49}$$

   For $b = 1$, this implies that $\mathbf{z}_k = \mathbf{h}_k$.

   • Select the bth user as follows:

   $$\pi_b = \arg\max_{k \in \mathscr{R}_b} \mathbf{z}_k \tag{6.50}$$

   $$\mathscr{A} = \mathscr{A} \bigcup \{\pi_b\} \tag{6.51}$$

   $$\mathbf{o}_{(b)} = \mathbf{o}_{\pi_b} \tag{6.52}$$

   $$b \leftarrow b + 1 \tag{6.53}$$

- If $|\mathscr{A}| < N_t$, then calculate $\mathscr{R}_b$, the set of users semiorthogonal to $\mathbf{o}_{(b)}$,

$$\mathscr{R}_{b+1} = \left\{ k \in \mathscr{R}_b, k \neq \pi_b \,\Big|\, \frac{\mathbf{o}_{(b)}^H \mathbf{h}_k}{\|\mathbf{o}_{(b)}\|^2 \|\mathbf{h}_k\|^2} < \alpha \right\} \tag{6.54}$$

where α is a small positive constant.

- If $\mathscr{R}_{b+1}$ is nonempty and the number of elements in the set $|\mathscr{A}| < N_t$ satisfies, then continue to algorithm. Otherwise, the algorithm is finished.

3. *Beamforming:*

The ZF transmit beamforming is performed as previously described by using the selected user set $\mathscr{A}$.

Simulation Results

We consider a multiuser MIMO system with $N_t = 2, 4$ and $N_r = 1$ antennas. The channel coefficients are modelled as random complex Gaussian variable with unit variance. We compare the system performance by employing different user scheduling algorithms and using RVQ-based codebooks that have been explained in Chap. 4. The quantization has been performed on CDI and CGI and their effects on the performances have been observed for full set, capacity and semi-orthogonal-based user selection algorithms.

Firstly of all, the performance results are obtained assuming that all users have the same average SNR. As shown in Fig. 6.6, the performance of sum rate is not affected by the quality of CGI for both in i.i.d where users have the same average SNR and non i.d.d channel where users have different average SNRs. Consequently, CGI is not required at the BS to perform user scheduling based on full user space set. However, the consideration of full user space set is a computationally complex procedure. In order to reduce the complexity, capacity and semi-orthogonal-based user selection algorithms can be used at the BS. The full feedback case has been illustrated in Fig. 6.7 for different number of transmit antennas. As shown in the simulation results, capacity-based greedy user scheduling gives better performance than SUS algorithm providing more gain especially when the number of antennas is increased. When limited feedback link is considered, both the quantization of CGI and CDI is required for these low complexity scheduling schemes. According to the performance results, it is observed that the effect of CDI is more dominant compared to CGI for SUS algorithm as illustrated in Fig. 6.8 and capacity-based user scheduling as shown in Fig. 6.9.

In the previous result, we consider uncorrelated flat fading channels. Under the case of correlated flat fading wireless channel, the throughput performances are shown in Fig. 6.10 [28] for $N_t = 4$ and $K = 10$ using ZF-BF with full user space set.

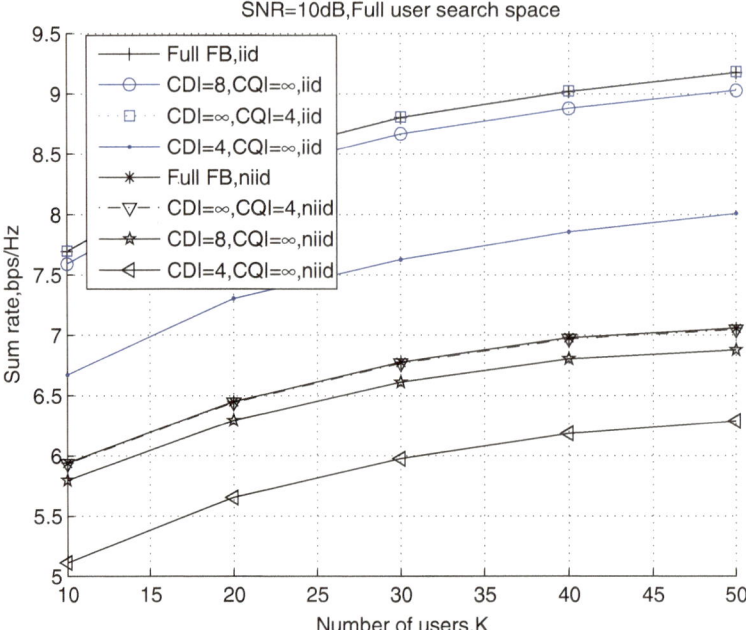

Fig. 6.6 Sum rate versus K with limited feedback for $N_t = 2$ using full user space set in i.i.d and non i.i.d channels

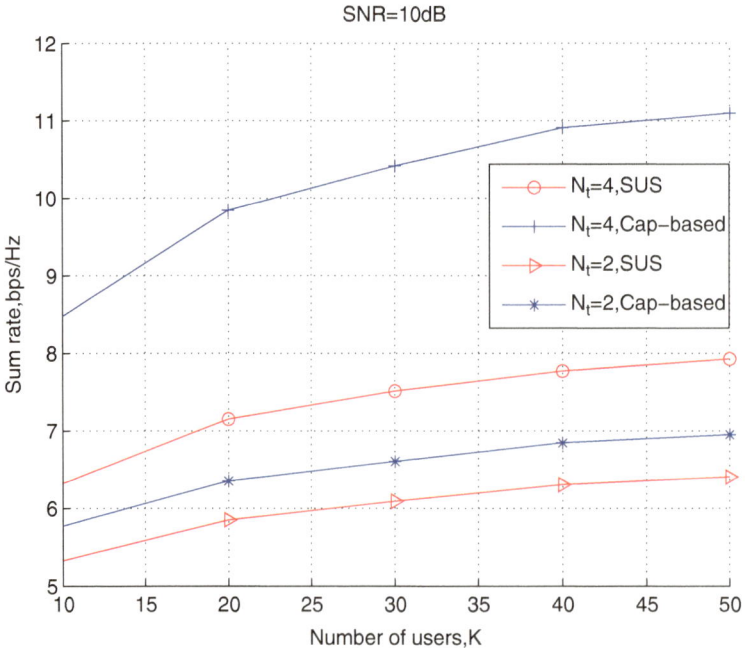

Fig. 6.7 Sum rate versus K for $N_t = 2$ and $N_t = 4$ for different scheduling algorithms

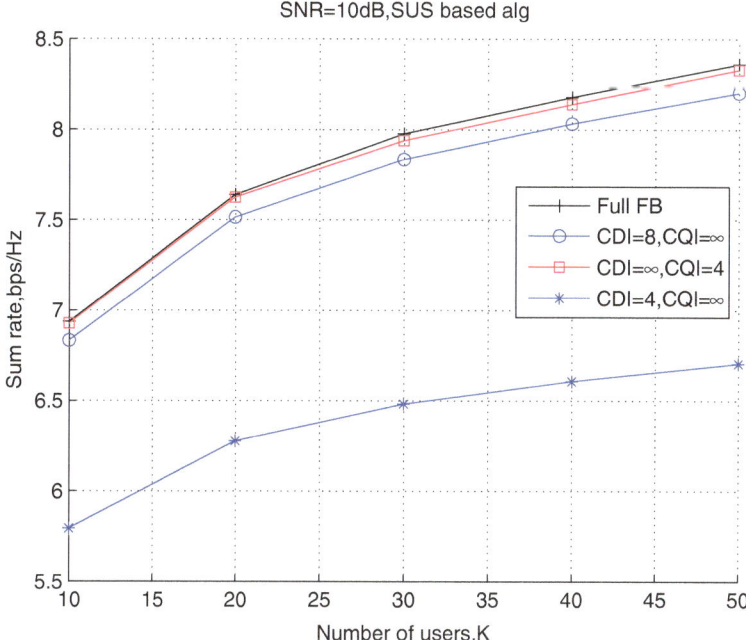

Fig. 6.8 Sum rate versus SNR with limited feedback for $N_t = 2$ using SUS algorithm

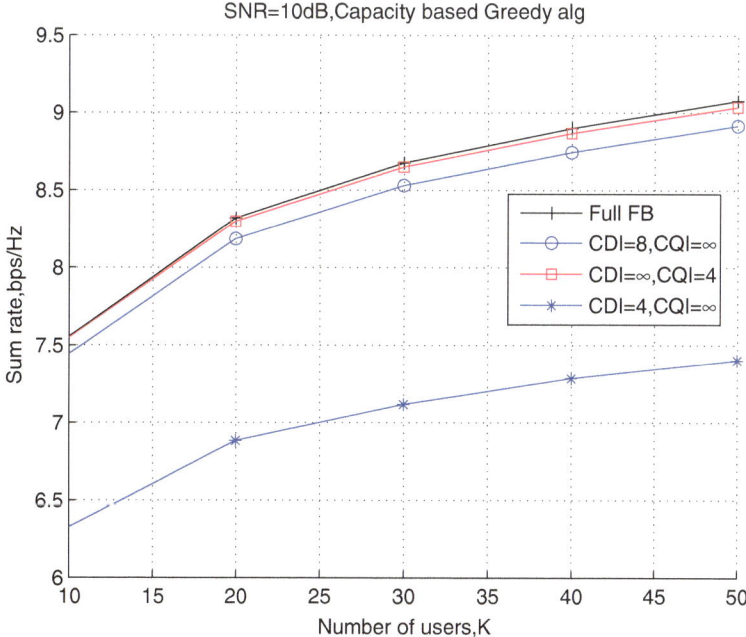

Fig. 6.9 Sum rate versus K with limited feedback for $N_t = 2$ using capacity-based algorithm

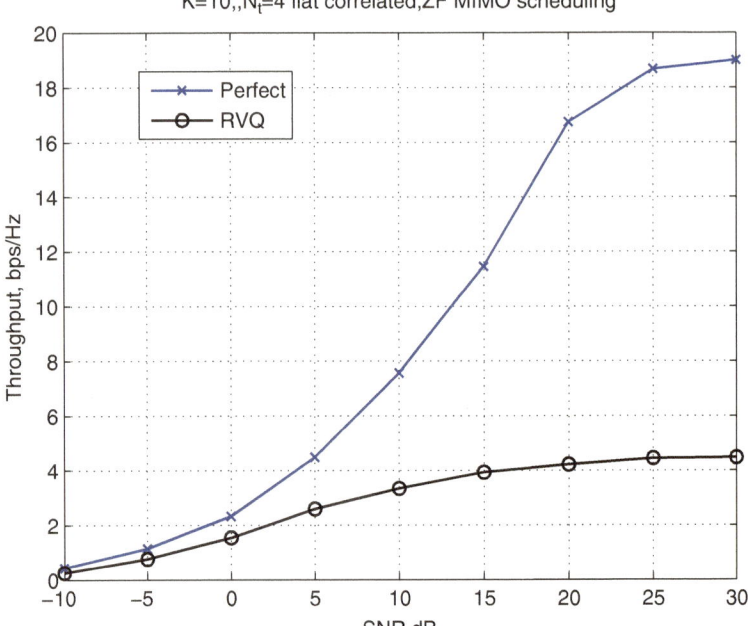

Fig. 6.10 Performance of MU-MISO systems in correlated flat fading channels for $N_t = 4$ and $K = 10$ [28]

6.2.3 User Selection at Receiver Side

As mentioned in the previous section, the number of active users is limited by the number of transmit antennas at the BS in the precoding scheme. The exhaustive search which consists in evaluating $\binom{K}{N_t}$ combinations quickly becomes prohibitive. However, the users having a poor channel (low norm or/and interfering with other good users) should not take part in the user selection algorithm, nor feed back their channel information.

By using a self discrimination criterion at the receiver side, it will be possible to reduce the feedback load and the complexity of the user selection algorithm at BS. In order to describe the less complex user selection and reduced feedback load, we propose three user selection criteria considering different users sets [24,25].

Let $\mathcal{R}$ be the set of all the users in the cell. From $\mathcal{R}$, we define three sets of users corresponding to three different self discrimination criteria in the following.

6.2.3.1 Norm-Based Selection

The set $\mathcal{U}_1$ is constructed by selecting the users that satisfy the norm-only selection which is also named as T1 criterion [11]:

$$\mathcal{U}_1 = \{k \in \mathcal{R} : \|\mathbf{h}_k\|^2 > \gamma_{th}\} \tag{6.55}$$

The threshold adjustment will be obtained either analytically or by simulation in order to guarantee an average number of users $\bar{K}$ that feedback their instantaneous CSI to the BS.

$$\bar{K} = K Pr\{k \in \mathcal{U}_1\}$$
$$= K Pr\{\|\mathbf{h}_k\|^2 > \gamma_{th}\} \tag{6.56}$$

The set $\mathcal{U}_1$ is determined by the incomplete gamma distribution $\Gamma(N_t, 1)$ which can be bounded by [15, 16]

$$[1 - \exp(-\beta\gamma)]^{N_t} \leq \int_0^{\gamma_{th}} f_\gamma(\gamma)\, d\gamma \leq [1 - \exp(-\gamma)]^{N_t} \tag{6.57}$$

where $\beta = (N_t!)^{-\frac{1}{N_t}}$ and $f_\gamma(\gamma)$ is the probability density function with $\chi_2^2(N_t)$.
Then, we can obtain

$$Pr\{k \in \mathcal{U}_1\} = \sum_{b=0}^{N_t-1} \frac{\exp(-\gamma_{th})(\gamma_{th})^b}{b!} \tag{6.58}$$

The average number of users that feed back their CSI to BS can be determined by

$$\bar{K} = K \sum_{b=0}^{N_t-1} \frac{\exp(-\gamma_{th})(\gamma_{th})^b}{b!} \tag{6.59}$$

6.2.3.2 Semi-orthogonal-Based Selection

The set $\mathcal{U}_2$ is constructed by selecting the users that satisfy the semi-orthogonal selection which is also named as T2 criterion [44].

We define the set of N_t-dimensional complex random orthogonal vectors $\boldsymbol{\phi}_b$ where $b=1,\ldots,N_t$. The $\boldsymbol{\phi}_b$ vectors are generated according to an isotropic distribution [15] and are equally likely pointed to any direction in the complex space. Based on the method of Heiberger, for the construction of $\boldsymbol{\phi}_b$, a $N_t \times N_t$ matrix $\mathbf{X}$ is first generated with entries $x_{ij} \sim \mathcal{N}(0, 1)$, then a QR factorization is computed. The method provides a matrix $\mathbf{Q}$ and $\mathbf{R}$ where the matrix $\mathbf{Q}$ composed of N_t orthogonal column vectors that represents the $\boldsymbol{\phi}_b$ vector.

Each user generates the same N_t random orthonormal vectors $\boldsymbol{\phi}_b$ which are also known by the BS.

Using a chordal distance which measures distances between subspaces of Euclidean N_t-dimensional space, the users measure the orthogonality between their normalized channel vector $\mathbf{g}_k = \frac{\mathbf{h}_k}{\|\mathbf{h}_k\|}$ and orthonormal vectors $\boldsymbol{\phi}_b$ as follows:

$$d^2(\mathbf{g}_k, \boldsymbol{\phi}_b) = 1 - |\mathbf{g}_k^H \boldsymbol{\phi}_b|^2 = \sin^2(\theta) \tag{6.60}$$

where θ is the angle between the lines generated by the column spaces of $\mathbf{g}_k$ and $\boldsymbol{\phi}_b$.

Let $\mathbb{O}^{N_t}$ be the unit sphere lying in $\mathbb{C}^{N_t}$ and centered at the origin. For any $0 < \epsilon_{th} < 1$, we can define a spherical cap on $\mathbb{O}^{N_t}$ with center $\mathbf{o}$ and square radius ϵ_{th} as the open set $\mathscr{B}_{\epsilon_{th}}(\mathbf{o}) = \{\mathbf{f} \in \mathbb{O}^{N_t} : d^2(\mathbf{f}, \mathbf{o}) \leq \epsilon_{th}\}$. Then, we have

$$\mathscr{U}_2 = \{k \in \mathscr{R} : \mathbf{g}_k \in \bigcup_{b=1}^{N_t} \mathscr{B}_{\epsilon_{th}}(\boldsymbol{\phi}_b)\} \tag{6.61}$$

According to [32], we have

$$Pr\{k \in \mathscr{R} : \mathbf{g}_k \in \bigcup_{i=1}^{N_t} \mathscr{B}_{\epsilon_{th}}(\boldsymbol{\phi}_b)\} = N_t \epsilon^{N_t-1} \tag{6.62}$$

6.2.3.3   Combined Criterion-Based Selection

The set of $\mathscr{U}_3$ is constructed according to combined selection by selecting the users that satisfy the *two* previous selections which is also named as T3 criterion [25]:

$$\mathscr{U}_3 = \{k \in \mathscr{R} : \mathbf{g}_k \in \bigcup_{b=1}^{N_t} \mathscr{B}_{\epsilon}(\boldsymbol{\phi}_b) \text{ and } \|\mathbf{h}_k\|^2 \geq \gamma_{th}\} \tag{6.63}$$

According to the user' channel independency, the channel norm and the channel direction becomes independent and Eq. (6.63) can be written as

$$\bar{K} = K Pr\{k \in \mathscr{U}_3\}$$
$$= K Pr\{k \in \mathscr{U}_1\} \times Pr\{k \in \mathscr{U}_2\}$$
$$= K Pr\{\|\mathbf{h}_k\|^2 > \gamma_{th}\} \times Pr\{k \in \mathscr{R} : \mathbf{g}_k \in \bigcup_{b=1}^{N_t} \mathscr{B}_{\epsilon_{th}}(\boldsymbol{\phi}_b)\} \tag{6.64}$$

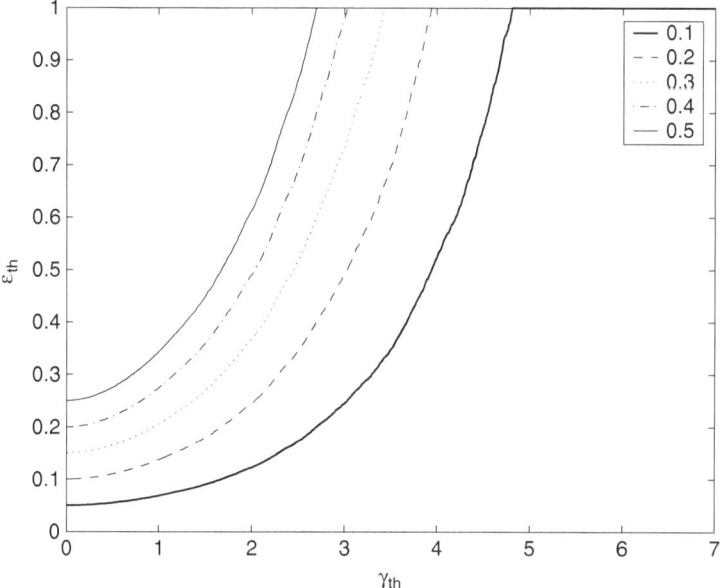

Fig. 6.11 ε_{th} and γ_{th} for $Pr\{k \in \mathcal{U}_3\} = 0.1$ to 0.5

Then, the average number of users for the combined criterion

$$\bar{K} = K N_t \sum_{b=0}^{N_t-1} \frac{e^{-\gamma_{th}}(\gamma_{th})^b}{b!} \epsilon^{N_t-1} \tag{6.65}$$

Therefore, the ϵ_{th} can be calculated as

$$\epsilon_{th} \geq \left(\frac{\bar{K}}{K} \frac{1}{N_t \sum_{b=0}^{N_t-1} \frac{e^{-\gamma_{th}}(\gamma_{th})^b}{b!}} \right)^{1/(N_t-1)} \tag{6.66}$$

In Fig. 6.11, we present the curves of the pair $(\gamma_{th}, \epsilon_{th})$ allowing predetermined probability $Pr\{k \in \mathcal{U}_3\}$ (10–50%). For each probability, it will be necessary to choose a pair by privileging either the criterion on the norm or the criterion on orthogonality.

6.2.3.4 Combined Criterion with Limited Feedback

In the quantized feedback scheme as explained in Chap. 4, the precoding vector **w** is taken from a set of 2^B vectors where B is the number of feedback bits. In contrast to the normalized i.i.d. channel isotropically distributed in $\mathbb{O}^{N_t}$, an important component of a limited feedback codebook tailored to a spherical cap region is the quantization of the localized region or local packing. A local Grassmannian packing with parameters N_t, N, $\mathbf{o}$, ϵ_{th} is defined in Chap. 4. From this spherical cap $\mathscr{B}_{\epsilon_{th}}(\mathbf{o})$, it is possible to compute the rotated spherical $\mathscr{B}_{\epsilon_{th}}(\mathbf{o}_{rot})$ by applying the following rotation map:

$$r(\mathbf{o}) = \mathbf{U}_{rot}\mathbf{o} \triangleq \mathbf{o}_{rot} \tag{6.67}$$

where $\mathbf{U}_{rot}$ is the rotation unitary matrix [37].

As in the i.i.d. case, we use vector quantization to design these local packings. For the T2 and T3 criterion, the codebook must be adapted according to orthogonal vectors $\boldsymbol{\phi}_b$. From the local packing, it is possible to compute the local packing associated to a rotation using the rotation matrix. When the user CDI is inside the spherical cap region, the user will feed back $\log_2(N)$ bits corresponding to the codebook index. In addition to that, it will be necessary to feed back $\log_2(N_t)$ bits corresponding to the index of the vector $\boldsymbol{\phi}_b$. Consequently, for a codebook size N, $B = \log_2(N \times N_t)$ bits will be necessary to quantify the CDI.

6.2.3.5   Performance Evaluation

We consider multiuser MISO systems with $N_t = 2$ antennas at the BS. Threshold γ_{th} for norm only criterion and the threshold pair $(\gamma_{th}, \epsilon_{th})$ for combined criterion are theoretically calculated in order to have an average number of users in the cell $\bar{K} = 4$. Only these users feedback their CGI bits relative to each channel gain $\|\mathbf{h}_k\|^2$ and CDI bits corresponding to the codebook index of their quantized CDI. Exploiting this feedback information, the base station will select the N_t users in order to maximize the sum data rate.

In all simulations, we consider the number of active users of $K = [10, 20, 30, 40, 50]$ and γ_{th} and the pair $(\gamma_{th}, \epsilon_{th})$ should be chosen such that $\gamma_{th} = [2 \ 3 \ 3.5 \ 3.9 \ 4.15]$ for norm only criterion to obtain $\mathscr{U}_1$ and the pair $(\gamma_{th}, \epsilon_{th}) = [(0, 1) \ (1, 0.35) \ (2, 0.23) \ (2.5, 0.18) \ (3, 0.1) \ (3.8, 0.09)]$ for combined criterion to get $\mathscr{U}_3$. The performance results of T1 and T3 algorithms are illustrated respectively in Figs. 6.12 and 6.13

6.2.4   Opportunistic Beamforming

As already presented in Sect. 6.1, when there are many users with experiencing different channel quality at different times, multi-user diversity can be achieved by scheduling a user with favorable channel conditions. The more users are present, the more likely one user has a very good channel at any time; hence, the total throughput of such a system tends to increase with the number of users. Multiuser diversity

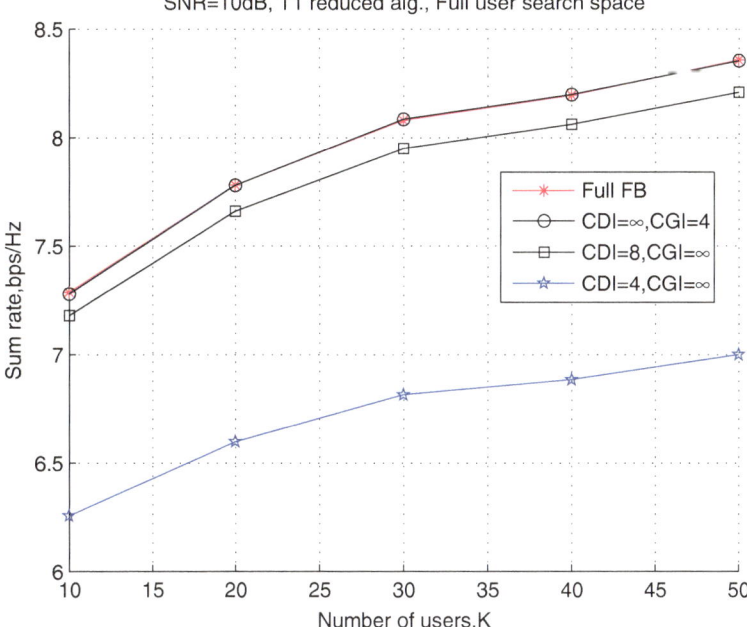

Fig. 6.12 Sum rate versus SNR with user selection based on T1 criterion with $N_t = 2$ using full user set

underlies opportunistic downlink scheduling [54] when the scheduler knows each user's fading level. In addition to that, by modulating in a controlled fashion the amplitude and phase of multiple transmit antennas, the fading rate and dynamic range can be increased, leading to higher multiuser diversity gains. This technique is called opportunistic beamforming (OBF). The primary studies on OBF or random beamforming (RBF) have been presented in [15, 54].

6.2.4.1   One Beam

For the OBF with one beam, only one beam is created and only one user transmits at each time [54]. Therefore, the training signal from each antenna is multiplied by $\theta_b; b = 1, \ldots, N_t$. Then, user k measures its SNR (or equivalently the data rate that the channel can support at the considered time slot) as

$$\gamma_k = \frac{\| P \sum_{b=1}^{N_t} \theta_b h_{b,k} \|^2}{N_0 B_W} \tag{6.68}$$

where θ_b is the phase generated according to an uniform distribution.

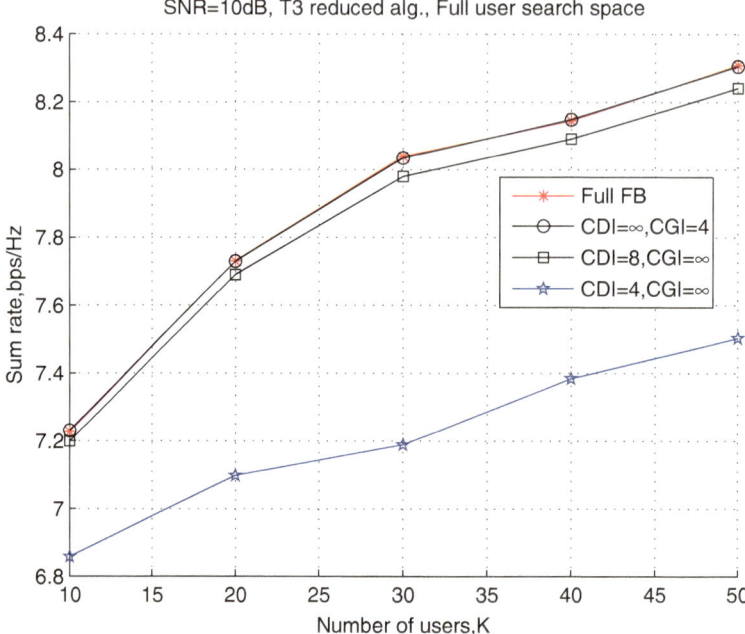

Fig. 6.13 Sum rate versus SNR with user selection based on T3 criterion with $N_t = 2$ using full user set

The users send back their SNR corresponding to this beam and the base station schedules the user with the highest SNR (or uses another scheduling rule) for transmission. There is no need to measure the individual channel gains (phase and/or amplitude) and the existence of multiple antennas is completely transparent to the receiver. In this context, the technique can be interpreted as OBF: phases and power allocated at the transmit antennas are varying in a pseudo-random manner, and the user which is currently closest to being in the generated beamforming configuration is scheduled at any time transmission.

This scheme approaches the performance of the optimal beamforming strategy for a large number of users. When there are fewer users, the generalization of OBF by using multiple random weighting vectors at each time slot has been proposed in [22], namely, as OBF with selection (OBF-S). The base station chooses the best weighting vector and performs the OBF according to this optimum vector. The main difference in OBF-S is that the users respond with the best beam and corresponding SNR for that beam; thus, OBF-S converges faster to the optimal beamforming strategy. With the best beam index strategy [9], each user only feeds back the index of its best beam. After receiving the best beam information of all users, the base station allocates a beam to a randomly selected user among all users requesting that beam, i.e., who feed back the corresponding beam index. When no user requests one or several beams, the base station will assign that beam to a randomly selected user. Note that with this strategy, each user only needs to feed back $log_2(N_t)$ bits of information.

Since both OBF and OBF-S send only one data stream, they do not take full advantage of the multiple antenna BC capacity gains.

6.2.4.2 Multiple Beams

In order to extend the OBF technique to create more than one orthogonal beam [15]. In this case, each user reports its best beam and its corresponding SINR to the BS (the number of feedback information is increased proportionally to the number of beams). The base station then schedules transmissions to multiple users based on the received SINR. As with OBF, this strategy requires a large number of users to approach full-CSI capacity gains.

The base station generates N_t random orthonormal vectors $\boldsymbol{\phi}_b \in \mathbb{C}^{1 \times N_t}$ at each coherence interval. The N_t vectors $\boldsymbol{\phi}_b$ are generated according to an isotropic distribution. It is obvious to use uniform power $P_b = \frac{P}{N_t}$ since the base station has no knowledge about the users' channels.

Each user k can compute its SINR for each beam and chooses the beam which gives it the maximal SINR:

$$\max_{1 \leq b \leq N_t} \gamma_{k,m} = \frac{P_b |\boldsymbol{\phi}_b \mathbf{h}_k|^2}{N_0 B_W + P_b \sum_{j \neq b}^{N_t} |\boldsymbol{\phi}_j \mathbf{h}_k|^2} \tag{6.69}$$

After that, each user feeds back its best SINR and corresponding beam index to the transmitter. The maximum SINR values contain information about the maximal channel gain and the minimal interference from other beams; therefore it is the best information combining multiuser diversity and interference. The transmitter can assign each beam to the user who has the best SINR for that beam.

It is interesting to state that in a large number of users regime ($K \rightarrow \infty$), the opportunistic RBF has no loss compared to a scheme that has full CSI at the transmitter [15]. This is explained by the fact that when the number of users is large, there exists almost surely a user well aligned to each beam and with very little interference from other beams.

In the case of quantized SINR strategy, there is a common quantization codebook with size N. The quantization index of SINR and the beam index is fed back to the base station. The BS selects the user with the largest quantized SINR on each of the N_t beams and beamforming is performed along the orthogonal beam vectors. The amount of feedback per user is equal to $\log_2(N_t N)$.

In [6], the authors present an approach that generates randomly multiple sets of orthonormal weight vectors during the training period which is divided into T minislots. Each user responds with its best beamforming weight and SINR for that beamforming vector in the corresponding orthogonal set. Then, the base station selects the set of users and beamforming weights, i.e., beam selection, that maximizes the sum capacity. This approach requires a little more feedback (a beam index and SINR value at each minislot). This technique generalizes OBF-S [22] by enabling parallel data transmissions to multiple users and improves [15] by using beam selection to achieve capacity benefits for a low number of users in the system.

6.2.4.3 PU2RC

Due to the distributed beam selection, numerous iterations of broadcast and feedback are required for implementing OBF-S, which incurs significant downlink overhead and feedback delay. As a result, the throughput gains of OBF-S over OBF are marginal. Alternatively, per unitary basis stream user and rate control (PU2RC), proposed in [13, 23], is a generalization of RBF. For PU2RC, there is a common quantization codebook consisting of $2^{B-\log_2 N_t}$ sets of orthogonal codebooks, where each orthogonal codebook consists of N_t orthogonal unit vectors. This allows each user to specify a particular orthogonal set using $B - \log_2 N_t$ bits and specify a particular beam within this set using $\log_2 N_t$. This codebook is common to all users, and the information fed back is the same as RBF (i.e., quantization index as well as a real number representing the SINR). User selection is performed as follows: for each of the orthogonal sets, the BS repeats the RBF user selection procedure and then selects the orthogonal set which yields the highest rate. If $B = \log_2 N_t$, there is only a single orthogonal set, and this scheme reduces to ordinary RBF.

In PU2RC, users whose channel vectors are close to the beamforming weight vectors are selected so as to create a user group for the maximum sum rate through the feedback information of the beam index. PU2RC requires limited feedback, such as the SINR and a vector index. Even if the performance of PU2RC is superior to that of RBF, the computational overhead to find a feasible vector index is higher than RBF and additional bits are required for the feedback.

6.2.5   OFDMA-Based Wireless Systems

In this section, we focus on a wideband systems. We examine reduced and limited feedback algorithms for multiuser multiantenna orthogonal frequency division multiple-access (OFDMA)-based wireless networks.

6.2.5.1 System Model

A MISO-OFDMA system with N_t transmit antennas, Q clusters and K users is considered. A cluster structure where the correlation is high among the subcarriers is employed so that the feedback of only one value is sufficient, e.g., the CSI value belonging to the subcarrier of minimum channel vector gain. For this model, the channel vector between the BS and the kth user for the qth cluster is described by

$$\mathbf{H}_{k,q} = \begin{bmatrix} H_{k,q,1} & H_{k,q,2} & \dots & H_{k,q,t} & \dots & H_{k,q,N_t} \end{bmatrix}^T \tag{6.70}$$

where $H_{k,q,t}$ is the frequency domain channel coefficient from the tth transmit antenna to the kth user for the qth cluster.
The channel coefficient of a cluster is determined by

$$\mathbf{H}_{k,q} = \bar{\mathbf{H}}_{k,n^*} \tag{6.71}$$

where $\bar{\mathbf{H}}_{k,n^*}$ is the $N_t \times 1$ channel vector associated to the kth user and the n^*th subcarrier and n^* is

$$n^* = (q-1)N_Q + \arg \min_{0 \le i \le N_Q - 1} \{||\bar{\mathbf{H}}_{k,(q-1)N_Q+i}||^2\}, \quad q = 1, 2, \dots, Q \tag{6.72}$$

where N_Q is the number of subcarriers in one cluster and calculated as $N_Q = N/Q$ with N is the total number of subcarriers in a OFDM symbol.
For a system where $N_t \le K$, let $\mathbb{S}_q$ be the set of assigned users for cluster q:

$$\mathbb{S}_q = \{A_{k,q,b} | A_{k,q,b} = 1 \quad \forall k, \forall b\}$$

where $A_{k,q,b}$ is a binary variable that indicates that cluster q is allocated to user k for beam b.

Let's denote $\mathbf{H}(\mathbb{S}_q)$ as a matrix consisting of N_t channel vectors of the selected users for cluster q. Then, the associated users' data is transmitted by using ZF precoding [60]. The ZF transmit beamforming matrix is calculated by

$$\mathbf{F}(\mathbb{S}_q) = \eta (\mathbf{H}(\mathbb{S}_q))^H [(\mathbf{H}(\mathbb{S}_q))\mathbf{H}(\mathbb{S}_q)^H]^{-1} \tag{6.73}$$

which includes N_t elements as $\mathbf{F}(\mathbb{S}_q) = \begin{bmatrix} \mathbf{F}_1(\mathbb{S}_q) & \mathbf{F}_2(\mathbb{S}_q) & \dots & \mathbf{F}_{N_t}(\mathbb{S}_q) \end{bmatrix}^T$ where $\mathbf{f}_b(\mathbb{S}_q)$ is the precoding vector for bth beam and qth subcarrier with the dimension of $N_t \times 1$.

In order to keep the short-term power constraint, we determine η as

$$\eta = \frac{1}{\sqrt{\operatorname{tr}[((\mathbf{H}(\mathbb{S}_q))\mathbf{H}(\mathbb{S}_q)^H)^{-1}]}} \tag{6.74}$$

For the kth user and the qth cluster, the relation between the data vector $\mathbf{S}(\mathbb{S}_q)$ and the received signal can be written as

$$\mathbf{Y}(\mathbb{S}_q) = \mathbf{H}(\mathbb{S}_q)^H \mathbf{X}(\mathbb{S}_q) + \mathbf{N}(\mathbb{S}_q)$$
$$= \mathbf{H}(\mathbb{S}_q)^H \mathbf{F}(\mathbb{S}_q)\mathbf{P}_q \mathbf{S}(\mathbb{S}_q) + \mathbf{N}(\mathbb{S}_q) \tag{6.75}$$

where $\mathbf{P}_q = \mathrm{diag}\left(\sqrt{P_T/(QN_t)}, \cdots, \sqrt{P_T/(QN_t)}\right)$ by sharing the total transmit power P_T equally between the clusters and beams and $\mathbf{X}(\mathbb{S}_q)$ is the transmitted vector from the BS.

Our objective is to maximize the average sum rate by optimizing both the cluster and beam allocation, $\mathbf{A} = [\mathbf{A}_1, \mathbf{A}_2, \ldots, \mathbf{A}_Q]$. To construct $\mathbf{A}_q$, all vectors of $\mathbf{A}_{k,q} = \left[A_{k,q,1}\ A_{k,q,2}\ \ldots\ A_{k,q,N_t} \right]^T$ are stacked column by column.

By performing ZF-BF, the optimization problem is formulated as

$$\mathbf{A}^* = \arg\max_{\mathbf{A}} \sum_{k=1}^{K} R_k \tag{6.76}$$

subject to

$$\sum_{k=1}^{K} A_{k,q,b} \leq 1; \quad \forall q, \forall b \tag{6.77}$$

$$A_{k,q,b} \in \{0,1\}; \quad \forall k, \forall q, \forall b \tag{6.78}$$

The instantaneous data rate R_k is calculated as

$$R_k = B_W \sum_{q=1}^{Q} \sum_{b=1}^{N_t} A_{k,q,b} \log_2(1 + \gamma_{k,q,b}) \tag{6.79}$$

where the SINR is

$$\gamma_{k,q,b} = \frac{\frac{P_T}{N_t}\left|\mathbf{H}_{k,q}^H \mathbf{F}_{k,q,b}\right|^2}{N_0 B_W + \frac{P_T}{N_t} \sum_{j=1; j\neq b}^{N_t} \left|\mathbf{H}_{k,q}^H \mathbf{F}_{k,q,j}\right|^2} \tag{6.80}$$

In order to simplify the optimization problem, the allocation is performed at a cluster level as

$$(\mathbf{A}_q)^* = \arg\max_{\mathbf{A}_q} \sum_{k=1}^{K} R_{k,q} \quad \forall q \tag{6.81}$$

The instantaneous data rate for each cluster is

$$R_{k,q} = B_W \sum_{b=1}^{N_t} A_{k,q,b} \log_2(1 + \gamma_{k,q,b}) \tag{6.82}$$

6.2.5.2 Reduced Rate Feedback Link

In order to exploit multiuser diversity, the users need to feed back their CSI to the BS. The feedback rate can be reduced by applying a selection a subset of clusters associated to CSI at receiver side. We present two different algorithms to perform user selection at the receiver side for MISO-OFDMA systems and a quantization method, thanks to the properties of the semi-orthogonal criterion described in Sect. 6.2.3.4.

P1: The Clustered S-Best Criterion

For the clustered S-best criterion which is also called P1 algorithm, each user selects independently a set $\mathbb{S}_k$ composed of the S clusters with the highest channel norm $\|\mathbf{H}_{k,q}\|$. Then, each user feds back only its CSI associated to the selected clusters to the BS. This criterion is the extension of the single antenna OFDMA systems [46] to the MISO-OFDMA case.

Let $\mathbb{T}_q$ be the set of users that feed back their CSI associated to the cluster q as,

$$\mathbb{T}_q = \{k \in \{1, 2, \ldots, K\} : q \in \mathbb{S}_q\} \tag{6.83}$$

Then, for each cluster q, the BS selects the set $\mathbb{S}_q$ from the set $\mathbb{T}_q$ to maximize the rate in Eq. (6.81).

Since the total feedback rate is proportional to KS, it is reasonable to adjust S in function of K according to a desired function $S = f(K)$.

The *objective* is defined in terms of a fraction of all clusters, η, which for L users send their CSI to the BS. κ can be adjusted to guarantee κQ clusters can allocate at least L users with a probability higher than $1 - P_{\text{obj}}$ as follows.

$$\Pr\left(\sum_{q=1}^{Q} I_q^L \leq \kappa Q | S, K\right) \leq P_{\text{obj}} \tag{6.84}$$

with

$$I_q^L = \begin{cases} 1 & \text{if } |\mathbb{T}_q| \geq L \\ 0 & \text{else} \end{cases}$$

where L is called a layer and takes values in $1 \leq L \leq N_t$.

The case of $L = N_t = 1$ has been examined in [46] for OFDMA systems with single antenna and the case for MISO-OFDMA systems considering the case of $L = N_t = 2$ to allocate more than one beam for each cluster has been presented in [35].

P2: The Combined Criterion

In the combined criterion which is also called P2 algorithm, The clusters are selected according to the information on both their channel norm/quality and channel direction [34]. Therefore, for each cluster, N_t random orthonormal vectors $\boldsymbol{\phi}_{b,q}$ with dimension $N_t \times 1$ for $b = 1, \ldots, N_t$ are generated.

The users measure the orthogonality between their channels and the random vectors $\boldsymbol{\phi}_{b,q}$ for each cluster using the chordal distance as

$$d^2 \left(\overset{=}{\underset{k,q}{\mathbf{H}}}, \boldsymbol{\phi}_{b,q} \right) = 1 - |(\overset{=}{\underset{k,q}{\mathbf{H}}})^H \boldsymbol{\phi}_{b,q}|^2 \tag{6.85}$$

where $\overset{=}{\underset{k,q}{\mathbf{H}}} = \frac{\mathbf{H}_{k,q}}{\|\mathbf{H}_{k,q}\|}$ is the normalized channel vector of the user k and cluster q.

For the semi-orthogonal criterion with any $0 < \epsilon < 1$, we have

$$\mathbb{T}'_q = \{k \in \{1, 2, \ldots, K\} : \mathbf{H}_{k,q} \in \bigcup_{b=1}^{N_t} \mathscr{B}_\epsilon(\boldsymbol{\phi}_{b,q})\} \tag{6.86}$$

where $\mathscr{B}_\epsilon(\mathbf{o})$ is a spherical cap on $\mathscr{O}^{N_t}$ with center $\mathbf{o}$ and square radius ϵ and $\mathbb{T}'_q$ is the set of semi-orthogonal users for cluster q.

The set of the selected clusters by user k is given by

$$\mathbb{T}'_k = \{q \in \{1, 2, \ldots, Q\} : k \in \mathbb{T}'_q\} \tag{6.87}$$

According to [32], the number of clusters which satisfy ϵ criterion for each user is calculated approximately as $N_t \epsilon_{th}^{N_t-1}$. Therefore, the choice of ϵ is critical since it is directly relative to the total number of clusters per user. Consequently, it should be guaranteed that at least S clusters which satisfy the semi-orthogonal criterion for each user are selected by adjusting the ϵ parameter properly.

Then, each user selects S clusters in terms of channel quality/norm from the set $\mathbb{T}'_k$ and constructs the set $\mathbb{S}_k$. From this set, for each q, we can obtain the set $\mathbb{T}_q$ using Eq. (6.83).

6.2.5.3 Performance Results

We obtain the performance results to illustrate the benefits of the reduced rate feedback channels in a single-cell MISO-OFDMA system with two transmit and one receive antennas through frequency selective wireless channels using. The number of clusters are 48 and the clusters are grouped into 18 subcarriers.

Firstly, we choose the reduced rate feedback channel parameters of S and ϵ as a function of K, P_{obj} and M. In order to increase multiuser diversity, the ϵ values are adjusted to have $1.3S$ clusters after performing the semi-orthogonal criterion.

Table 6.1 The parameters for reduced feedback algorithms	K	10	20	30	40	50
	S	30	15	10	8	6
	ϵ	0.4	0.2	0.15	0.125	0.1

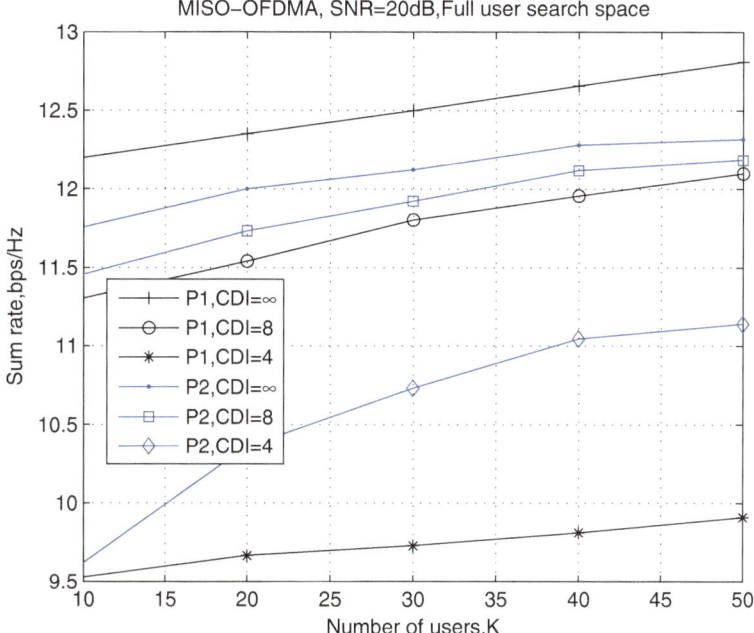

Fig. 6.14 Sum capacity performance of MISO OFDMA systems with different the user selection algorithms

In Table 6.1, the parameters are listed for a given target according to the number of users in the cell for the case of $L = N_t = 2$, $\kappa = 0.95$, $P_{\text{obj}} = 0.1$ at the spectral usage=0.99.

As illustrated in Fig. 6.14, we compare the sum capacity for reduced feedback based on T1 and T2 criteria by assuming full CSI is available at the BS and considering the quantization of CDI. According to the results, the combined criterion outperforms the clustered S-best criterion. Since the MISO-OFDMA systems are sensitive to interference, it is important to minimize the CDI quantization error. The quantization error of the CDI is reduced by using the proposed codebook design. Moreover, when the number of CDI quantization bits decreases, the performance difference between the two criteria increases.

6.2.6   MAT Algorithm

On the multi-antenna downlink systems, the ability to perform beamforming (i.e., linearly precoding) multiple data streams simultaneously to several users (up to N_t) comes nevertheless at a price in terms of requiring the base station transmitter to be informed of the channel coefficients of all served users. As already shown previous parts, transmitter knowledge of the CSI can be very important. For point-to-point wireless systems, the CSI only provides power gains using water-filling strategies. For multiuser wireless channels, CSI can also provide multiplexing and beamforming gains to send data along multiple beams to different users simultaneously.

In practice, it is not easy to achieve the theoretical gains of these techniques. In the high SNR regime, where the multiplexing gain offered by these techniques is particularly significant, the performance of these techniques is very sensitive to inaccuracies of the CSIT. However, it is hard to obtain accurate CSIT particularly in FDD systems where the channel state has to be measured at the receiver and fed back to the transmitter. This feedback process leads to two sources of inaccuracies:

- Quantization error: Since there is a limited feedback rate link the accuracy of the CSI at the BS is getting low.
- Delay issue: There is a delay between the time when the CSI is measured at the user side and the time when the CSI is used at the BS. The delay comes from the fact that the users require time to estimate CSI using pilots and then feed it back to the BS reliability. In time-varying wireless channels, when the CSI is started to use at BS, the channel has already changed at the user side.

Many works in the literature has focused on the first issue that has been examined in detail in the previous sections. The general conclusion is that the rate of the feedback channel needed to achieve the perfect CSIT multiplexing gain scales well with the SNR. Nowadays, the second issue that is feedback delay is paid more attention. The standard approach of dealing with feedback delay is to exploit the time correlation of the channel to predict the current CSI from the delayed version of the channel as seen in Chap. 4. However, for higher mobility case, the coherence time of the channel becomes shorter compared to the feedback delay, and the delayed feedback CSI reveals no information about the current CSI. Therefore, different transmission strategies have been presented in [29] to deal with the delayed CSI for multiantenna multiuser systems.

In this section, we focus on a wireless system, where the transmitter has N_t antennas and there are K receivers each with a single receive antenna. The transmitter wants to send an independent data stream to each receiver. To model completely outdated CSI, we allow the channel state to be independent from one symbol time to the next, and the CSI is available to both the transmitter and the receivers one symbol time later. This means that by the time the feedback reaches the transmitter, the current channel is already completely different. We also assume that the overall N_t-by-K channel matrix is full rank at each time.

In [53], the DoF attainable by ZF and MAT algorithm delayed CSI have been examined by assuming delayed CSI is available at the BS. The key performance metrics considered in this work are the DoF (also known as the multiplexing gain), the feedback overhead and the net DoF. The net DoF is the prelog capacity term remaining after subtracting the feedback DoF consumed depending on the feedback rate.

The DoF gain of single user transmission always achieves one DoF without requiring feedback and MISO BC with K users having partial CSIT and applying linear precoding achieves DoF up to K when the CSIT is sufficiently current and accurate.

The DoF with outdated CSIT for MAT algorithm but no quantization error is given by

$$DoF(K) = \frac{K}{1 + \frac{1}{2} + \ldots + \frac{1}{K}} \tag{6.88}$$

For example, MAT promises the $4/3$ DoF for $K = 2$.

The DoF of ZF case is given by

$$DoF_{ZF}(K) = K \left(1 - \frac{N_{fd} + (K-1)}{N_{ch}} \right) \tag{6.89}$$

where N_{fd} is the feedback delay and N_{ch} is coherence time.

6.2.6.1 The Scheme for $K = N_t = 2$

The Maddah Ali and Tse (MAT) algorithm [21, 29] for $N_t = 2$ transmit antennas and $K = 2$ users is explained in the following:

At $t = 1$, u_A and v_A which are the symbols for the receiver A are transmitted from antenna 1 and antenna 2, respectively. Then, the received signals at receiver A and receiver B are

$$y_A[1] = u_A h_{1A}[1] + v_A h_{2A}[1] + n_A[1] \tag{6.90}$$

$$y_B[1] = u_A h_{1B}[1] + v_A h_{2B}[1] + n_B[1] \tag{6.91}$$

At $t = 2$, u_B and v_B which are the symbols for the receiver B are transmitted from antenna 1 and antenna 2, respectively. Then, the received signals at receiver A and receiver B are

$$y_A[2] = u_B h_{1A}[2] + v_B h_{2A}[2] + n_A[2] \tag{6.92}$$

$$y_B[2] = u_B h_{1B}[2] + v_B h_{2B}[2] + n_B[2] \tag{6.93}$$

At $t = 3$, the $y_B[1] + y_A[2]$ and 0 are transmitted from antenna 1 and antenna 2, respectively, assuming the h_A and h_B are quantized and fed back to transmitter; the received signals at receiver A and receiver B are

$$y_A[3] = h_{1A}[3](y_B[1] + y_A[2]) + n_A[3] \tag{6.94}$$

$$y_B[3] = h_{1B}[3](y_B[1] + y_A[2]) + n_B[3] \tag{6.95}$$

The received signals can be rewritten at receiver A as

$$\begin{bmatrix} y_A[1] \\ y_A[3] - h_{1A}[3]y_A[2] \end{bmatrix} = \mathbf{H}_A \begin{bmatrix} u_A \\ v_A \end{bmatrix} + \begin{bmatrix} n_A[1] \\ n_A[3] - h_{1A}[3]n_A[2] \end{bmatrix} \tag{6.96}$$

where

$$\mathbf{H}_A = \begin{bmatrix} h_{1A}[1] & h_{2A}[1] \\ h_{1A}[3]h_{1B}[1] & h_{1A}[3]h_{2B}[1] \end{bmatrix} \tag{6.97}$$

Then, the transmitted symbols are encoded at receiver A as

$$\begin{bmatrix} \tilde{u}_A \\ \tilde{v}_A \end{bmatrix} = \mathbf{H}_A^{-1} \begin{bmatrix} y_A[1] \\ y_A[3] - h_{1A}[3]y_A[2] \end{bmatrix} \tag{6.98}$$

Similarly, the received signals can be rewritten at receiver B as

$$\begin{bmatrix} y_B[2] \\ y_B[3] - h_{1B}[3]y_B[1] \end{bmatrix} = \mathbf{H}_B \begin{bmatrix} u_B \\ v_B \end{bmatrix} + \begin{bmatrix} n_B[2] \\ n_B[3] - h_{1B}[3]n_B[1] \end{bmatrix} \tag{6.99}$$

where

$$H_B = \begin{bmatrix} h_{1B}[2] & h_{2B}[2] \\ h_{1B}[3]h_{1A}[2] & h_{1B}[3]h_{2A}[2] \end{bmatrix} \tag{6.100}$$

Then, the transmitted symbols are encoded at receiver B as

$$\begin{bmatrix} \tilde{u}_B \\ \tilde{v}_B \end{bmatrix} = H_B^{-1} \begin{bmatrix} y_B[2] \\ y_B[3] - h_{1B}[3]y_B[1] \end{bmatrix} \tag{6.101}$$

6.2.7 Massive MIMO

The concept of adding a large number of transmit antennas, often called massive MIMO systems, has been evolving over the past few years by scaling up MIMO by possibly orders of magnitude compared to current state of the art while each antenna unit uses extremely low power in the order of mW.

As presented in [30], adding more antennas at the base station is always beneficial even with very noisy channel estimation since the base station can recover information even with a low SNR once it has sufficiently many antennas. This motivates the concept of using a very large number of transmit antennas, where the number of antenna elements can be at least an order of magnitude more than the current cellular systems (10s–100s) [31]. Massive MIMO systems that use antenna arrays with a few hundred antennas, simultaneously serve many tens of terminals in the same time-frequency resource to boost the network capacity [17] and by increasing the range of transmission with improved energy efficiency [33]. Fundamental limits, optimal transmit precoding and receive strategies, and real channel measurement issues for massive MIMO systems have been summarized in [41]. For massive MIMO systems, there are many challenges subjects including channel estimation, cooperative communications, interference management etc.

Since the large number of degrees of freedom is available in the user channels for massive MIMO systems, a per-user beamforming such as matched beamforming can be implemented easily [31] unlike conventional MU-MIMO systems with a small number of transmit antennas. However, without accurate CSI, the sum rate of massive MU-MIMO systems would also be effected strongly [20]. Therefore, the challenge is to scale channel estimation and feedback strategies to effectively provide CSI at BS. It is difficult to scale the codebook-based techniques described in Chap. 4 to massive MIMO. In order to maintain the same level of channel quantization error, the feedback overhead must increase proportional to the number of transmit antennas. Therefore, efficient codebook designs are needed for massive MIMO systems.

6.2.8 PMI Feedback Schemes

The BS and users have some predefined common codebooks and each element of the codebook can be indexed as precoding matrix/vector index (PMI) and the user will be precoded or beamformed using the corresponding precoding matrix/vector if it is scheduled. Using channel quality indicator (CQI), preferred stream index or preferred matrix index (PMI), BS scheduling, resource allocation and rate adaptation decisions for closed-loop MU-MIMO are made. In addition to the PMI for the serving cell, the recommendation or restriction PMI for neighboring cells may be feedback to control the intercell interference and hence make better pairing among users or cells. For MU-MIMO, an important concern is how to deal with the interference in the same cell or neighboring cells. The other question is that given a codebook-based precoding schemes, what is the information the user can provide to the BS to pair the users more efficiently to control the interference level, i.e., besides the traditional preferred PMI?

6.2.8.1 Normal Feedback Scheme

In normal feedback scheme (NFS), the user only feeds back its preferred matrix maximizing its SNR. With the reported PMI and the help of other system information, the BS will allocate one PMI to the user if this user is scheduled. It should be noted that the allocated PMI may or may not be same as the reported PMI depending on usage of codebook or non-codebook-based precoding schemes at the BS. For NFS, the best PMI for a user can be selected from a codebook based on the following rule of maximizing the received signal power [61].

$$\mathbf{w}_{k*} = \max_{i \in [1,2,...,N]} |\mathbf{h}_k^H \mathbf{w}_i|^2 \tag{6.102}$$

For MU-MIMO, the corresponding CQI is calculated at the receiver side assuming that the interfering users are scheduled by the serving base station using rank-1 precoders for single stream orthogonal to each other. Hence it will orthogonal to the rank-1 precoder represented by the reported PMI. However, with this NFS feedback, BS can identify the best PMI for the user, but it does not know how to pair another user to minimize mutual interference. Therefore, BS can only assume the PMI represented CSI is accurate and then make random pairing. Correspondingly, the user will have to assume that the users that construct according to random pair to have the highest interference to obtain a surely workable CQI as

$$SINR_{k*} = \min_{l \in [1,2,...,N]} \frac{P|\mathbf{h}_{k*}^H \mathbf{w}_{k*}|^2}{N_0 B_W + P|\mathbf{h}_k^H \mathbf{w}_l|^2} \tag{6.103}$$

6.2.8.2 Best Companion Pairing

In [38,39], the best companion pairing (BCP) is suggested to coordinate the multiple users in both single cell and multicell scenarios. The basic principle is to maximize SINR and therefore maximize cell throughput. To achieve the target, additional feedback information of so-called Best Companion indexes are provided, which are actually the PMI of to-be-paired user with least interference to the target user, $(\mathbf{w}_{k*}, \mathbf{w}_{k+})$.

The best companion in terms of maximizing the gain:

$$\mathbf{w}_{k*} = \max_{i \in [1,2,...,N]} \|\mathbf{h}_k^H \mathbf{w}_i\|^2 \tag{6.104}$$

The best companion in terms of minimizing the interference:

$$\mathbf{w}_{k+} = \min_{l \in [1,2,...,N]} (\|\mathbf{h}_k^H \mathbf{w}_l\|^2) \tag{6.105}$$

Correspondingly, the CQI can be calculated as

$$SNR_{k*} = P \frac{\|\mathbf{h}_{k*}^H \mathbf{w}_{k*}\|^2}{N_0 B_W + P \|\mathbf{h}_k^H \mathbf{w}_{k+}\|^2} \qquad (6.106)$$

With this scheme, the intracell mutual MU-MIMO interference for single cell scenario or the intercell interference for multicell scenario can be effectively minimized with the extra codebook-based information of the to-be-paired user(s) with least interference. For multicell scenarios, serving BS will share all the PMI feedback information via backhaul to the interfering neighbor cells for coordinated scheduling.

6.3 Multiuser MIMO Systems with Multiple Receive Antennas

Multiuser MIMO systems with multiple receive antennas have proved their ability to achieve high bit rates in a scattering wireless network. In a point-to-point scenario, it has been shown that the capacity scales linearly with the minimum number of transmit and receive antennas, regardless of the availability of CSI at the transmitter. This linear increase is the multiplexing gain. In a system with a large number of users, the BS can increase the throughput by selecting the best users set communicate with achieving both multiplexing and multiuser diversity gains.

6.3.1 Precoding Strategies

It is known that selecting the user who has the best channel is optimal in terms of capacity in multiuser SISO systems in Chap. 5. However, the optimal case for multiuser (MU) MIMO systems is to serve multiple users simultaneously. Since data are transmitted to multiple users in the same frequency simultaneously, inter-user interference needs to be dealt with in MU-MIMO systems. DPC is the optimal technique to remove the inter-user interference. It has been shown that the interference which is known a priori to the transmitter does not affect the channel capacity. DPC is a nonlinear scheme that achieves the channel capacity in MIMO broadcast channel. However, it is impractical for implementation since its complexity is prohibitive. In order to reduce the complexity, many precoding algorithms for MU-MIMO systems have been proposed such as ZF, MMSE and block diagonalization (BD) [4, 27, 43]. The ZF scheme removes inter-user interference among the selected users by designing a precoding matrix which is the pseudo inverse of the selected users channels. On the other hand, the BD scheme uses null space which is computed by SVD. For a given user, a matrix is constructed by stacking all the users channels except its own channel and finds the null space by SVD. The number of users who can be simultaneously supported with BD is limited by the number of transmit antennas and the number of receive antennas since each user precoding matrix must lie in the null space of all the other users channels.

In the single stream per user case, only one stream can be allocated to each user and the algorithm chooses the best precoding vector for each user. In the multiple streams per user case, more than one stream can be allocated to each user respecting two main constraints: $N_{s_k} \leq \min(N_r, N_t)$ that represents the number of streams allocated of each user and $N_s = \sum_{k=1}^{K} N_{s_k} \leq \min(\sum_{k=1}^{K} N_r, N_t)$ that represents the total number of streams allocated by the BS. The maximal number of streams is determined with the rank of matrix the overall matrix $\mathbf{H} = [\mathbf{H}_1^T, \ldots, \mathbf{H}_K^T]^T$ of size $K N_r \times N_t$ where the individual channel matrix $\mathbf{H}_k; k = 1, 2, \ldots, K$ of size $N_r \times N_t$.

The overall received signal for downlink multiuser MIMO systems with multiple receive antennas is given by

$$\mathbf{y} = \mathbf{HFs} + \mathbf{n} \tag{6.107}$$

where $\mathbf{F}$ is the precoding matrix with $N_t \times N_s$, $\mathbf{s}$ is the data stream vector with $N_s \times 1$.

6.3.1.1 Zero Forcing

ZF is nothing else but a pseudo inverse of the channel vector for multiuser MIMO systems with only one receive antennas. This simple approach could be easily extended to the multiuser MIMO system with more than one receive antennas by applying the pseudo inverse to the overall matrix $\mathbf{H}$ for the case of $N_t \geq K N_r$, given by

$$\mathbf{F}_{ZF} = \eta \mathbf{H}^H (\mathbf{HH}^H)^{-1} \tag{6.108}$$

where $\eta = \frac{1}{\sqrt{\mathrm{tr}((\mathbf{HH}^H)^{-1})}}$.

6.3.1.2 MMSE

A main problem with ZF is that the sum rate does not scale linearly with the number of antennas, since each stream is weighted by the corresponding singular value and that the singular values of the channel matrix are largely spread. In order to overcome this limitation, MMSE beamforming (also called regularized channel inversion) has been used by regularizing the channel inversion to improve the condition of the inverse [45]. The MMSE beamformer is given by

$$\mathbf{F}_{MMSE} = \eta \mathbf{H}^H (\alpha \mathbf{I} + \mathbf{HH}^H)^{-1} \tag{6.109}$$

where η is chosen such that $\text{tr}(\mathbf{F}_{MMSE}\mathbf{F}_{MMSE}^H) = 1$ and α is the power allocation term which can be set according to equal power sharing.

6.3.1.3 Block Diagonalization

The block diagonalization (BD) involves a precoding strategy the signals to be transmitted to suppress interference at each user due to all other users (but not due to different antennas for the same user).

For the kth user, the received signal of BD can be given by

$$\mathbf{y}_k = \mathbf{H}_k \mathbf{F}_k \mathbf{s}_k + \mathbf{H}_k \sum_{j=1, j \neq k}^{K} \mathbf{F}_j \mathbf{s}_k + \mathbf{n}_k \qquad (6.110)$$

The first term is the desired signal for the kth user, and the second term is the interference term. BD eliminates the interference term by finding the null space of the channel matrices of the other users. Therefore, the precoding matrix $\mathbf{F}_k$ is designed to be in the null space of each $\mathbf{G}_k = \begin{bmatrix} \mathbf{H}_1^T & \dots & \mathbf{H}_{k-1}^T & \mathbf{H}_{k+1}^T & \dots & \mathbf{H}_K^T \end{bmatrix}^T$ to eliminate the interference term, where $\text{tr}[\mathbf{FF}^H] = 1$ and

In order to find the precoding matrix, $\mathbf{G}_k$ can be computed by SVD

$$\mathbf{G}_k = \mathbf{U}_k \Sigma_k [\tilde{\mathbf{V}}_k \quad \mathbf{V}_k]^H \qquad (6.111)$$

where $\tilde{\mathbf{V}}_k$ holds the first $\tilde{N}_s = \text{rank}(\mathbf{G}_k)$ right singular vectors and $\mathbf{V}_k$ holds the last $N_t - \tilde{N}_s$ right singular vectors. Then, we obtain the precoding vector as the following:

$$\mathbf{F}_k = \mathbf{V}_k \qquad (6.112)$$

A MU-MIMO system is decomposed into K independent SU-MIMO systems by BD. The number of the transmit antennas should be larger than the sum of the number of receive antennas of any $K - 1$ users for the existence of the null space of all $\mathbf{G}_j; 1 \leq j \leq K, j \neq k$. The condition to be satisfied is then given by $N_t > (K - 1)N_r$.

6.3.1.4 Single-Stream SJNR Based Precoding

The objective is to design transmit precoding under the total transmit power constraint $\sum_{k=1}^{K} P_k = P$ and transmitted power per user is $\text{tr}(\mathbf{f}_k \mathbf{f}_k^H) = 1$ with $\text{tr}(\mathbf{s}_k \mathbf{s}_k^H) = P_k$. The signal-to-jamming-plus-noise ratio (SJNR) [57] defined as the signal power of user k over the noise plus total power of interference caused by the user k for the other receivers is given as

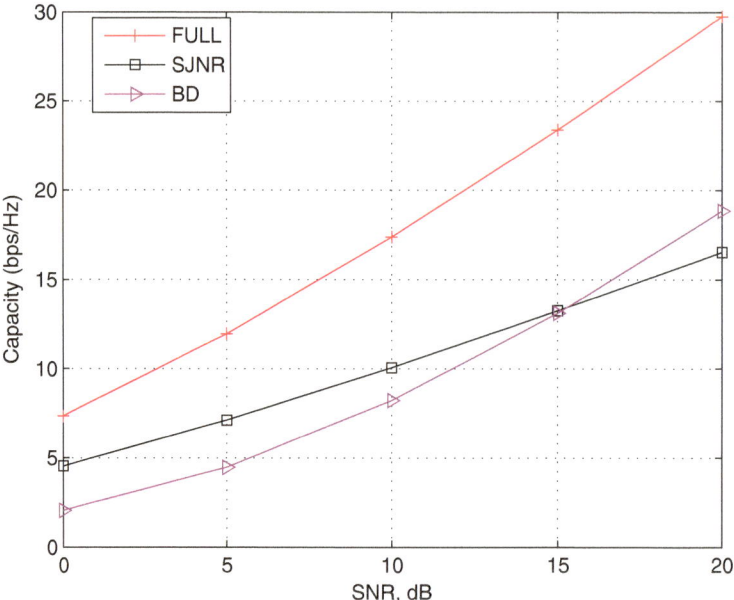

Fig. 6.15 Comparison of different precoding strategies for multiuser MIMO systems with $N_t = 4$, $N_t = K = 2$

$$SJNR_k = \frac{\mathbf{f}_k^H \mathbf{H}_k^H \mathbf{H}_k \mathbf{f}_k}{\sum_{j=1, j \neq k}^{K} \mathbf{f}_k^H \mathbf{H}_j^H \mathbf{H}_j \mathbf{f}_k + (N_0 B_w / P_k)} \quad (6.113)$$

The precoder vector that will maximize the SJNR is obtained as

$$\mathbf{f}_k = \xi_m \left[\left(\sum_{j=1, j \neq k}^{K} \mathbf{H}_j^H \mathbf{H}_j + \frac{N_0 B_W}{P_k} \mathbf{I} \right)^{-1} \mathbf{H}_k^H \mathbf{H}_k \right] \quad (6.114)$$

where $\xi_m[\mathbf{X}]$ represents the largest eigenvector of $\mathbf{X}$.

For this scheme, the decoding vector is defined as,

$$\mathbf{z}_k = \frac{\mathbf{f}_k^H \mathbf{H}_k^H}{\|\mathbf{f}_k \mathbf{H}_k\|} \quad (6.115)$$

The comparison results for MIMO-SJNR and MIMO-BD precoding strategies are illustrated in Fig. 6.15 for multiuser MIMO systems with multiple receive antennas. As shown in the performance results, the SJNR outperforms BD at low SNRs and then BD is started to provide better capacity.

6.3.1.5   Multistream Iterative Weighted MMSE-Based Precoding

The objective for the multistream case is to design the precoding matrices $\mathbf{F}_k$ under the total transmit power constraint $\sum_{k=1}^{K} P_k = P$ and transmitted power per user is $\mathrm{tr}(\mathbf{F}_k \mathbf{F}_k^H) = 1$ with $\mathrm{tr}(\mathbf{s}_k \mathbf{s}_k^H) = P_k$. Therefore, inspired by the idea adopted for the single-stream case, we will extend it to the multistream case.

In multistream iterative-weighted MMSE-based precoding [5], the precoding and decoding matrices can be optimized based on MMSE criterion iteratively as described in the following [5]:

- For each k, initialize the precoders $\mathbf{F}_k$ such that $\mathbf{H}_k^H$ and compute decoding matrices $\mathbf{Z}_k$ based on MMSE filtering:

$$\mathbf{Z}_k = \mathbf{F}_k^H \mathbf{H}_k^H \left(\mathbf{I}_{N_r} + \sum_{i=1}^{K} \mathbf{H}_k \mathbf{F}_i \mathbf{F}_i^H \mathbf{H}_k^H \right)^{-1} \tag{6.116}$$

- Compute the weighting matrix $\mathbf{S}_k$ for each k:

$$\mathbf{S}_k = \mathbf{I}_k + \mathbf{F}_k^H \mathbf{H}_k^H \left(\mathbf{I}_{N_r} + \sum_{i=1, i \neq k}^{K} \mathbf{H}_k \mathbf{F}_i \mathbf{F}_i^H \mathbf{H}_k^H \right)^{-1} \mathbf{H}_k \mathbf{F}_k \tag{6.117}$$

- Compute the new precoders $\mathbf{F}_k$ for each k as

$$\mathbf{F}_k = \beta \left(\sum_{j=1}^{K} \mathbf{H}_j^H \mathbf{Z}_j^H \mathbf{S}_j \mathbf{Z}_j \mathbf{H}_j + \frac{\mathrm{tr}(\mathbf{S}_j \mathbf{Z}_j \mathbf{Z}_j^H) N_0 B_W}{P_k} \mathbf{I}_{N_t} \right)^{-1} \tag{6.118}$$

where β is a scalar factor to respect total power constraint.
- Repeat second and third steps until it converges.

6.3.2   User Scheduling Methods

In MU-MIMO systems where there are many users in a cell, the optimal scheduling (user selection) is computationally prohibitive. The optimal strategy is a brute-force approach to find the best subset of users exhaustively. The exhaustive search method needs to consider the cases with a complexity roughly $O(K^{N_s})$. Many suboptimal user selection schemes have been proposed to reduce the computational complexity of the optimal (exhaustive) user selection scheme. A suboptimal user selection scheme by using GSO in MU-MISO systems was proposed in [60]. However, it may be difficult to use GSO in MU-MIMO systems because the channel is a matrix while GSO can only be used for vectors. For MU-MIMO systems, two low complexity user scheduling algorithms, the capacity-based algorithm and the Frobenius norm-based algorithm, were proposed in [47]. These algorithms are based on a greedy method and achieve performance close to the optimal scheduling algorithm. However, their computational complexity is still relatively high. The capacity-based algorithm needs frequent computation of SVD, and the Frobenius norm-based algorithm also needs heavy GSO computation. In [26], a low complexity scheduling algorithm with the BD scheme was proposed to maximize the total throughput by introducing chordal distance as a new distance metric. The scheduling algorithm selects a new user who has the maximum chordal distance with previously selected users channels.

6.3.2.1 Capacity-Based Greedy User Scheduling Algorithm

This algorithm first selects a single user that has the highest capacity. Then, from the remaining unselected users, it finds the user that provides the highest total capacity together with those selected users. The algorithm terminates when N_s users are selected or when the total capacity reduces if more users are scheduled (the total capacity may decrease with an additional user because the size of the null space for every user also reduces). Clearly, this algorithm needs to search over user sets, which greatly reduces the complexity compared to the exhaustive search method. Since the user scheduling criterion is based on the sum capacity, it is called the capacity-based greedy suboptimal user scheduling algorithm.

- Initially, let $\Omega = 1, 2, \ldots, K$ and $\Upsilon = \emptyset$.
- Find $s_1 = \arg\max_{k \in \Omega} \log_2 \det\left(\mathbf{I} + \frac{P}{N_0 B_w} \mathbf{H}_k \mathbf{H}_k^H\right)$. Let $\Omega = \Omega \cap s_1$ and $\Upsilon = \Upsilon \cup s_1$.
- Calculate initial capacity as

$$C_t = \log_2 \det\left(\mathbf{I}_{N_r} + \frac{P}{N_0 B_w} \mathbf{H}_k \mathbf{H}_k^H\right) \tag{6.119}$$

- $i = 2$ to N_s
  - For every $k \in \Omega$:
    - Let $\bar{\Upsilon} = \Upsilon \cup k$.
    - Find the precoding matrix $\mathbf{F}_j$ for each $j \in \bar{\Upsilon}$ by using BD algorithm.
    - Obtain the effective channel $\bar{\mathbf{H}}_j = \mathbf{H}_j \mathbf{F}_j$ for each $j \in \bar{\Upsilon}$.
    - Find the total capacity for the user set $\bar{\Upsilon}$ and denoted as $C_{temp,k}$.
  - Let $s_i = \arg\max_{k \in \Omega} C_{temp,k}$.
  - If $\max_{k \in \Omega} C_{temp,s_i} < C_t$, algorithm terminated and the selected user set becomes Υ.
  - Otherwise, let $\Omega = \Omega \cap s_i$ and $\Upsilon = \Upsilon \cup s_i$ and $C_t = \max_{k \in \Omega} C_{temp,s_i}$.

6.3.2.2 Frobenius Norm-Based Algorithm

Although the capacity-based suboptimal user selection algorithm greatly reduces the size of the search set, the algorithm still may not be cost-effective for real-time implementation because SVD, which is computationally intensive, is required for each user in each iteration to find the total throughput. In this section, we examine another suboptimal user selection algorithm which is based on the channel Frobenius norm. The motivation is that the capacity is closely related to eigenvalues of the effective channel after precoding. Although the channel Frobenius norm cannot characterize the capacity completely, it is related to the capacity because the Frobenius norm indicates the overall power of the channel, i.e., the sum of the eigenvalues of $\mathbf{HH}^H$ equals $\|\mathbf{H}\|_F$.

- Let $\Omega = 1, 2, \ldots, K$ and $\Upsilon = \emptyset$.
- Find $s_1 = \arg \max_{k \in \Omega} \|\mathbf{H}\|_F$.
- Let $\Omega = \Omega \cap s_1$ and $\Upsilon = \Upsilon \cup s_1$. Assign $\mathbf{V} = \mathbf{V}_{s_1}$ where $\mathbf{H}_{s_1} = \mathbf{U}_{s_1} \Sigma_{s_1} \mathbf{V}_{s_1}^H$.
- $i = 2$ to N_s

  - For every $k \in \Omega$, let $\tilde{\mathbf{H}}_k = \mathbf{H}_k - \mathbf{H}_k \mathbf{V}^H \mathbf{V}$. Then $\tilde{\mathbf{H}}_k$ is in the null space of $\mathbf{V}$.
  - For $j = 1$ to $i - 1$

    · Let $\tilde{\mathbf{H}}_{s_j,k} = [\mathbf{H}_{s_1}^T \ldots \mathbf{H}_{s_{j-1}}^T \mathbf{H}_{s_{j+1}}^T \ldots \mathbf{H}_{s_{j-1}}^T \mathbf{H}_{s_j}^T]$.
    · Let $\mathbf{W}_{s_j,k}$ be the row basis for $\tilde{\mathbf{H}}_{s_j,k}$ after GSO.

  - For each $s \in \Upsilon$, let $\hat{\mathbf{H}}_s = \mathbf{H}_s - \mathbf{H}_s \mathbf{W}_{s,k}^H \mathbf{W}_{s,k}$. Then $\tilde{\mathbf{H}}_s$ is in the null space of $\hat{\mathbf{H}}_{s,k}$.

$$s_i = \arg \max_{k \in \Omega} (\sum_{s \in \Upsilon} \|\tilde{\mathbf{H}}_s\|_F^2 + \|\tilde{\mathbf{H}}_k\|_F^2) \qquad (6.120)$$

  - $\Omega = \Omega \cap s_i$ and $\Upsilon = \Upsilon \cup s_i$.
  - Apply GSO procedure to $\hat{\mathbf{H}}_{s_i}$ and obtain $\tilde{\mathbf{V}}_{s_i}$.
  - Set $\mathbf{V} = [\tilde{\mathbf{V}} \quad \tilde{\mathbf{V}}_{s_i}]^T$.

6.3.2.3 ZFBF-SUS Algorithm

Another linear precoder that is the so-called ZFBF-SUS (zero forcing beam forming with successive user selection) has been proposed in [50, 60]. This recursive precoder design technique is based on the selection of semi-orthogonal users. In the case where each user has more than one receive antennas, the selection is not done on users but on their available streams. In order to select the best candidate, the SVD of all users are considered. The selection process in [50] is a recursive solution where at each recursion the user who generates the least interference to the previously selected ones is chosen. Once the selection procedure accomplished, either a ZFBF or a DPC precoding on the selected streams can be applied to eliminate the residual part of the interference.

- Selecting the required streams:

    - Perform a SVD over the channel matrices of all users, $\mathbf{H}_k$.
    - Select the best available stream based on the highest singular value.
    - Construct a set of streams minimizing the interference part based on a criterion such as Interference $< \delta$ where δ is a predefined threshold taken equal to $\frac{1}{log(K)}$. Select among the set the stream with the biggest singular value.
    - Construct a set of streams in the previous step until the maximum number of allowed streams N_s is reached recursively.

- Construct precoders by applying a ZF-BF over the selected streams.

6.3.3   Limited Feedback Strategies

As already mentioned in detail, in order to fully exploit multiuser diversity, user scheduling should be performed by exploiting the feedback information of the users at the transmitter. This feedback information can be provided based on a finite channel codebook as described in Chap. 4. Each user quantizes its channel based on the codebook and feeds back the corresponding index together with a CGI based on approximate SINR value. The BS uses the quantized CSI to compute the precoder based on ZF criterion or BD and uses the available CGI to schedule the users by maximizing the rate.

We assume equal power allocation is used at the base station for all active users and hence each user only feeds back its strongest right eigenvector. Specifically, assuming K active users in the system and denoting the strongest right singular vector of the kth user by $\hat{\mathbf{v}}_k$, the regularized ZF beamformer for user k, which we denote by $\mathbf{f}_k$, is given by the normalized kth column of the matrix given by

$$\mathbf{F} = \Gamma^H \left(\Gamma \Gamma^H + \frac{K N_0 B_W}{P} \mathbf{I}_K \right)^{-1} \tag{6.121}$$

where $\Gamma = [\hat{\mathbf{v}}_1, \hat{\mathbf{v}}_2, \dots, \hat{\mathbf{v}}_K]^T$.

The SINR per user becomes

$$\gamma_k = \frac{P}{K N_0 B_W} \mathbf{f}_k^H \mathbf{H}_k^H \left(\mathbf{I} + \sum_{i=1, i \neq k}^{K} \frac{P}{K N_0 B_W} \mathbf{H}_k \mathbf{f}_i \mathbf{f}_i^H \mathbf{H}_k^H \right)^{-1} \mathbf{H}_k \mathbf{f}_k \tag{6.122}$$

Then the sum capacity is determined as

$$R = B_w \sum_{k=1}^{K} \log_2(1 + \gamma_k) \tag{6.123}$$

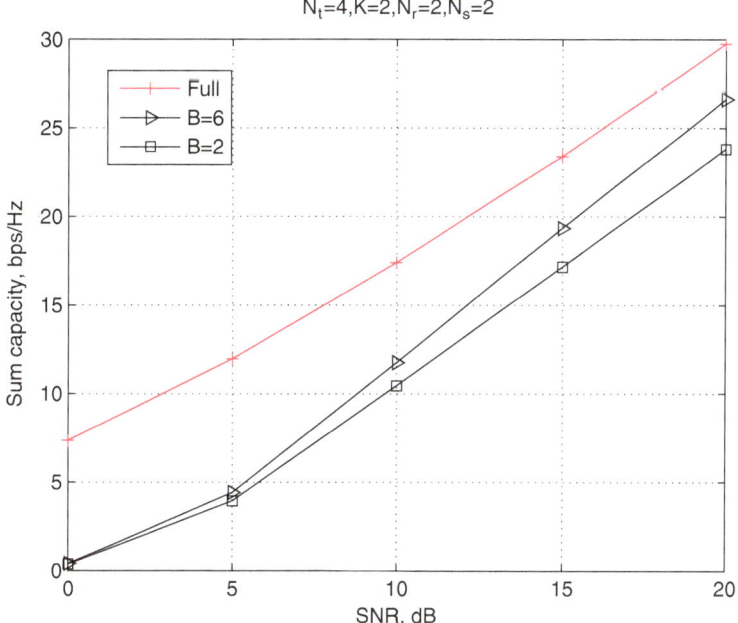

Fig. 6.16 Sum capacity of multiuser MIMO under limited feedback link

The sum capacity performance of multiuser MIMO system with $N_t = 4, K = 2, N_s = 2$, and $N_r = 2$ is illustrated in Fig. 6.16 by employing ZF precoding by considering full and quantized CSI.

6.3.4 Opportunistic Beamforming

In [8], it has been shown that the RBF converges to the eigen beamforming when many users are in a cell. In order to build two or more parallel channels when the number of antenna increases, the eigen beamformer technique is utilized using the subspace method. Thus, it is possible to transmit the signal through subspaces that can be considered as parallel channels with SVD approach.

To derive the metric, which is simple but can show the channel status, we first rewrite the received signal of the eigen beamforming by

$$\hat{\mathbf{y}} = \mathbf{U}_k^H \mathbf{H}_k \mathbf{V}_b \mathbf{x}_b + \mathbf{N}$$
$$= \Lambda_k \mathbf{V}_k^H \mathbf{V}_b \mathbf{x}_b + \mathbf{N}$$

where $\mathbf{H}_k$ denotes the channel matrix of the kth user and can be decomposed as $\mathbf{H}_k = \mathbf{U}_k \Lambda_k \mathbf{V}_k^H$ and $\mathbf{V}_b$ is the eigen beamformer that can be generated from an arbitrary random matrix $\mathbf{H}_b$. If the eigen beamformer $\mathbf{V}_b$ is equal to $\mathbf{V}_k$, which is obtained from the kth user channel $\mathbf{H}_k$, the term $\mathbf{V}_k^H\mathbf{V}_b$ represents the identity matrix. Thus, checking how close the $\mathbf{V}_k^H\mathbf{V}_b$ is to the identity matrix can be a criterion for denoting the accuracy of beamforming to user k. In addition, we can obtain multiuser diversity if we can choose the user who has the largest eigenvalues. Considering both the accuracy of beamforming and eigenvalues of the channel matrix $\mathbf{H}_k$, we develop a metric for feedback, effective SNR (ESNR), which includes eigenvalues of the kth user channel as

$$ESNR_k^a = \frac{\lambda_k^a \Psi_{aa}}{\lambda_k^a \displaystyle\sum_{j,j \neq a} \Psi_{aj} + N_0 B_w}$$

where a denotes the subspace index, Ψ_{aj} is the ath row and the jth column value of the matrix $\Psi = \mathbf{V}_k^H \mathbf{V}_b$, λ_k^a the singular value of the kth user channel. The $ESNR_k^a$ is SNR of subspace when equal power is allocated to each antenna at the transmitter.

The procedure is as follows:

- At the BS, a random unitary matrix, $\mathbf{V}_b$ is applied to the pilot symbols.
- Data is transmitted using N_t transmit antennas.
- At the kth receiver, the received signal is multiplied by $\mathbf{U}_k^H$, which is obtained by the SVD of $\mathbf{H}_k = \mathbf{U}_k \Lambda_k \mathbf{V}_k^H$.
- $ESNR_k^a$ for $a = 1, 2, \ldots, A$ is calculated and fed back them to the BS.
- At the BS, ESNRs from all the users are collected and the corresponding capacities are calculated. Then, one user according to a scheduling algorithm is selected.

6.3.4.1 Opportunistic Feedback

In [48], the opportunistic feedback is designed to improve the sum-rate capacity with a minimal amount of feedback by adjusting multiple parameters such as thresholds, random access probabilities, and the number of feedback minislots. First of all the SNR for MIMO channel is calculated as

$$\gamma_{m,k} = \frac{P}{N_t N_0 B_w [\mathbf{H}_k \mathbf{H}_k^H]_{k,k}} \tag{6.124}$$

In contention-based feedback channels, users report their CSI via shared feedback channels if their channel state is over a threshold. The feedback period consists of one static-based (SB) feedback channel and N_t contention-based feedback channels. The SB feedback channel is defined as the feedback channel dedicated to a selected user, whereas the CB feedback channel is defined as the feedback channel that all users share. The SB feedback channel consists of N_t minislots. The BS assigns the mth minislot of the SB feedback channel to a randomly selected user to receive a feedback message without a collision. Users share N_t CB feedback channels, where the mth feedback channel out of the N_t contention-based feedback channels is used to transmit the feedback information of the mth spatial channel. Each contention-based feedback channel is divided into two feedback periods, i.e., $\eta_{m,1}$ minislots and $\eta_{m,2}$ minislots, where $\eta_{m,1} + \eta_{m,2} = \eta$ for $m = 1, 2, \ldots, N_t$.

The operation of the mth contention-based feedback channel is described here. On each spatial channel, users are classified into two classes on the basis of their scheduling metric. If the scheduling metric of user k on the mth spatial channel, i.e., $\gamma_{m,k}$, is over threshold ν_1, the user belongs to class 1, and if $\nu_1 < \gamma_{m,k} < \nu_2$, the user belongs to class 2, where $\nu_1 > \nu_2$. Class-1 users access $\beta_{m,1}$ minislots with access probability $p_{m,1}$ in the first feedback period, and they also access $\beta_{m,2}$ minislots with access probability $p_{m,2}$ in the second feedback period. However, Class-2 users access only $\beta_{m,2}$ minislots with access probability $p_{m,2}$ in the second feedback period. If two or more users simultaneously transmit feedback messages, their messages are assumed to be un-recoverable at the BS because the BS has difficulty separating the multiple received messages without information of collaboration. When a BS successfully receives a feedback message in the mth contention-based feedback channel, the BS selects the user and terminates the remaining feedback process of the mth contention-based feedback channel by broadcasting a termination message . Accordingly, the number of actually used minislots in each contention-based feedback channel is less than or equal to β. If there is no successful feedback in the mth contention-based feedback channel, the BS schedules the user that reports its feedback information through the mth minislot of the SB feedback channel. The design of the proposed feedback protocol is to choose the following parameters on the mth spatial channel $m = 1, 2, \ldots, N_t$:

- Random access probabilities: $p_m = \{p_m, p_{m,1}, p_{m,2}\}$
- Scheduling metric thresholds; $\nu_m = \nu_{m,1}, \nu_{m,2}$
- Number of minislots in each feedback channel: $\beta_m = \beta_{m,1}, \beta_{m,2}$, where $\beta_{m,1} + \beta_{m,2} = \beta$

If the spatial channels are assumed to be i.i.d., the aforementioned parameters are identically used for all spatial channels. However, if the spatial channels are non-i.i.d., the BS should calculate the parameters for every spatial channel, respectively.

6.4 Conclusion

In this chapter, we have analyzed feedback strategies for multiple antenna for multiuser systems in detail. In the first part, we have focused on the achievable capacity by uplink and downlink multiuser transmission schemes assuming full CSI is available at both the transmitter and receiver sides. In the second part, we have mainly focused on the downlink transmission by considering multiuser MIMO systems with only one receive antenna. The precoding and user selection algorithms are examined and the performance results for multiuser schemes which allocate more than one user have been illustrated with full and quantized feedback. In order to further reduce feedback load, the user selection at the receiver side has been performed for OFDMA-based multiuser MIMO systems. The CDI codebooks have been designed using a local packing by taking into account the users' direction.

As a future works for multiuser multiple antenna systems, the massive MIMO systems and the delay issues have been examined by introducing MAT algorithm. In the last part, we have considered multiuser MIMO systems with multiple receive antennas with precoding and user scheduling algorithms and the performance results have been illustrated for quantized feedback. The performance results have indicated that multiuser MIMO systems with multiple receive antennas improve the performance compared to multiuser MISO systems.

References

1. Caire G, Shamai S (2003) On the achievable throughput of a multi antenna gaussian broadcast channel. IEEE Trans. on Information Theory. 43: 1691–1706.
2. Costa M H R (1983) Writing on Dirty Paper. IEEE Trans. on Info. Theory.29: 439–441.
3. Yu W, Cioffi J (2004). Trellis and convolutional precoding for transmitter based interference pre-subtraction. IEEE Trans. on Communications. 53: 1220–1230.
4. Choi L U, Murch R D (2004). A transmit preprocessing technique for multiuser MIMO systems using a decomposition approach. IEEE Trans. Wireless Commun. 3: 20–24.
5. Christensen S S, Agarwal R, de Carvalho E, J.M. Cioffi (2009). Weighted sum-rate maximization using weighted mmse for MIMO-BC beamforming design. In Proc. IEEE International Conference on Communications, Germany.
6. Choi W, Forenza A, Andrews J G , Heath R W (2007). Opportunistic space-division multiple access with beam selection. IEEE Trans. Commun., vol. 55: 2371–2380.
7. Castro P M, Joham M, Castedo L, Utschick W (2007). Robust Precoding for Multi-User MISO Systems with Limited-Feedback Channels. In International ITG Workshop in Smart Antennas, Germany.
8. Chung J, Hwang C-S, Kim K, Kim Y K (2003). A Random Beamforming Technique in MIMO Systems Exploiting Multiuser Diversity. IEEE Journal on Selected Areas in Communication. 21: 848–855.
9. Diaz J, Simeone O, Bar-Ness Y (2006). Sum-Rate of MIMO Broadcast Channels with One Bit Feedback. In Proc. of Int. Symp. Inform. Theory (ISIT), USA, 1944–1948.
10. Dimic G, Sidiropoulos N D (2005). On Downlink Beamforming With Greedy User Selection: Performance Analysis and a Simple New Algorithm. IEEE Trans. on Signal Processing. 53: 3857–3868.
11. Gesbert D, Alouini M S (2004). How much feedback is multi-user diversity really worth? In Proc. IEEE International Conference on Communications, France.
12. Harashima H, Miyakawa H (1972). Matched-transmission technique for channels with inter-symbol interference. IEEE Trans. Communication. 20: 774 –780.
13. Huang K, Andrews J G, Heath R W (2009). Performance of Orthogonal Beamforming for SDMA with Limited Feedback. IEEE Trans. on Vehicular Technology, 58: 152–164.
14. Hochwald B M, Peel C B, Swindlehurst A L (2005). A vector-perturbation technique for near-capacity multiantenna multiuser communication - Part II: Perturbation. IEEE Trans. Communication. 53: 537–544.
15. Sharif M, Hassibi B (2005). On the capacity of MIMO Broadcast channels with partial side information. IEEE Trans. on Information Theory. 51: 506–522.
16. Huang K, Heath R W, Andrews J. (2007). Multi-user aware limited feedback for MIMO systems. submitted to IEEE Transactions on Signal Processing.

17. Hoydis J, ten Brink S, Debbah M (2011). Massive MIMO: how many antennas do we need? Proceedings of Allerton Conference on Communication, Control, and Computing.
18. Jindal N, Vishwanath S, Goldsmith A (2004). On the duality of Gaussian multiple-access and broadcast channels. IEEE Trans.on Information Theory. 50: 768–783.
19. Jindal N, Goldsmith A (2005). Dirty paper coding vs. TDMA for MIMO broadcast channel. IEEE Trans. on Information Theory. 51: 1783–1794.
20. Jindal N (2006). MIMO broadcast channels with finite-rate feedback. IEEE Transactions on Information Theory. 52: 5045–5060.
21. Cadambe V R, Jafar S A (2008) Interference Alignment and Degrees of Freedom of the K-User Interference Channel.IEEE Transactions on Information Theory. 54:3425–3441.
22. Kim I, Hong S, Chassemzadeh S, Tarokh V (2005). Optimum opportunistic beamforming based on multiple weighting vectors. IEEE Trans. on Wireless Communication. 4: 2683–2687.
23. Kim J S, Kim H, Park CS, Lee K B (2006). On the performance of multiuser MIMO systems in WCDMA/HSDPA: Beamforming, feedback and user diversity. IEICE Trans. on Communication, 89: 2161–2169.
24. Khanfir H, Le Ruyet D, Ozbek B (2007). Semi-orthogonal user selection for MIMO systems with quantized feedback. ITG/IEEE workshop on smart antennas. Germany.
25. Khanfir H, Le Ruyet D, Ozbek B (2012). Reduced feedback load using User Selection Algorithms for the Multiuser MISO Systems. Wiley Transactions on Emerging Telecommunications. 63: 295–305.
26. Ko K, Lee J (2012). Multiuser MIMO User Selection Based on Chordal Distance. IEEE Trans. on Communication. 60: 649–654.
27. Lee J, Jindal N (2006). Dirty paper coding vs. linear precoding for MIMO broadcast channels. In Proc. 2006 Asilomar Conf. Signal, Syst. Comp.
28. Long Term Evolution Advanced (LTE-A) Link Level simulator.[Online]. Available: http://www.nt.tuwien.ac.at/ltesimulator/
29. Maddah-Ali M A, Motahari S A, Khandani, A K (2008) Communication Over MIMO X Channels: Interference Alignment,Decomposition, and Performance Analysis. IEEE Trans. Information Theory.54: 595–602.
30. Marzetta T L (2006). How much training is required for multiuser MIMO? In proceedings of IEEE Asilomar Conference on Signals, Systems, and Computers.
31. Marzetta T L (2010). Noncooperative cellular wireless with unlimited numbers of base station antennas. IEEE Transactions on Wireless Communications, 9 : 3590–3600.
32. Mukkavilli K K, Sabarwal A, Erkip E, Aazhang B (2003). On beamforming with finite rate feedback in multiple antenna systems. IEEE Trans. on Information Theory. 49: 2562–2579.
33. Ngo H Q, Larsson E G, Marzetta T L (2011). Uplink power efficiency of multiuser MIMO with very large antenna arrays. In Proceedings of Allerton Conference on Communication, Control, and Computing.
34. Ozbek B, Le Ruyet D (2009). Reduced feedback designs for SDMA-OFDMA systems. In Proc. of IEEE International Communication Conference (ICC),Dresden.
35. Ozbek B, Le Ruyet D (2012). Feedback channel designs for fair scheduling in MISO-OFDMA systems. EURASIP Journal on Wireless Communications and Networking 2012, 2012:220 doi:10.1186/1687-1499-2012-220.
36. Peel C, Hochwald B, Lee Swindlehurst A (2003). A vector perturbation technique for near capacity multi-antenna multi-user communication. IEEE Trans. on Communications. 53: 537–544.
37. Raghavan V, Jr Heath R W, Sayeed A M (2007). Systematic Codebook Designs for Quantized Beamforming in Correlated MIMO Channels. IEEE Journal on Selected Areas in Communications. 25: 1298–1310.
38. UE PMI Feedback Signalling for User Pairing-Coordination Alcatel-Lucent, R1-083759, R1-084141, R1-090051, R1-090777.
39. Best Companion Reporting for Improved Single-Cell MU-MIMO Pairing Alcatel-Lucent, R1-090926, R1-091307, R1-092031,R1-092546, R1-093333.

40. Rhee W, Yu W, Cioffi J M (2004). The optimality of beamforming in uplink multiuser wireless systems. IEEE Trans. on Wireless Communications. 3: 86–96.
41. Rusek F, Persson D, Lau B K, Larsson E G, Marzetta T L, Edfors O, Tufvesson F (2013). Scaling up MIMO: opportunities and challenges with very large arrays. IEEE Signal Processing Magazine, 30 : 40–60.
42. Sharif M, Hassibi B (2007). A comparison of time-sharing, DPC, and beamforming for MIMO broadcast channels with many users. IEEE Trans. Communication. 55: 11–15.
43. Spencer Q H, Swindlehurst A L, Haardt M (2004). Zero-forcing methods for downlink spatial multiplexing in multiuser MIMO channels. IEEE Trans. Signal Processing. 52: 461–471.
44. Swannack C, Uysal-Biyikoglu E, Wornell G W (2005). Finding NEMO: Near Orthogonal Sets and Applications to MIMO Broadcast Scheduling. In Int Conf on Wireless Networks, Communications and Mobile Computing. 2: 1035–1040.
45. Peel C B, Hochwald B M, Swindlehurst A L (2005). A vector perturbation technique for near-capacity multiantenna multiuser communication-part i: channel inversion and regularization. IEEE Transactions on Communications. 53: 195–202.
46. Svedman P, Wilson S K, Cimini L J, Ottersten B (2007). Opportunistic Beamforming and Scheduling for OFDMA systems. IEEE Transaction on Communications. 55: 941–952.
47. Shen Z, Chen R, Andrews J G, Heath J R W, Evans B L (2006). Low complexity user selection algorithms for multiuser MIMO systems with block diagonalization. IEEE Trans. Signal Process. 54: 3658–3663.
48. So J, Cioffi J M (2009). Multiuser Diversity in a MIMO System with Opportunistic Feedback. IEEE Trans. on Vehicular Technology. 58: 4909–4918.
49. Slim I, Mezghani A, Nossek J A (2011). Quantized CDI Based Tomlinson Harashima Precoding for Broadcast Channels. In Proceedings of IEEE Intern. Conf. On Communications (ICC).
50. Sun L, Mckay M (2010). Eigen-based transceivers for the MIMO broadcast channel with semi-orthogonal user selection. IEEE Trans. on Signal Processing. 58:5246–5261.
51. Tu Z, Blum R S (2003). Multiuser diversity for a dirty paper approach. IEEE Commun. Lett. 7:370–372.
52. Tomlinson M (1971). New automatic equalizer employing modulo arithmetic. Electronic Letters 7: 138–139.
53. Xu J, Andrews JG, Jafar S A (2012). Broadcast Channels with Delayed Finite-Rate Feedback: Predict or Observe? IEEE Trans. on Wireless Communications. 11: 1456–1467.
54. Viswanath P, Tse DNC, Laroia R (2002). Opportunistic beamforming using dumb antennas. IEEE Trans. on Inform. Theory. 48:1277–1294.
55. Viswanath P, Tse D (2003). Sum capacity of the vector gaussian broadcast channel and uplink downlink duality. IEEE Trans. on Information Theory. 49:1912–1921.
56. Vishwanath S, Jindal N, Goldsmith A (2003). Duality, achievable rates and sum rate capacity of Gaussian MIMO broadcast channel. IEEE Trans.on Information Theory. 49: 2658–2668.
57. Wu Y, Zhang J, Xu M, Zhou S, Xu X (2005). Multiuser MIMO downlink precoder design based on the maximal SJNR criterion. In Proc. IEEE Global Telecommunications Conference. 5 : 2698–2703.
58. Yu W, Cioffi J (2002). Sum Capacity of Gaussian Vector Broadcast Channels. In Proc. of Int. Symp. Information Theory.
59. Yu W, Cioffi J (2004) Sum capacity of gaussian vector broadcast channels. IEEE Trans. on Info. Theory. 50:1875–1892.
60. Yoo T, Goldsmith A (2006). On the optimality of multiantenna broadcast scheduling using zero-forcing beamforming. IEEE J. on Select. Areas in Commun. 24:528–541.
61. Zhou H, Zhang J, Xue J, Zhang Y, Tian J, Power K, Vadgama S, Nakatsugawa K (2008). Proposal for User Grouping Scheme for MU-MIMO. IEEE 802.16m-08392.
62. Zamir R, Shamai S, Erez U (2002). Nested Linear Lattice Codes for Structured Multiterminal Binning. IEEE Trans. on Info. Theory. 19:1250–1276.

Part II
Advanced Issues and Standard

Chapter 7
Feedback Strategies for Multicell Systems

7.1 Introduction

The main objective of next-generation wireless networks is to accommodate increasing user demand and to achieve a ubiquitous high-data-rate coverage so that mobile broadband services comparable to those of the wirelines are realized in a cost-efficient manner. The increasing demand for wireless multimedia has led to coordinated multicell transmission which can increase data rate and reduces outage in cellular systems by mitigating intercell interference (ICI). In order to mitigate ICI, adaptive power allocation and multiple antenna techniques are employed in the multicell networks.

In this chapter, we address the ICI mitigation problem considering two different frameworks. In the first framework, we will perform user scheduling and power allocation for networks. In order to perform distributed power allocation in multicell orthogonal frequency division multiple access (OFDMA) networks, the channel state information (CSI) belonging to all users is shared between base station(BS)s. However, the amount of feedback increases with the number of users, base stations and subcarriers. Therefore, it is important to perform a selection at the user side for OFDMA-based multicell networks. In the second framework, we consider multiple antenna strategies to mitigate ICI in cooperative networks. The cooperative networks are quite sensitive to the quality of the CSI of serving and interfering base stations. Therefore, the bit partitioning strategies affect the overall performance of the cooperative networks.

Section 7.2 introduces cooperative networks and different resource allocation (RA) problem assuming that perfect and full feedback is available at all BSs. Section 7.3 considers reduced feedback strategies for orthogonal frequency-division multiplexing (OFDM)-based multicell networks by performing user selection at the receiver side for distributed networks. Section 7.4 will examine the limited feedback strategies for multiantenna multicell systems. The purpose is to quantize CSI belongs to serving and interfering BSs by applying bit partitioning strategies based on different criteria.

B. Özbek and D. Le Ruyet, *Feedback Strategies for Wireless Communication*, DOI 10.1007/978-1-4614-7741-9_7, © Springer Science+Business Media New York 2014

7.2   Cooperative Networks

The increasing demand for wireless multimedia and interactive Internet services foster intensive research efforts on the design of novel wireless communication systems architectures for high-speed, reliable and cost-effective transmission solutions. Upcoming cellular standards like the 3GPP are targeting universal frequency reuse in a bid to increase peak data rates. This could, however, lead to high levels of ICI due to simultaneous transmissions on the same frequency by neighboring base stations. ICI can significantly reduce data rates and cause outages in cellular systems, especially at cell edges.

Next-generation wireless networks will go beyond the point-to-point or point-to-multipoint paradigms of classical cellular networks. While the multiple input multiple output (MIMO) system is now a key technology to improve the performance and capacity of wireless communications over conventional single antenna systems, the concept of cooperative communications has more recently emerged as a solution to exploit the potential MIMO gains on a distributed scale [14]. In a cooperative communication environment, different nodes can share resources to distribute the phases of transmission and/or processing, and each terminal can be considered as a potential relay.

The cellular standards like the 3GPP LTE Advanced [32] are targeting universal frequency reuse in a bid to increase peak data rates. This could, however, lead to high levels of ICI due to simultaneous transmissions on the same frequency by neighboring base stations and can significantly reduce data rates and cause outages in cellular systems, especially at cell edges. Multicell cooperation is one solution to manage ICI in future commercial wireless standards.

Base station entails sharing control signals, transmit data, user propagation CSI and/or precoders via high-capacity wired backhaul links to coordinate transmissions. In practice, however, backhaul will be bandwidth-limited due to the prohibitive costs involved in establishing high-capacity links. This restricts the amount of information that can be exchanged among base stations, which in turn determines the level of cooperation and the performance gains obtained. Multicell base station cooperation can be broadly divided into three different levels of cooperation.

1. *Control-level cooperation.* These cooperative strategies exchange only control-level information among base stations, leading to small load on the backhaul link. They usually involve some form of joint allocation of available resources to orthogonalize user transmissions in adjacent cells, by assigning different frequency bands of operation and/or timing cycles. While these techniques yield higher sum rates than static transmission algorithms, they do not utilize all the available frequency and time resources and hence, do not realize the performance gains that can be potentially obtained using base station cooperation.

2. *Partial cooperation.* Partial cooperative strategies, where base stations exchange only the CSI of active users, offer a fair balance between ensuring a reasonable load on the backhaul and attaining the performance gains using cooperation.

The shared CSI can be used by base stations to design individual precoding matrices (or beamforming vectors, for single-stream transmission) on site to transmit exclusively to users within their own cell. This is known as coordinated beamforming in 3GPP LTE Advanced.

3. *Full cooperation.* Full cooperation leads to the highest sum rates at the cost of increased overhead due to global CSI requirements and the exchange of a greater amount of information among base stations, including CSI and transmitted data. Full cooperation is typically high complexity and imposes a large load on backhaul links. Examples of these techniques include multicell dirty paper coding, multicell zero forcing, and minimum-mean squared error precoding. These are known as joint transmission in 3GPP LTE Advanced.

The overheads related to cooperative multicell processing can be divided into two main categories.

Signaling Overheads

- CSI estimation: Users estimate a greater number of channel coefficients than a multiuser MIMO system since it is proportional to the number of cooperative BSs.
- CSI feedback: Feedback of high number of estimated channel coefficients from users to BSs.
- Time synchronization: Collaborating BSs need to be tightly synchronized in time.

Infrastructural Overheads

- Central unit: The central unit gathers CSI from the BSs, performs scheduling and designs the transmission parameters according to the chosen transmission strategy.
- Low-latency backhaul links: Collaborating BSs are connected to the central unit via low-latency links in order to exchange CSI, scheduling decisions, and transmission parameters.

7.3 Resource Allocation for Cooperative Multicell Networks

In this section, the RA which includes power allocation and user scheduling are presented for multicell networks having one transmit and one receive antenna based on BS coordination for the downlink of a cellular OFDMA system. The RA problem can be classified into two categories. One is margin adaptive which minimizes the total power consumption subject to prescribed rate requirements for users. When there are no such constraints, proportional rate constraints can be imposed

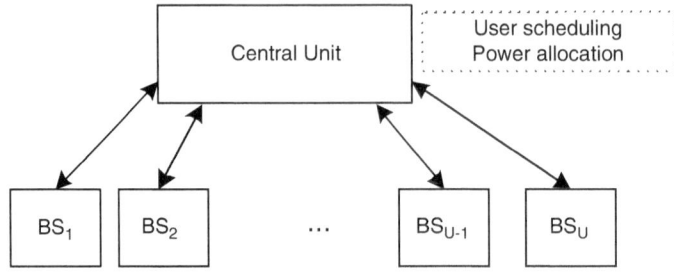

Fig. 7.1 A centralized cooperative multicell system

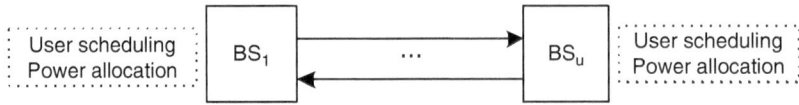

Fig. 7.2 A decentralized or distributed cooperative multicell system

to guarantee fairness among users. The second one is to maximize the sum rate over multicell networks for a given power constraint.

The optimal resource allocation for multicell OFDMA systems requires to solve the problem of power and subcarrier allocation jointly in all the considered cells, accounting into interaction between users of different cells via the multicell interference. Unfortunately, in most of the practical cases, this global optimization problem is not convex and does not have, therefore, simple closed-form solution. BSs in a cellular network are connected via backhaul links to a central processing unit (e.g., a dedicated control station or a preassigned BS), which has the global knowledge of transmit messages for all the MSs in the network and transmission channels from each BS to all the MSs shown in Fig. 7.1. Thereby, the central processing unit is able to jointly design the downlink transmissions for all BSs and provide with the appropriate signals to transmit. In the centralized cooperative multicell networks, the user scheduling and power allocation can be performed jointly or iteratively by a central unit. A joint RA over all subcarriers is preferred in order to better exploit the frequency and multiuser diversity inherent in OFDMA systems.

A decentralized or distributed multicell network as in Fig. 7.2 is considered to reduce the necessary infrastructural overheads and costs. In these multicell networks, it is assumed that each BS collects local and nonlocal CSI where each MS sends its CSI estimate belonging to all cooperating BSs to its own serving BS. In this case, each BS can perform scheduling and design the transmission parameters independently without the need of any CSI exchange with a central unit. In the decentralized or distributed cooperative multicell networks, the user scheduling and power allocation can be performed iteratively.

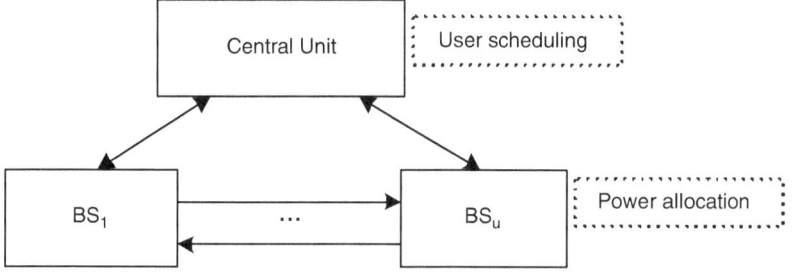

Fig. 7.3 A semi-distributed cooperative multicell system

It is also possible to perform user scheduling in the centralized unit and power allocation in each BS distributively and iteratively as shown in Fig. 7.3. In this case, the overheads can be reduced partially.

7.3.1 System Model for Multicell Network

As shown in Fig. 7.4, we consider a downlink cellular OFDMA system composed of U base stations, each of which has K_u users. Using OFDMA systems with N subcarriers, BS u serves a group of users belonging to the user set $\mathcal{U}_u$ for cell u assuming that the number of users at each cell is smaller than N. Throughout the section, it is assumed that each cell has an equal number of users, such that $K_u = K$ for $u = 1, 2, \ldots, U$.

The channel power gain of subcarrier n from BS v to a user k in cell u is denoted by $G_{v,u,k,n}$ which includes path loss, shadowing and multipath effect of the wireless channels. Note that if $u \neq v$, $G_{v,u,k,n}$ represents the power gain of the cochannel interfering link from BS v to the user k in $\mathcal{U}_u$ for subcarrier n, otherwise ($u = v$) it denotes the power gain of the communication channel from (serving) BS u to the user k in $\mathcal{U}_u$ for subcarrier n. It is assumed that the channel between each BS-user pair remains static over a sufficiently long duration, so that resource allocation can be carried out for that duration.

The power-allocation-related notations are defined in the following. $p_u^{\max}$ denotes the total available power at BS u, $p_{u,n}$ represents the allocated power to subcarrier n by BS u, and $p_{u,q}^{\max}$ is the maximum allowed power level to subcarrier n by BS u. $\{p_{u,n}\}_{u=1}^{U}$ is stacked into a $U \times 1$ vector $\mathbf{p}_n = [p_{1,n}, p_{2,n}, \ldots, p_{U,n}]^T$ and then the power allocation matrix $\mathbf{P}$ of size $U \times N$ is constructed.

The subcarrier-allocation-related notations are defined in the following. A binary variable $A_{u,k,n}$, which indicates that subcarrier n is allocated to user k in cell u if $A_{u,k,n} = 1$. $\{A_{u,k,n}\}_{n=1}^{N}$ is stacked into the vector $\mathbf{A}_{u,k} = [A_{u,k,1}, A_{u,k,2}, \ldots, A_{u,k,N}]^T$ and then $\{\mathbf{A}_u^k\}_{k=1}^{K}$ is stacked into a matrix $\mathbf{A}_k$ column by column. Finally, a matrix $\mathbf{A} = [\mathbf{A}_1, \ldots, \mathbf{A}_U]$ of size $N \times (\sum_{u=1}^{U} |U_u|)$ is defined and it indicates how subcarriers are allocated to all users.

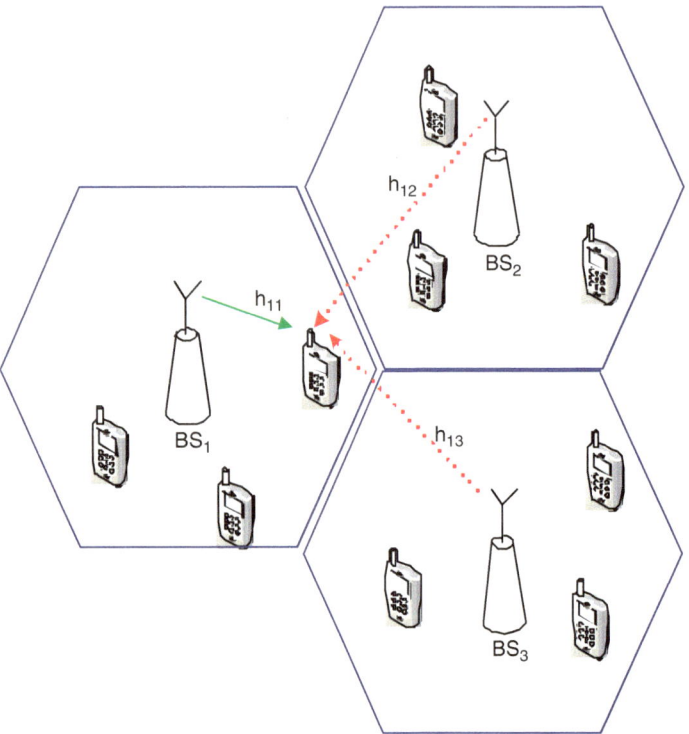

Fig. 7.4 A multiuser multicell system

The capacity of the user k in cell u as a function of $\mathbf{A}_{u,k}$ and $\mathbf{P}$ is expressed as

$$R_{u,k}(\mathbf{A}_{u,k}, \mathbf{P}) = \sum_{n=1}^{N} A_{u,k,n} R_{u,k,n}(\mathbf{p}_n) \tag{7.1}$$

where $R_{u,k,n}(\mathbf{p}_n)$ is the rate of subcarrier n when allocated to user k in cell u and it is calculated by

$$R_{u,k,n}(\mathbf{p}_n) = \log_2(1 + \gamma_{u,k,n}(\mathbf{p}_n)) \tag{7.2}$$

where $\gamma_{u,k,n}(\mathbf{p}_n))$ is the signal-to-interference-noise ratio (SINR) and determined by

$$\gamma_{u,k,n}(\mathbf{p}_n) = \frac{P_{u,n} G_{u,u,k,n}}{\sigma^2 + \sum_{v=1, v \neq u}^{U} P_{v,n} G_{v,u,k,n}} \tag{7.3}$$

The rate-adaptive resource allocation problem can be defined to maximize the function $\mathbf{f}(\mathbf{A}, \mathbf{P})$ which can be sum cell capacity, weighted sum cell capacity or sum cell of minimum users capacity subject to total power constraints for each BS.

$$\max_{\mathbf{A}, \mathbf{P}} \mathbf{f}(\mathbf{A}, \mathbf{P}) \tag{7.4}$$

subject to

$$\sum_{n=1}^{N} P_{u,n} \leq P_u^{\max}, \quad \forall u \quad (C_1)$$

$$P_{u,n} \geq 0 \quad (C_2)$$

$$\sum_{k \in \mathscr{U}_u} A_{u,k,n} \leq 1, \quad \forall u, \forall n \quad (C_3)$$

$$A_{u,k,n} \in \{0, 1\}, \quad k \in \mathscr{U}_u, \forall u, \forall n \quad (C_4)$$

where (C_1) is the sum power constraint per BS with (C_2) and (C_3) is the OFDMA subcarrier allocation orthogonality constraint with (C_4). The optimization variables are the set of subcarriers $\mathbf{A}$ per BS and the set of power values $\mathbf{P}$ per BS and per subcarrier.

For the maximization, the optimization function is:

$$\mathbf{f}(\mathbf{A}, \mathbf{P}) = \sum_{u=1}^{U} \sum_{k \in \mathscr{U}_u} R_{u,k} \tag{7.5}$$

For maximization, the optimization function is

$$\mathbf{f}(\mathbf{A}, \mathbf{P}) = \sum_{u=1}^{U} \sum_{k \in \mathscr{U}_u} w_{u,k} R_{u,k} \tag{7.6}$$

where $w_{u,k}$ is a weight for each user at the cell u.

For maximization, the optimization function is

$$\mathbf{f}(\mathbf{A}, \mathbf{P}) = \sum_{u=1}^{U} w_u \min_{k \in \mathscr{U}_u} R_{u,k} \tag{7.7}$$

where w_u represents a weight factor of the users in uth cell.

The resource allocation problem that minimizes the function $\mathbf{f}(\mathbf{A}, \mathbf{P})$ for power consumption subject to user rate constraints can be considered as follows:

$$\mathbf{f}(\mathbf{A}, \mathbf{P}) = \min_{\mathbf{A}, \mathbf{P}} \sum_{u=1}^{U} \sum_{n=1}^{N} P_{u,n} \tag{7.8}$$

subject to

$$R_{u,k} \geq R_{u,k}^{tt}, \quad \forall u, \forall k \quad (C_1)$$

$$P_{u,n} \geq 0, \quad \forall u, \forall n \quad (C_2)$$

$$\sum_{k \in \mathcal{U}_u} A_{u,k,n} \leq 1, \quad \forall u, \forall n \quad (C_3)$$

$$A_{u,k,n} \in \{0, 1\}, \quad k \in \mathcal{U}_u, \forall u, \forall n \quad (C_4)$$

where (C_1) is the target rate of each user at each BS.

7.3.2   State of the Art

For single-cell OFDMA systems, a margin-adaptive resource allocation subject to prescribed rate requirements for users is examined [61]. Application-dependent user rate constraints are given in [66]. When there are no such constraints, proportional rate constraints can be imposed to guarantee proportional fairness among users [50]. To achieve the maximum fairness among users, a max-min problem can be solved to provide similar rates to users [44].

Rate-Adaptive RA

No exact solution exists for Problem (7.5) and (7.6) that represents multicell OFDMA case in contrast to the single-cell OFDMA case where the exact solution to the problem of sum capacity maximization has been identified. An approach to tackle a variant of this problem with per subcarrier peak power constraint has been proposed in [15]. The approach consists in performing a decentralized algorithm that maximizes an upperbound on the network sum rate. Interestingly, this upperbound is proved to be tight in the asymptotic regime when the number of users per cell is allowed to grow to infinity. However, the algorithm does not guarantee fairness among the different users. A numerical approach to solve this problem has been adopted in [30]. The authors have proposed a centralized iterative allocation scheme allowing to adjust the number of cells reusing each subcarrier. This algorithm promotes allocating subcarriers which are reused by a small number of cells to users with bad channel conditions. It also provides an interference limitation procedure in order to reduce the number of users whose rate requirements are unsatisfied.

In single-carrier-based multicell networks, the optimization Problems (7.5) and (7.6) can be reduced to a power allocation problem. The optimal power allocation for Problem (7.5) has been derived analytically in [10,16] when $U = 2$ by studying the capacity region. The optimal power allocation is remarkably shown in a simple solution that is achieved by transmitting at full power only at base station 1 or only base station 2 or both. The single-carrier case of the Problems (7.5) and (7.6) is not convex and the distributed iterative algorithm presented in [56] may only lead to local optima. In high SINR regime, when $\log_2(1 + \text{SINR}) \approx \log_2(\text{SINR})$, the single-carrier sum cell capacity maximization problem belongs to the class of geometric programming (GP). In [9], the GP becomes a convex optimization problem and thus tractable for centralized power control scenario. In [7, 20], the global optimum of this optimization problem is obtained in a distributed way. The analytical derivation for the single-carrier weighted sum data rate optimization problem is fully characterized in [41] for $U = 2$. Unfortunately, these techniques are mainly intended for single-carrier and are not directly suitable for general cellular OFDMA contexts.

In the high SINR regime, the weighted sum rate in Problem (7.6) is also solved for single user OFDM through a decomposition in dual space [20]. Joint allocation of subcarriers and power has been studied in [63] for discrete multitone systems. Their conclusions directly apply to uplink multicell OFDMA and have been extended to downlink multicell OFDMA in [60]. They show that the duality gap of the weighted sum rate problem tends to zero when the number of subcarriers goes to infinity. Consequently, under this assumption, the joint resource allocation problem can be solved via Lagrange dual decomposition although it is very complex and not distributed. In single-cell downlink OFDMA, this property is verified with a finite number of subcarriers [51]. However, this conclusion no longer stands in multicell downlink OFDMA.

Since no exact solution has yet been found for Problem (7.6), only suboptimal (with respect to the optimization criterion) approaches to tackle it exist. A rate-adaptive joint RA algorithm has been proposed in [59]. This algorithm optimizes subcarrier and power allocation iteratively, such that the weighted sum rate in Problem (7.6) keeps increasing until convergence. Specifically, the algorithm adopts duality-based methods to optimize power allocation. Another approach presented in [41] consists in performing resource allocation via two phases: First, the users and subcarriers are identified with the simplifying assumption of uniform power allocation. In the second phase, an iterative distributed algorithm called Dual Asynchronous Distributed Pricing (DADP) [20] is applied for the remaining users under high SINR assumption. In [40], instead of performing subcarrier allocation by assuming uniform power allocation, a graph-based subcarrier allocation is performed with the combination of distributed power allocation where the weight of each user is proportional to its queue length. As shown in semi-distributed multicell networks, even the power allocation performs distributively by sharing only the "price" information belonging to each user; the graph-based subcarrier allocation requires a centralized unit to the build interference graph.

In [60], three iterative suboptimal algorithms are proposed to solve the joint user scheduling and power allocation problem by employing a central unit to process global channel information. In [11], a semi-distributed dynamic resource allocation scheme is proposed to suppress, which requires a central unit to perform power allocation. In [26], a novel-distributed algorithm of binary power allocation and scheduling for capacity maximization is presented. In [21], a distributed algorithm based on non-cooperative game theory and binary power allocation is presented. The distributed coordinated power allocation and user scheduling algorithm has been investigated for weighted sum throughput maximization for downlink single-carrier multicell networks in [31]. This resource allocation is separated into two suboptimal problems: user scheduling that has been performed individually at each cell for a fixed power value and power control problem based on exchanging price messages among the cells as in [41].

The optimization Problem (7.7) has been considered in [62] by optimizing subcarrier and power allocation among coordinated BSs subject to total power constraints at each BS. The purpose of this algorithm is to guarantee that the users in each cell have similar rates using an iterative algorithm which optimizes the subcarrier and the power allocation alternatively until reaching the convergence. At each iteration, the power allocation is updated by fixing the subcarrier allocation and solving a successive set of convex optimization problems with a duality-based algorithm. After that, the subcarrier is updated by fixing the power allocation and solving a mixed integer problem for each cell.

Margin-Adaptive RA

No exact solution of the Problem in (7.8) has yet been provided. Some margin-adaptive joint RA algorithms have been proposed for the downlink of multicell OFDMA systems in [1, 39]. When the RA is optimized for each subcarrier individually, the rate-adaptive algorithms in [26, 27] can be adopted for margin-adaptive RA. In the approach in [42], subcarrier assignment is performed before power allocation to tackle the problem. Once the subcarrier assignment is determined, multicell power allocation, i.e., the determination of power level for each user in each cell, is performed via an iterative allocation algorithm. Each iteration of this algorithm consists in solving the power allocation problem separately in each cell and on each assigned subcarrier based on the current level of multicell interference. The interference values experienced on each subcarrier are then updated for the next iteration of the algorithm. Necessary conditions for the convergence of this iterative algorithm are also provided.

Other performance metrics for the optimization of OFDMA resource allocation have been considered in the literature. For example, authors of [22] considered the problem of subcarrier assignment and power control that minimize the percentage of unsatisfied users under rate and power constraints.

7.3.3 Centralized Joint User Scheduling and Power Allocation

In order to maximize the network throughput optimally, power allocation should be jointly optimized with scheduling. The joint power allocation and scheduling problem consists in finding the power allocation vector $\mathbf{P}$ and scheduling vector $\mathbf{A}$ that will maximize the chosen utility function. A straightforward approach to achieve the optimal solution of the Problems in (7.5) and (7.6) would be an exhaustive search over all possible combinations of $(\mathbf{P}, \mathbf{A})$ under the given constraints. This approach entails a significant computational cost as well as feedback overhead.

7.3.4 Semi-Distributed User Scheduling and Power Allocation

Since the fully centralized coordinated resource allocation algorithms for multicell OFDMA systems require a large number of global channel information, they are not scalable and suitable for practical networks. Therefore, it is possible to decentralized the systems by employing a distributed and semi-distributed algorithms to maximize or minimize the target function for given constraints. In order to obtain distributed solution, the optimization problem can be divided into two suboptimal problems: (1) user scheduling and (2) power allocation.

First, user scheduling can be performed by fixing the power values in a distributed or centralized manner. After, for a fixed user scheduling, the power allocation is performed.

7.3.4.1 Single-Carrier Multicell Systems

For single-user single-carrier multicell systems, only power allocation among the BSs is performed iteratively in a distributed way. In this scheme, the feedback information is each user's interference prices and is shared among all the users. Then, the maximization of the optimization problem on sum cell capacity defined in (7.6) is simplified as follows:

$$\max_{\mathbf{P}_u} \sum_{u=1}^{U} w_u \log_2(1 + \gamma_u(\mathbf{P}_u)) \tag{7.9}$$

$$0 \le P_u \le P_u^{\max}, \quad \forall u \tag{7.10}$$

For the case of $U = 2$ and $w_1 = w_2 = 1$, the optimal solution is obtained in [16] as $(P_1, P_2) = (P^{\max}, P^{\max})$ or $(0, P^{\max})$ or $(P^{\max}, 0)$. For maximizing the sum throughput, the binary approach also leads to good performance in terms of spectral efficiency with $U > 2$ cells, although it is no longer optimal. In this centralized approach, each BS can only transmit at full power $P^{\max}$, or not transmit at all.

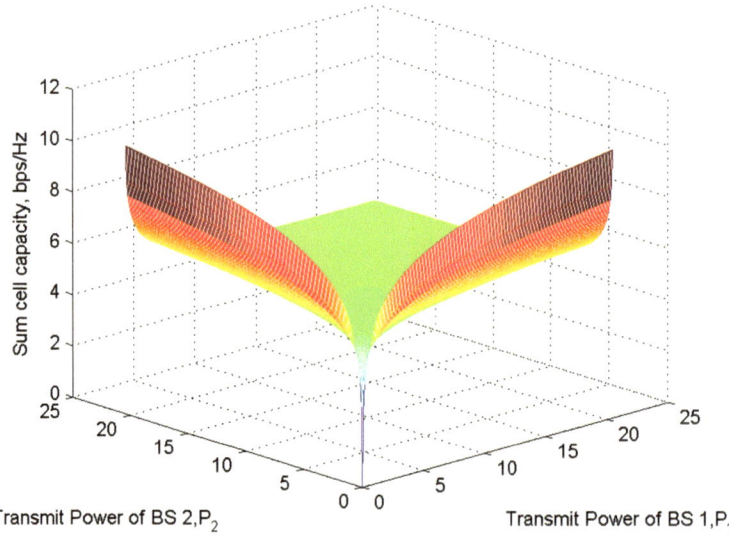

Fig. 7.5 Sum cell capacity versus different transmit power values for $U = 2$

As shown in Fig. 7.5, the maximum capacity is obtained by turning one of the BS off and transmitting at full power from the other considering the case of symmetric user locations in the 2-cell network.

When the number of BSs is higher than 2, the iterative power allocation algorithms become more attractive because of their reduced complexity. The maximization of the weighted sum rate problem, a centralized iterative algorithm, may have several local optima and consequently, it is not certain that this process will converge to a good solution. This will depend on the set of initial power values. For high SINR values, the optimization problem has a unique global maximum since it belongs to GP and consequently it is solved in a distributed way [20] in which the power values belonging to each BSs are updated iteratively and distributively. The users cooperate between them by exchanging interference information which is called interference prices. The interference price could also be periodically broadcasted through this beacon. Since each user announces only a single price, the number of prices scales linearly with the size of the network.

The algorithm is described as follows:

- For $i = 1$ to $i^{\max}$:

  - Iterative power allocation:

$$P_u(i + 1) = 0 \quad \text{if} \quad W_u(P_u(i)) \leq 0$$
$$P_u(i + 1) = P_u^{\max} \quad \text{if} \quad W_u(P_u(i)) \geq P_u^{\max}$$
$$P_u(i + 1) = W_u(P_u(i)) \tag{7.11}$$

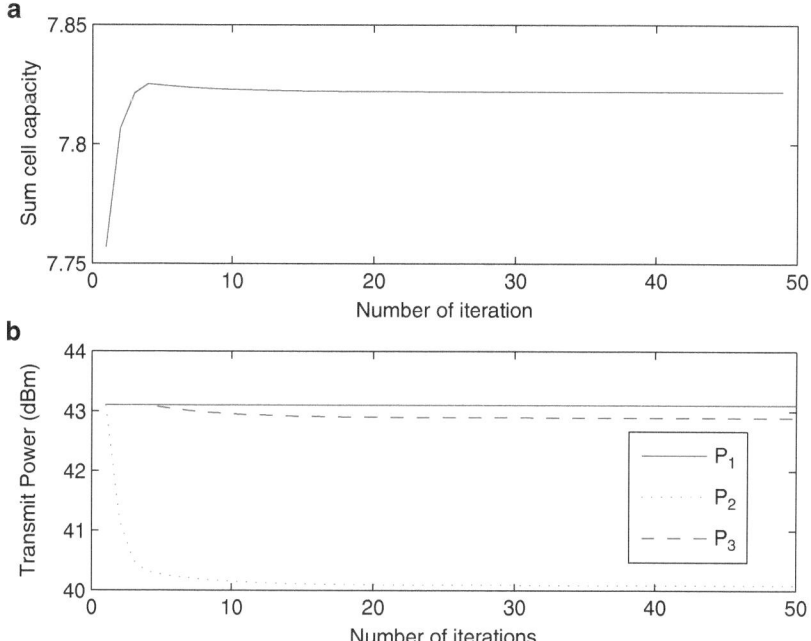

Fig. 7.6 (**a**) Sum cell capacity and (**b**) power values versus number of iteration for equal weights

where $W_u(P_u(i)) = \frac{w_u}{\sum_{v=1, v \neq u}^{U} G_{v,u} m_v(i)}$ with the price definition $m_v(i) = \frac{w_v}{I_v(i)}$ and interference term $I_v(i) = \sigma^2 + \sum_{z=1, z \neq v}^{U} P_z(i) G_{z,v}$.

If the messages $m_v(i)$ are transmitted at each iteration, then the power update for the user in the cell u at the next iteration only depends on $m_v(i)$ and on distributed information of user in cell u. The channel coefficients $G_{v,u}$ for $u, v = 1, 2, \ldots, U$ are estimated through training sequences. For $P^{\max} = 43\text{dBm}$ and $P_{\min} = 0$, the weighted sum cell capacity and power values results are obtained considering different weight factors for 3-cell networks as illustrated in Figs. 7.6 and 7.7.

7.3.4.2   Single-Carrier Multicell Systems with Scheduling

For multiuser single-channel multicell systems, as in the semi-distributed scheme, the user allocation is performed at the central unit and the power allocation can be performed iteratively in a distributed way. In this scheme, the feedback information can be divided into two groups. In the first group, the feedback information which is gathered at the central unit is channel gain belonging to all users in all cells. In the second group, the feedback information is each user's interference prices and is shared among all the users.

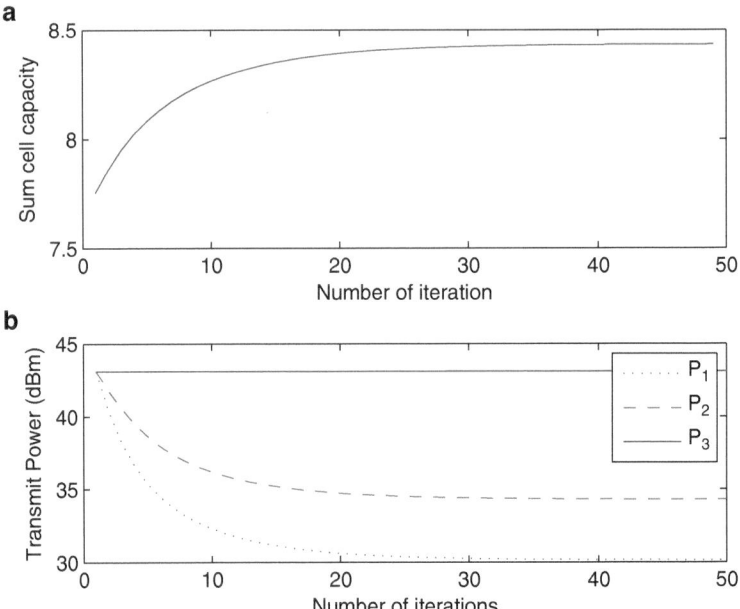

Fig. 7.7 (**a**) Weighted sum cell capacity and (**b**) power versus number of iterations and w $= [0.2, 0.3, 0.5]$

In order to perform both user scheduling and power allocation among BSs for single-carrier multicell systems, the optimization problem for weighted sum cell capacity maximization is written as follows:

$$\max_{\mathbf{P}_u, \mathbf{A}_u} \sum_{u=1}^{U} \sum_{k=1}^{K} w_{u,k} A_{u,k} \log_2(1 + \gamma_{u,k}(\mathbf{P}_u)) \tag{7.12}$$

$$0 \leq P_u \leq P_u^{\max}, \quad \forall u$$

$$\sum_{k \in \mathcal{U}_u} A_{u,k} \leq 1, \quad \forall u$$

$$A_{u,k} \in \{0, 1\}, \quad k \in \mathcal{U}_u, \forall u$$

The exhaustive search is too complex for a practical implementation. However, the maximization of sum cell capacity allows us to obtain the optimum combination of power allocation and scheduling. The suboptimal solution is performed by separating user scheduling and power allocation.

Overall Iterative RA Optimization Algorithm

- ***Initialize:***
  Set $P_u(1)^* - \Gamma_u^{\max}, \forall u$.
- ***Iteration:***
- For $j = 1$ to $j^{\max}$:

  – Allocate users at each BS u separately to maximize weighted sum capacity based on the power values obtained at jth iteration.
  For $u = 1$ to U:

  $$\mathbf{A}_{u,k}^*(j+1) = \arg \max_{\forall k \in \mathscr{U}_u} \sum_{k=1}^{K} w_{u,k} A_{u,k} \log_2(1 + \gamma_{u,k}(P_u(j)^*)) \qquad (7.13)$$

  – Allocate the power values of each subcarrier among BSs distributively under the power constraint on subcarrier to maximize weighted sum cell capacity:
  For $i = 1$ to $i = i^{\max}$:

  $$P_u(i+1) = 0 \quad \text{if} \quad W_u(P_u(i)) \leq 0$$
  $$P_u(i+1) = P_u^{\max} \quad \text{if} \quad W_u(P_u(i)) \geq P_u^{\max}$$
  $$P_u(i+1) = W_u(P_u(i)) \qquad (7.14)$$

  – Allocate the power for each BS as,

  $$P_u(j+1) = P_u(i^{\max}) \qquad (7.15)$$

The power allocation is performed distributively and iteratively and global iteration between user scheduling and power allocation continues until it converges. The performance results that maximize sum cell capacity and power values are shown in Fig. 7.8.

7.3.4.3 OFDM-Based Multicell Systems with Scheduling

For multiuser OFDM multicell systems, the feedback information can be grouped similarly to single-carrier multicell systems with scheduling case. The only difference is the complexity of the user scheduling algorithm since it should be performed for all subcarriers. However, suboptimal techniques can be applied by separating the user allocation at subcarrier level. In this case, the user allocation is performed at each subcarrier individually subject to power restriction on subcarrier. Then, the optimization problem becomes

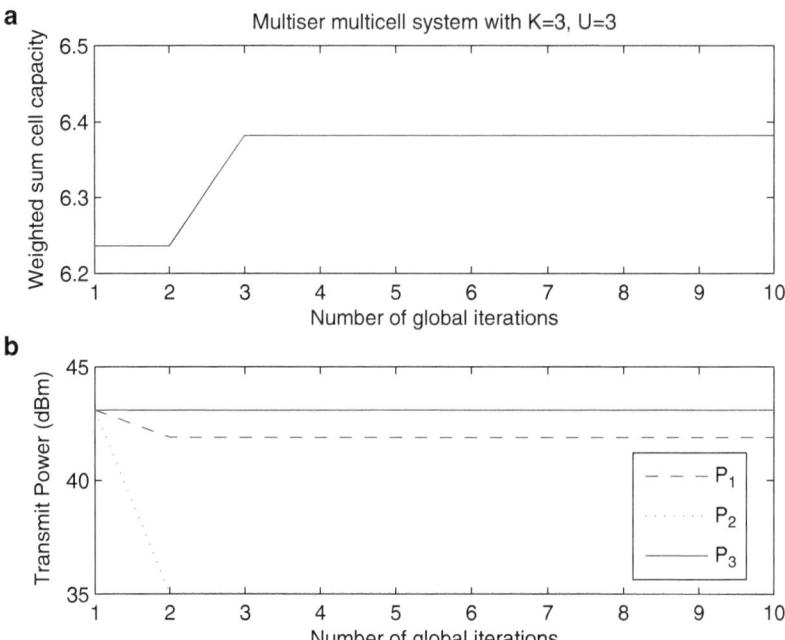

Fig. 7.8 (**a**) Sum cell capacity and (**b**) transmitted power versus the number of global iterations for $K = 3$ and $U = 3$

$$\max_{\mathbf{P,A}} \sum_{u=1}^{U} \sum_{k=1}^{K} \sum_{n=1}^{N} w_{u,k} A_{u,k,n} \log_2(1 + \gamma_{u,k,n}(\mathbf{P}))$$

$$0 \leq P_{u,n} \leq P_{u,n}^{\max}, \quad \forall u, \forall n$$

$$\sum_{k \in \mathscr{U}_u} A_{u,k,n} \leq 1, \quad \forall u, \forall n$$

$$A_{u,k,n} \in \{0, 1\}, \quad k \in \mathscr{U}_u, \forall u, \forall n \tag{7.16}$$

Overall Iterative RA Optimization Algorithm

- *Initialize:*
  Set $P_{u,n}(1)^* = P_{u,n}^{\max}, \forall u, \forall n$.
  $P_{u,n}^{\max}$ can be chosen as $P_u^{\max}/N$ by assuming equal power sharing among subcarriers.
- *Iteration:*
- For $j = 1$ to $j^{\max}$:

    – Allocate users at each BS u and each subcarrier n separately to maximize weighted sum capacity based on the power values obtained at jth iteration.

For $u = 1$ to U:

- For $n = 1$ to N:

$$\mathbf{A}^*_{u,k,n}(j+1) = \arg\max \sum_{k=1}^{K} w_{u,k} A_{u,k,n} \log_2(1 + \gamma_{u,k,n}(P_{u,n}(j)^*)) \quad (7.17)$$

– Allocate the power values of each subcarrier among BSs distributively according to user scheduling obtained as $\mathbf{A}^*_{u,k,n}(j+1)$ and under the power constraint on subcarrier to maximize weighted sum cell capacity:
For $n = 1$ to N:

- For $i = 1$ to $i = i^{\max}$:

$$P_{u,n}(i+1) = 0 \quad \text{if} \quad W_{u,n}(P_{u,n}(i)) \leq 0$$
$$P_{u,n}(i+1) = P^{\max}_{u,n} \quad \text{if} \quad W_{u,n}(P_{u,n}(i)) \geq P^{\max}_{u,n}$$
$$P_{u,n}(i+1) = W_{u,n}(P_{u,n}(i)) \quad (7.18)$$

– Allocate the power for each BS and for each subcarrier as

$$P_{u,n}(j+1) = P_{u,n}(i^{\max}), \quad \forall u, \forall n \quad (7.19)$$

As output of the overall iterative RA optimization algorithm, the converged values for user scheduling and power allocation are obtained as shown in Fig. 7.9.

7.4   Reduced Feedback Links for Multi-user OFDM Multicell Systems

Base station (BS) cooperation entails sharing control signals, transmit data, CSI and/or precoders via high-capacity wired backhaul links to establish coordinated transmission. In practice, however, the backhaul will be bandwidth-limited due to the prohibitive costs involved in establishing high-capacity links. Therefore, the amount of information exchanged among BSs should be restricted, which in turn determines the level of cooperation and the performance gains. In order to reduce the backhaul load, partial cooperative strategies have been considered where the BSs share only the users' CSIs [3].

In coordinated multicell multiuser OFDM systems with only shared CSI, the optimal resource allocation (RA) requires to solve the problem of both the power and subcarrier allocations jointly in all considered cells, taking into account inter-actions between users of different cells via the multicell interference. The existing RA algorithms have assumed that the quantized CSI belonging to the serving and interfering BSs of all users is available at the BSs, which causes a high feedback

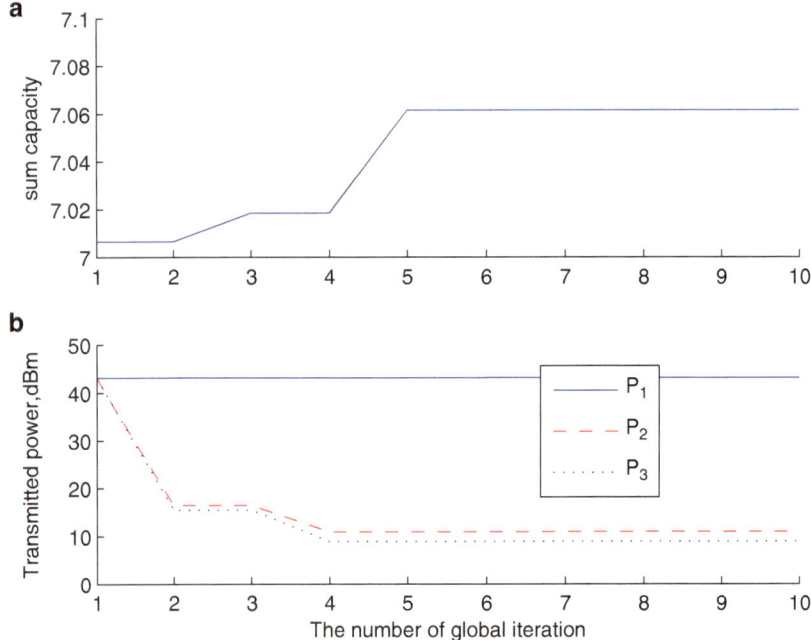

Fig. 7.9 Weighted sum cell capacity versus global iterations for $K = 3$ and $U = 3$ with $w = [0.6, 0.3, 0.1]$

load and a sophisticated resource allocation algorithm at the BSs. In this section, we will examine the multicell OFDMA systems while reducing the power consumption and satisfying users' data rate constraints.

7.4.1   System Model

We consider a downlink cellular OFDMA system where BS is equipped with one transmit and one receive antenna, composed of U base stations, each of which has K users. Using OFDMA system with Q clusters, BS u serves a group of users belonging to the user set $\mathcal{U}_u$ for cell u assuming that the number of users at each cell is smaller than Q.

Firstly, the subcarriers are grouped into clusters where the correlation is high among the adjacent subcarriers. Then, only one value for each cluster is sufficient to represent it. In multicell networks, in order to determine the representative value of each cluster, we distinguish the channel coefficients belonging to the serving BS and interfering BSs. For the serving BS, the CSI value belonging to the subcarrier of minimum channel gain is chosen as

$$\overline{\overline{G}}_{u,u,k,q} = G_{u,u,k,x}, \quad u = 1, 2, \ldots, U \tag{7.20}$$

where

$$x = (q-1)N_Q + \arg \min_{1 \leq i \leq N_Q} \{G_{u,u,k,(q-1)N_Q+i}\}, \quad \forall q$$

where $N_Q = N/Q$ is the number of subcarriers in one cluster, and N is the number of subcarriers in an OFDM symbol.

For interfering BSs, the CSI value belonging to the subcarrier of maximum channel gain is selected as

$$\bar{G}_{v,u,k,q} = G_{v,u,k,y}, \quad v, u = 1, 2, \ldots, U \quad v \neq u \tag{7.21}$$

where

$$y = (q-1)N_Q + \arg \max_{1 \leq j \leq N_Q} \{G_{v,u,k,(q-1)N_Q+j}\}, \quad \forall q$$

Then, by assuming full and equal power transmission from, respectively, all BSs and all clusters, the *approximate SINR* value of each user in multicell network is calculated as

$$\tilde{\gamma}_{u,k,q} = \frac{(P_u^{\max}/Q) \, \overline{\overline{G}}_{u,u,k,q}}{(\sigma^2/Q) + (P_u^{\max}/Q) \sum_{v=1,v \neq u}^{U} \bar{G}_{v,u,k,q}}$$

$$= \frac{P_u^{\max} \, \overline{\overline{G}}_{u,u,k,q}}{\sigma^2 + P_u^{\max} \sum_{v=1,v \neq u}^{U} \bar{G}_{v,u,k,q}} \tag{7.22}$$

where $\sigma^2 = N_0 B_W$ is the noise power.

The reason of selecting minimum channel gain for the serving BS and maximum channel gain for interfering BSs is to avoid outage by calculating the lowest *approximate SINR* value for each cluster.

The margin-adaptive resource allocation problem is defined as

$$\min_{\mathbf{A},\mathbf{P}} \sum_{u=1}^{U} \sum_{k=1}^{K} \sum_{q=1}^{Q} A_{u,k,q} P_{u,q} \tag{7.23}$$

subject to

$$R_{u,k} \geq R_{u,k}^{tt}, \quad k \in \mathscr{U}_u, \forall u \quad (C_1)$$

$$\sum_{q=1}^{Q} P_{u,q} \leq P_u^{max}, \quad \forall u \quad (C_2)$$

$$P_{u,q} \geq 0, \quad \forall u, \forall q \quad (C_3)$$

$$\sum_{k \in \mathscr{U}_u} A_{u,k,q} \leq 1, \quad \forall u, \forall q \quad (C_4)$$

$$A_{u,k,q} \in \{0,1\}, \quad k \in \mathscr{U}_u, \forall u, \forall q \quad (C_5)$$

(C_1) is the target rate constraint for each user at each cell, (C_2) and (C_3) are the power constraints and (C_4) and (C_5) are the OFDMA cluster allocation constraints. The optimization variables are the set of clusters $\mathbf{A}$ per BS and the set of power values $\mathbf{P}$ per BS and per cluster.

7.4.2 The Reduced Feedback Link Designs

For single-cell OFDMA systems, in order to reduce the feedback load, adjacent subcarriers are grouped into clusters and only the CSI related to the S-best clusters for each user are fed back to the transmitter [52]. We perform clustering for multicell networks by considering the criterion based on approximated SINR values instead of channel gain only [55].

7.4.2.1 Clustered S-Best for Fixed Feedback

For fixed feedback criterion, each user k at each cell u builds independently a set $\mathbb{S}_{u,k}$ composed of the S-best clusters based on $\tilde{\gamma}_{u,k,q}$ values. Then, the CSIs of serving and interfering BSs of $\mathbb{S}_{u,k}$ are fed back and shared among BSs.

Let $\mathbb{T}_{u,q}$ be the set of users that have fed back their CSI associated to the cluster q at cell u as

$$\mathbb{T}_{u,q} = \{k \in \{1, 2, \ldots, K\} : q \in \mathbb{S}_{u,k}\} \quad (7.24)$$

The S parameter is fixed to satisfy all users' rate constraints according to the number of users in the cell as a function of K and $R_{u,k}^{tt}$ such that $S = f(K, R_{u,k}^{tt})$. Total feedback is proportional to KS at each cell.

Then, the set $\mathbb{T}_{u,q}$ is constructed as in Eq. (7.24).

7.4.3 *Distributed User Scheduling and Power Allocation*

In order to illustrate the performance of the proposed reduced feedback links for multicell networks, we examine the optimal solution for $U = 2$ and $U = 3$ cells in a distributed manner by considering the clustered structure and selected user set which have been introduced in the previous section. In order to solve this optimization problem in a distributed manner, the power values are calculated for any combination of the allocation of users to each cluster at each cell. Then, the user pair which requires the minimum sum power satisfying the constraints is selected in each BS separately.

7.4.3.1 The Distributed Solution for U = 2

A simple scenario with only 2 cells is considered with the two following assumptions:

- *Assumption 1*: The users' target rates are the same by having the same SINR value, γ^{tt}, per cluster.
- *Assumption 2*: The channel gains $\bar{\bar{G}}_{1,1,k_1,q}$, $\bar{G}_{1,2,k_1,q}$, $\bar{\bar{G}}_{2,2,k_2,q}$ and $\bar{G}_{2,1,k_2,q}$ for $k_1 \in \mathbb{T}_{1,q}$ and $k_2 \in \mathbb{T}_{2,q}$ for each cluster are perfectly known at $\mathbf{BS}_1$ and $\mathbf{BS}_2$.

  Then, the optimization problem can be simplified as

$$\min \sum_{k=1}^{K} \sum_{q=1}^{Q} (A_{1,k,q} P_{1,q} + A_{2,k,q} P_{2,q}) \qquad (7.25)$$

All constraints remain the same, except the first constraint (C_1) which can be simplified to

$$\sum_{q=1}^{Q} A_{u,k,q} \geq Q^{tt}$$

where $Q^{tt} = \frac{R^{tt} Q}{B \log_2(1 + \Gamma \gamma^{tt})}$ with Γ is a constant gap that is chosen depending on the required bit error rate (BER).

The joint user scheduling and power minimization algorithm is given by

- For $q = 1$ to Q:
  - *Step 1*: For $\mathbf{BS}_1$, the required power for each user pair is calculated:
    - For any (k_1,k_2) where $k_1 \in \mathbb{T}_{1,q}$ and $k_2 \in \mathbb{T}_{2,q}$, by solving (7.25), the power can be determined as

$$\tilde{P}^{k_1,k_2}_{1,q} = \frac{a_1 + a_2 b_{12}}{1 - b_{12} b_{21}}$$

$$\tilde{P}^{k_1,k_2}_{2,q} = \frac{a_2 + a_1 b_{21}}{1 - b_{21} b_{12}}$$

where

$$a_u = \frac{\sigma^2_q \gamma^{tt}}{\overline{\overline{G}}_{u,u,k_u,q}}, \quad u = 1, 2$$

$$b_{12} = \frac{\overline{G}_{1,2,k_1,q} \gamma^{tt}}{\overline{\overline{G}}_{1,1,k_1,q}}$$

$$b_{21} = \frac{\overline{\overline{G}}_{2,1,k_2,q} \gamma^{tt}}{\overline{G}_{2,2,k_2,q}}$$

where $\sigma^2_q = \sigma^2/Q$.
- End.

- *Step 2:* Choose the feasible power pairs $(\tilde{P}^{k_1,k_2}_{1,q}, \tilde{P}^{k_1,k_2}_{2,q})$ that satisfy the power constraints in (C_2) and (C_3) and the conditions of $b_{12} < 1$ and $b_{21} < 1$.
- *Step 3:* Among the feasible power pairs, the user pair that requires the minimum total transmitted power is selected as follows:

$$(k^*_1, k^*_2) = \arg \min_{k_1, k_2} (\tilde{P}^{k_1,k_2}_{1,q} + \tilde{P}^{k_1,k_2}_{2,q})$$

- *Step 4:* At **BS**$_1$, the user k^*_1 is scheduled as $A_{1,k^*_1,q} = 1$ and its power is assigned as $P_{1,q} = \tilde{P}^{k^*_1,k^*_2}_{1,q}$.
- *Step 5:* Check that the user k^*_1 satisfies the rate constraint in (C_1). If so, remove this user from the set $\mathbb{T}_{1,q}, \forall q$. Moreover, this user does not continue to feed back any information in the OFDM frame.

- End.

The same algorithm is separately performed at **BS**$_2$ and the user is scheduled and the power is assigned as $P_{2,q} = \tilde{P}^{k^*_1,k^*_2}_{2,q}$ and $A_{2,k^*_2,q} = 1$.

7.4.3.2   The Distributed Solution for U = 3

We extend the same solution to $U = 3$ cells with *Assumption 1* and *Assumption 2* by considering 3 BSs in the network.

In order to obtain feasible power values, firstly we select the users which fulfil the following condition [42]:

$$\frac{\gamma^{tt}}{E_{u,k,q}} < 1 \quad k_u \in \mathbb{T}_{u,q}, \forall u, \forall q \tag{7.26}$$

where

$$E_{u,k,q} = \frac{\overline{\overline{G}}_{u,u,k,q}}{\sum\limits_{v=1,v\neq u}^{U} \bar{G}_{v,u,k,q}} \tag{7.27}$$

Then, we perform the joint user scheduling and power allocation algorithm by taking into account these users $k'_u \in \mathbb{T}'_{u,q}, \forall u, \forall q$.

The joint user scheduling and power minimization is given by

- For $q = 1$ to Q:

  - For $\mathbf{BS}_1$, the required power for each user pair is calculated:

    - For any (k'_1, k'_2, k'_3) where $k'_1 \in \mathbb{T}'_{1,q}$, $k'_2 \in \mathbb{T}'_{2,q}$, $k'_3 \in \mathbb{T}'_{3,q}$, the power can be determined by

      $$\mathbf{P} = \mathbf{B}^{-1}\mathbf{A}$$

    where $\mathbf{P}$, $\mathbf{A}$ and $\mathbf{B}$ are given in the following:

    $$\mathbf{P} = [\tilde{P}_{1,q}^{k'_1,k'_2,k'_3} \quad \tilde{P}_{2,q}^{k'_1,k'_2,k'_3} \quad \tilde{P}_{3,q}^{k'_1,k'_2,k'_3}]^T$$

    $$\mathbf{A} = [a_1 \quad a_2 \quad a_3]^T$$

    $$\mathbf{B} = \begin{bmatrix} 1 & -b_{12} & -b_{13} \\ -b_{21} & 1 & -b_{23} \\ -b_{31} & -b_{32} & 1 \end{bmatrix}$$

    with

    $$a_u = \frac{\sigma_q^2 \gamma^{tt}}{\overline{\overline{G}}_{u,u,k'_u,q}}, u = 1, 2, 3$$

    $$b_{1v} = \frac{\bar{G}_{v,1,k'_1,q}\gamma^{tt}}{\overline{\overline{G}}_{1,1,k'_1,q}}, v = 2, 3$$

    $$b_{2w} = \frac{\bar{G}_{w,2,k'_2,q}\gamma^{tt}}{\overline{\overline{G}}_{2,2,k'_2,q}}, w = 1; w = 3$$

    $$b_{3y} = \frac{\bar{G}_{y,3,k'_3,q}\gamma^{tt}}{\overline{\overline{G}}_{3,3,k'_3,q}}, y = 1, 2$$

- End.

– Considering the constraints of (C_2) and (C_3), the feasible power levels at each BS for each pair of users are calculated and then, *Step 2* to *Step 5* described in the previous algorithm are performed to schedule the user for $\mathbf{BS}_1$.

• End.

The same algorithm is separately performed at $\mathbf{BS}_2$ and $\mathbf{BS}_3$ and the user is scheduled and the power is assigned as $P_{2,q} = \tilde{P}_{2,q}^{k_1^*,k_2^*,k_3^*}$, $A_{2,k_2^*,q} = 1$ at the $\mathbf{BS}_2$ and $P_{3,q} = \tilde{P}_{3,q}^{k_1^*,k_2^*,k_3^*}$, $A_{3,k_3^*,q} = 1$ at the $\mathbf{BS}_3$.

7.4.4   Performance Results

We obtain the performance results to illustrate the benefits of the reduced rate feedback links in a multicell OFDMA networks for $U = 2$ and $U = 3$. It is assumed that the users are uniformly distributed in multicell network with a cell radius of 500 m. The transmitted power and the noise power density are chosen as 43.10 dBm and -174 dBm/Hz, respectively. The channel model is 3GPP TU and the path loss is $L_p = 128.1 + 37.6 \log_{10}(d)$ dB. The target SINR values are chosen as $\gamma^{tt} = 7.45$ dB and $\gamma^{tt} = 4.65$ dB at BER=10^{-6} [17] to perform link adaptation. The number of subcarriers per cluster is fixed at 12 and the number of clusters is equal to 40. The frame duration is 10 ms with a feedback time of 1 ms.

In Fig. 7.10, the percentage of satisfied users according to the number of feedback clusters is demonstrated based on the proposed SINR criterion for optimal solution when $U = 2$. Based on the user satisfaction criterion, the minimum number of required clusters S is listed in Table 7.1 for different rate constraints. As seen in the table, when the number of users in the cell increases, the required number of feedback clusters also increases to satisfy all users' constraints. Besides, the required number of feedback clusters increases when the required target rate gets higher.

The total transmitted power that satisfies all users' rate constraints is drawn, respectively, for $U = 2$ and $U = 3$ in Figs. 7.11 and 7.12. The performance results indicate that the proposed fixed feedback channel reduces the feedback load significantly as shown in Figs. 7.13 and 7.14 while requiring almost the same average transmitted power compared to full feedback schemes.

7.5   Limited Feedback Designs for Multiantenna Multicell Systems

The use of multiple antenna techniques in downlink wireless networks increases the overall throughput by exploiting the degrees of freedom in the spatial domain. It is possible to accommodate up to $N_t - 1$ interference signals in a coordinated multicell

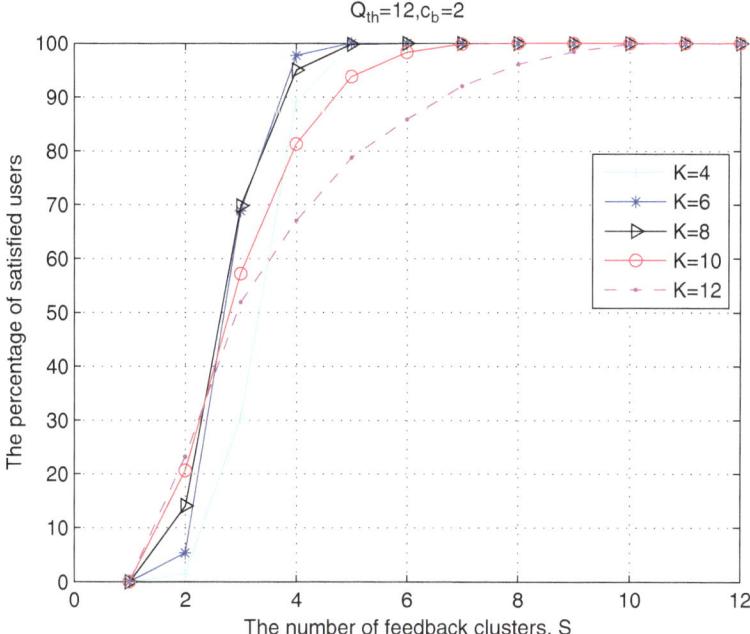

Fig. 7.10 The percentage of satisfied users in each cell versus the number of feedback clusters for $c = 2, Q^{target} = 12, U = 2$

Table 7.1 The number of required feedback clusters

Case	K				
	4	**8**	**12**	**16**	**20**
$R^{tt} = 1\,M, \gamma^{tt} = 7.45dB, U = 2$	3	3	3	6	12
$R^{tt} = 0.4\,M, \gamma^{tt} = 4.65dB, U = 2$	2	2	2	2	2
$R^{tt} = 0.4\,M, \gamma^{tt} = 4.65dB, U = 3$	5	5	5	7	11

transmission when a base station (BS) is equipped with N_t antennas by using linear transceiver processing techniques. Full coordinated multicell transmission requires to exchange all the users' CSI as well as their data information. In order to reduce the backhaul load, partial cooperative strategies have been considered where the BSs share only the users' CSIs [3]. In partial coordinated multicell networks, each base station designs its beamforming vector for communicating to its own user by employing different transmission strategies such as maximum ratio combining (MRC) and partial zero-forcing (PZF) [23]. It is important to develop cooperative techniques that maximize the performance while keeping the feedback load at a reasonable level. It is important to develop cooperative techniques with limited feedback that maximize the performance while keeping the feedback load at a reasonable level. The limited feedback systems have extensively been investigated from point-to-point MIMO channels to MIMO broadcast channels in [8, 24, 64, 65].

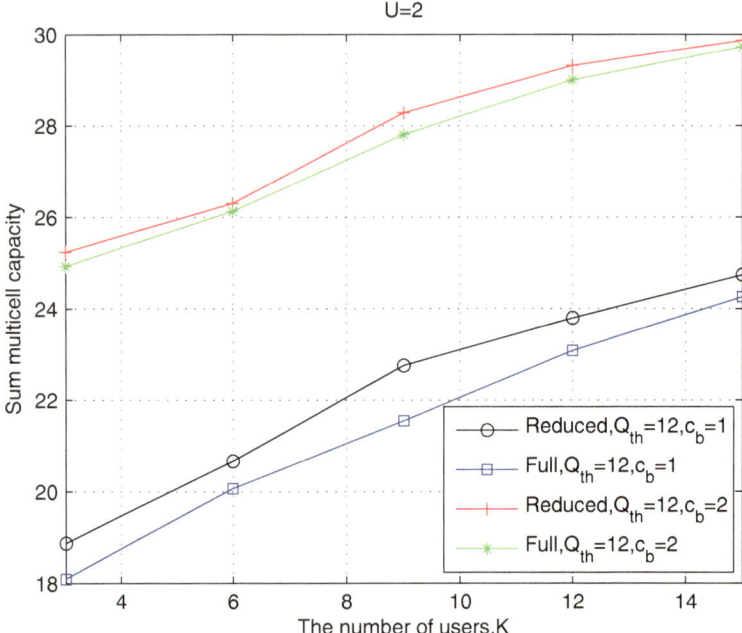

Fig. 7.11 The transmitted power versus the number of users for $c_b = 2$, $Q^{target} = 12$, and $U = 2$

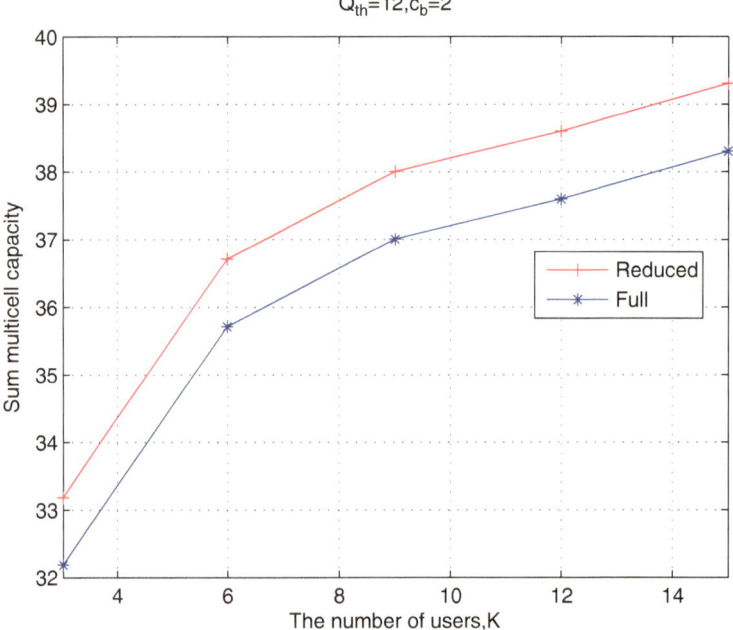

Fig. 7.12 The transmitted power versus the number of users for $c_b = 2$, $Q^{target} = 12$, and $U = 3$

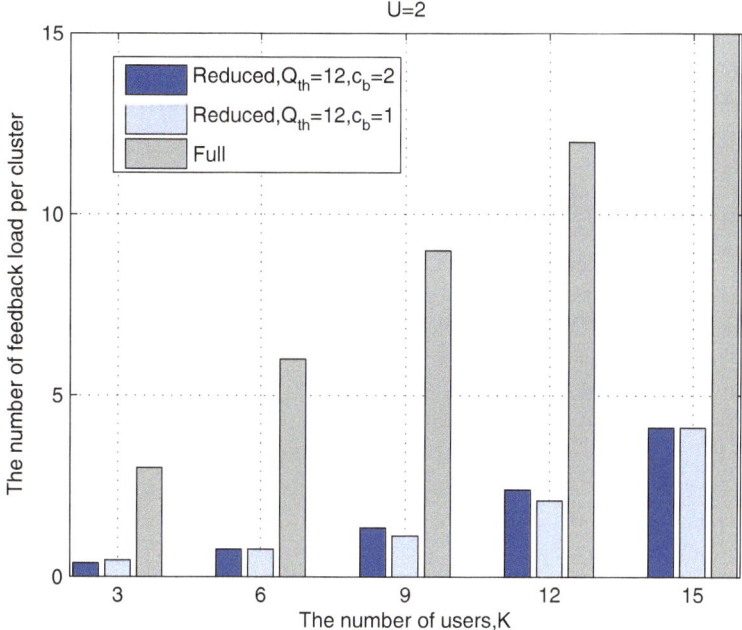

Fig. 7.13 The feedback load per cluster versus the number of users in the cell for $c_b = 2, 1$, $Q^{target} = 12$, and $U = 2$

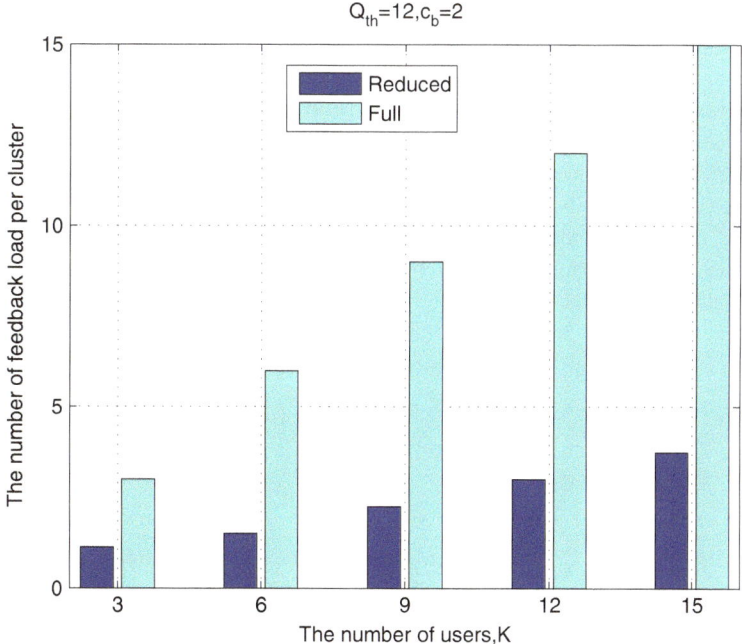

Fig. 7.14 The feedback load per cluster versus the number of users in the cell for $c_b = 2$, $Q^{target} = 12$, and $U = 3$

The performance of cooperative multicell networks is highly dependent on the quality of the CSIs of both serving and interfering BSs which is quantized and sent back by the users through a limited feedback link. The strategy of bit partitioning between the serving and interfering BSs affects the overall performance of cooperative networks. In [43], a robust decentralizing framework has been presented by evaluating the effect of feedback errors under a digital feedback model. In [67], the adaptive ICI method which still provides throughput gain with carefully designed feedback strategies has been presented where multiple BSs jointly select the transmission strategies. In [4], at high SINR, a feedback-allocation strategy has been presented to reduce the mean loss caused by random vector quantizer (RVQ) in sum capacity. The adaptive bit partitioning for delayed limited feedback channels to perform ICI by allocating more bits to quantize the stronger channels with smaller delays and fewer bits to weaker channels with larger delays has been presented in [5]. Adaptive feedback schemes for coordinated zero forcing (ZF) have been investigated by minimizing the expected quantization error to maintain the optimal multiplexing gain in [34]. In order to reduce the feedback and backhaul overheads, a selective feedback which prevents users whose channel quality does not exceed a threshold has been presented in [37]. Limited feedback schemes for cooperative multicell processing (CoMP) channel quantization with per-cell codebook have been examined including CoMP channel reconstruction, CoMP codebook generation and per-cell codewords selection in [46]. A scalable two-stage feedback mechanism which includes a first stage of individual per-cell feedback to support single-cell MU-MIMO and a second stage of multicell feedback to efficiently enable coordinated multicell transmission per-cell feedback has been described in [33]. Bit partitioning techniques have been proposed for cooperative multiantenna 2 cell networks with fixed and adaptive feedback rate in [53].

These limited feedback strategies assume that each user has a fixed number of bits to quantize CSIs of the serving and interfering BSs without taking into account its position in the cell and improve the average capacity while providing poor performance for cell-edge users. By employing more quantization bits for the cell-edge users, it is possible to further improve the capacity at cell edges. In [28] and [31], a precoding matrix index (PMI) restriction method by informing the other BSs about the precoding vector which causes large interference has been examined to mitigate for cell-edge users.

In this section, we consider a U-cell multicell network including a BS with N_t transmit antennas. Each BS communicates only one user which has a single antenna.

7.5.1   System Model

A downlink multicell network with U cells where each base station has N_t transmit antennas and each user is equipped with a single antenna is considered as shown in Fig. 7.15.

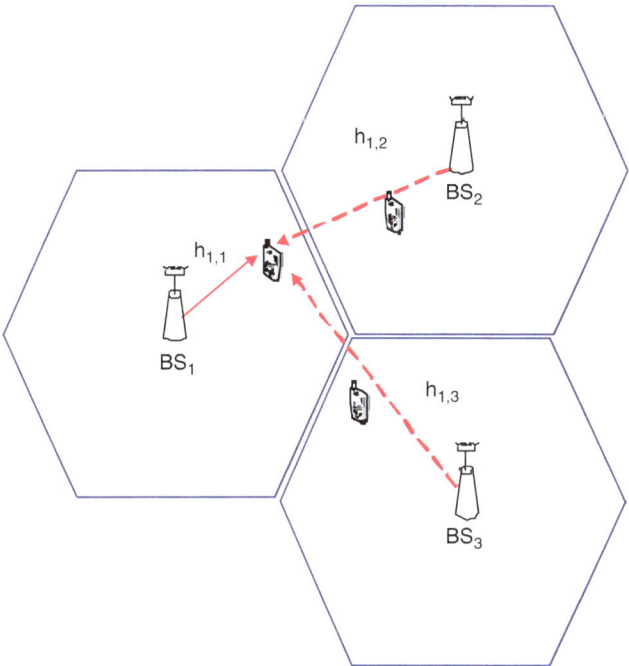

Fig. 7.15 A multicell network with N_t transmit antennas

Using a narrow band flat-fading model, the baseband received signal for the user in cell u is written as

$$y_u = \sqrt{P_{u,u}} \mathbf{h}_{u,u}^H \mathbf{w}_u x_u + \sum_{v \neq u} \sqrt{P_{u,v}} \mathbf{h}_{u,v}^H \mathbf{w}_v x_v + n_u \qquad (7.28)$$

where $P_{u,v}$ is the received power of user u from BS v, $\mathbf{h}_{u,v}$ is the channel between the user of cell u and BS v, $\mathbf{w}_u$ is the beamforming vector at BS u, and n_u is the complex additive white Gaussian noise at the receive antenna with zero mean and $\mathbb{E}[|n_u|^2] = 1$. It is assumed that each component of $\mathbf{h}_{u,v}$ has independent identically distributed random variable with $\mathscr{CN}(0,1)$ and $\mathbb{E}[\|x_u\|^2]$ is normalized to 1.

The associated user rate in each cell u is calculated as

$$R_u = \mathbb{E}[\log_2(1 + \mathrm{SINR}_u)] \qquad (7.29)$$

where the instantaneous SINR of the user in cell u is

$$\mathrm{SINR}_u = \frac{P_{u,u}|\mathbf{h}_{u,u}^H \mathbf{w}_u|^2}{1 + \sum_{v \neq u} P_{u,v}|\mathbf{h}_{u,v}^H \mathbf{w}_u|^2} \qquad (7.30)$$

The received power from serving and interfering BSs is calculated considering the following path loss model:

$$P_{u,v} = P_0 \left(\frac{d_{u,v}}{R} \right)^{-\alpha}$$

(7.31)

with R the radius of the cell, P_0 the received power at the cell edge, α the path loss exponent and $d_{u,v}$ the distance between user u and BS v.

7.5.1.1 Transmission Strategies

Two classical transmission strategies are considered to design the beamforming vectors: (1) MRC beamforming where each user serves its own user and is not required to cancel interference from other cells. (2) PZF beamforming which cancels interference from the other cells.

(1) MRC Beamforming

The other cell interference can be ignored since the user is far from the other BSs. Therefore, the precoding vector is designed according to the channel direction of the user itself. For the uth BS, the precoding vector is determined by

$$\mathbf{w}_u = \frac{\mathbf{h}_{u,u}}{\|\mathbf{h}_{u,u}\|}$$

(7.32)

For MRC, distribution of the received power becomes

$$2|\mathbf{h}_{u,u}^H \mathbf{w}_u|^2 \sim \chi^2_{2N_t}$$

(7.33)

where χ^2_n denotes the chi-square random variable with n degrees of freedom.

(2) PZF Beamforming

In order to maximize $|\mathbf{h}_{u,u}^H \mathbf{w}_u|^2$, some degrees of freedom are used for cancelation. This corresponds to the selection of $\mathbf{w}_u$ in the direction of the projection of the channel vector $\mathbf{h}_{u,u}$ on the nullspace of $\mathbf{H}_u = [\mathbf{h}_{1,u}, \mathbf{h}_{2,u}, \ldots, \mathbf{h}_{u-1,u}, \mathbf{h}_{u+1,u}, \ldots, \mathbf{h}_{U,u}]$.
  The precoding vector is given by

$$\mathbf{w}_u' = (\mathbf{I} - \mathbf{P})\mathbf{h}_{u,u}$$

(7.34)

where the projection matrix on $\mathbf{H}_u$ is $\mathbf{P} = \mathbf{H}_u(\mathbf{H}_u^* \mathbf{H}_u)^{-1}\mathbf{H}_u^*$ and $*$ is the transpose conjugate operator.
  Then, the PZF vector is obtained after normalization as

$$\mathbf{w}_u = \frac{\mathbf{w}_u'}{\|\mathbf{w}_u'\|}$$

(7.35)

For PZF precoding, distribution of the received power is [23]

$$2|\mathbf{h}_{u,u}^H \mathbf{w}_u|^2 \sim \chi^2_{2(N_t-(U-1))} \tag{7.36}$$

For 3-cell network coordination, each BS has four different strategies. For BS_1 as an example, these strategies are described as follows: (1) MRC beamforming, denoted by $t_1 = $ MRC. (2) PZF beamforming for user 2, denoted as $t_1 = $ PZF(2). (3) PZF beamforming for user 3, denoted as $t_1 = $ PZF(3). (4) PZF beamforming for user 2 and user 3. It is denoted by $t_1 = $ PZF(2, 3). The same kind of transmission strategies can be selected either for user 2 and user 3.

Then, the strategy sets can be defined for BS_1, BS_2, and BS_3, respectively,

$$\mathbb{T}_1 = \{MRC, PZF(2), PZF(3), PZF(2, 3)\},$$
$$\mathbb{T}_2 = \{MRC, PZF(1), PZF(3), PZF(1, 3)\},$$
$$\mathbb{T}_3 = \{MRC, PZF(1), PZF(2), PZF(1, 2)\}.$$

The strategies taken by BSs are $(t_1 \times t_2 \times t_3) \in \mathbb{T}_1 \times \mathbb{T}_2 \times \mathbb{T}_3$. Therefore, there are 4^3 different triples to be selected for an optimum solution.

Firstly, we assume that CSI of the serving and interfering BSs for each user in the cell is sent back perfectly to the serving BS and shared among BSs through the perfect backhaul link. Then, our objective in this section is to select the strategies $t_i; i = 1, 2, 3$ to maximize the sum multicell capacity as

$$(t_1^*, t_2^*, t_3^*) = \arg \max_{t_1 \in \mathbb{T}_1, t_2 \in \mathbb{T}_2, t_3 \in \mathbb{T}_2} R_1^{ffb} + R_2^{ffb} + R_3^{ffb} \tag{7.37}$$

7.5.1.2 Limited Feedback Channel

In order to address the lack of perfect CSI, a classical solution is to quantize the channel direction information (CDI) and the channel quality information (CQI) before transmission over the finite rate feedback link as already introduced in Sect. 7.4. It is assumed that each user has a perfect knowledge of CSI belonging to serving and interfering BSs and CQI is perfectly available at the BS and the users only fed back their CDI to serving BS associated to all these links. Therefore, before sending back the CDI of serving and interfering BSs, the user quantizes the direction of the channel vectors where the codebook is known by the users and the BSs.

According to the results derived in Chap. 4, it is possible to compute the average user rate in cell 1 under the quantized channel for 3-cell networks as follows:

$$R_1^{lfb}(B_{1,1}, B_{1,2}, B_{1,3}) = \mathbb{E}\left[\log_2 \left(1 + \frac{P_0(d_{1,1}/R)^{-\alpha} \gamma_{1,1} Z}{1 + P_0(d_{1,2}/R)^{-\alpha} \kappa_{1,2} Y + P_0(d_{1,3}/R)^{-\alpha} \kappa_{1,3} Y} \right) \right] \tag{7.38}$$

where $\gamma_{1,1}$ and $\kappa_{1,j}; j = 2, 3$ are the errors for a given number of quantization bits for serving and interfering BSs. These values are calculated by

$$\gamma_{1,1} = 1 - 2^{B_{1,1}} \beta\left(2^{B_{1,1}}, \frac{N_t}{N_t - 1}\right)$$

$$\kappa_{1,j} = 2^{B_{1,j}} \beta\left(2^{B_{1,j}}, \frac{N_t}{N_t - 1}\right); \quad j = 2, 3 \tag{7.39}$$

where $B_{1,1}$ and $B_{1,j}$ are the numbers of bits used to quantize the normalized channel vectors of $\mathbf{g}_{1,1}$ and $\mathbf{g}_{1,j}$ with $j = 2, 3$.

Since the quantization error is upper bounded by

$$\mathbb{E}[\sin^2(\theta_{u,v})] < 2^{-\frac{B_{u,v}}{N_t - 1}}; \quad u, v = 1, \dots, U \tag{7.40}$$

where θ is the angle between the exact and quantized channel direction.

Then, Eq. (7.39) is represented as

$$\gamma_{1,1} < 1 - 2^{-\frac{B_{1,1}}{N_t - 1}}$$

$$\kappa_{1,j} < 2^{-\frac{B_{1,j}}{N_t - 1}}; \quad j = 2, 3 \tag{7.41}$$

7.5.2 Bit Partitioning Strategies

Our objective is to select the bit partitioning among the serving and interfering BSs to maximize the sum multicell capacity under the constraint that the number of total feedback bits to quantize serving and interfering BSs is fixed for each user.

For a given position, $d_{u,u}$, the optimization problem for $U = 3$ is

$$\max_{B_{u,v}} R_T = \max_{B_{u,v}} \sum_{u=1}^{3} \mathbb{E}\left[\log_2\left(1 + \frac{P_0(d_{u,u}/R)^{-\alpha}\gamma_{u,u}Z}{1 + P_0 \sum_{v=1; v \neq u}^{3}(d_{u,v}/R)^{-\alpha}\kappa_{u,v}Y}\right)\right] \tag{7.42}$$

The constraints are

$$\sum_{v=1}^{3} B_{u,v} = B_{\max}, \quad u = 1, 2, 3.$$

$$B_{u,v} \geq 0; \quad u \neq v; u, v = 1, 2, 3.$$

$$B_{u,u} > 0; \quad u = 1, 2, 3. \tag{7.43}$$

In the multicell network, the cell area can be divided in two regions: (1) the non cooperative region (NCR) corresponding to the center of the cell and (2) The cooperative region (CR) corresponding to the edge of the cell. If the user, u, is in the

NCR, MRC beamforming is performed by the other BSs. For this user, the number of bits of interfering BSs, $B_{u,v}$, is selected as zero in this scenario. Otherwise, PZF beamforming is applied by 2 neighborhood cells or one of the neighborhood cells depending on the location of this user. In this case, the number of bits, $B_{u,v}$ is selected as higher than zero. Therefore, the selection of the number of quantization bits also corresponds to the selection of transmission strategies.

The separation of regions can be defined according to the power in the cell edge and the total number of quantization bits. For example, if the power in the cell edge is quite high depending on the transmitter power and the cell radius, only the CR region can be constructed for 3-cell networks. This will be demonstrated later.

Quantization of CSI leads to a loss in the sum rate. In [4], a feedback-bit partitioning strategy has been developed to reduce the mean loss in the sum rate given by

$$\mathbb{E}[\Delta R] \approx \mathbb{E}[R_{Full} - R_{Quan}] \tag{7.44}$$

where $R_{Full} = \sum_{u=1}^{U} R_u$ with R_u as given in Eq. (7.29) and $R_{Quan} = \sum_{u=1}^{U} \hat{R}_u$. The rate with quantized link $\hat{R}_u$ is calculated by replacing the quantized version of channel in Eq. (7.29).

For user i, depending on its location, an cancelation is performed by the other base station or not. In order to determine the required number of bits for the users in the cooperative region or not, we give to calculate the user rate at two extreme locations. The first location is the border between the non-cooperative and the cooperative region where the user does not require ICI cancelation. The second one is the cell edge where the user requires ICI cancelation. For a given normalized distance $d_{th} = d_{i,i}/R$ between the serving BS and the users, it is possible to improve the rate of the users within that region which is between d_{th} and 1.

Assuming MRC is performed by BS_i for user j, the rate at these two extreme positions is defined as $R_{i,bs}$ and $R_{i,cell}$, respectively:

$$R_{i,bs} = \mathbb{E}\left[\log_2 \left(1 + \frac{P_0 d_{th}^{-\alpha} \gamma'_{i,i} Z}{1 + P_0 (2 - d_{th})^{-\alpha} Y} \right) \right] \tag{7.45}$$

$$R_{i,cell} = \mathbb{E}\left[\log_2 \left(1 + \frac{P_0 \gamma_{i,i} Z}{1 + \kappa_{i,j} P_0 Y} \right) \right] \tag{7.46}$$

where $2Z \sim \chi^2_{2N_t}$ and $2Y \sim \chi^2_2$.

The criterion is based on the maximization of the sum rate at these two extreme locations to improve the rate of cell-edge users under the constraint that the average number of feedback bits within each cell is fixed.

The optimization problem for $U = 2$ is defined by

$$\max_{B'_{i,i}, B_{i,i}, B_{i,j}} R_w = R_{i,bs} + R_{i,cell} \tag{7.47}$$

The constraints

$$d_{th}B'_{i,i} + (1 - d_{th})(B_{i,i} + B_{i,j}) = B_{\text{avg}}$$

$$B'_{i,i} > 0$$

$$B_{i,i} > 0$$

$$B_{i,j} > 0 \qquad\qquad (7.48)$$

The first constraint which determines the average required feedback bits for 2-cell Wyner model satisfies that the average number of feedback bits within each cell is fixed. The other constraints imply that the CDIs of the user are always quantized either in the cooperative or in the non-cooperative region.

7.5.2.1 Extension to Wideband Channels

The bit partitioning schemes which have been presented for mitigation for single-carrier transmission schemes can be extent to the wideband channels such as OFDMA-based multicell systems. All these schemes can be applied to each sub-carrier or cluster individually and then power allocation methods can be employed as described in previous sections to further reduce ICI among BSs.

7.5.3 Performance Results

We illustrate the performance results to show the benefits of these strategies in a cooperative multicell network. For the simulations, the parameters are chosen as $N_t = 4$, $R = 1$ km and $\alpha = 3.7$.

7.5.3.1 Fixed Rate Limited Feedback Channel

The number of bits, $B_{u,v}$, determined by the proposed bit partitioning in 3-cell networks is shown considering two different occasions. In the first one, the user location changes only through the x-axis and the distance between the user and the interfering BSs is always equal. As it is shown in Fig. 7.16, the cooperative and non-cooperative regions are determined according to the power levels at cell edge and the total number of allocated bits. When both the power at cell edges and the number of bits are low, the NCR is quite large compared to CR. However, when the number of quantization bits is increased even the power is low, the non-cooperative region becomes absent. For the high power values at cell edge, there is no NCR anymore since the interference power is quite high and should be eliminated. In the second one, the user location changes both in the x- and the y-axis, and consequently

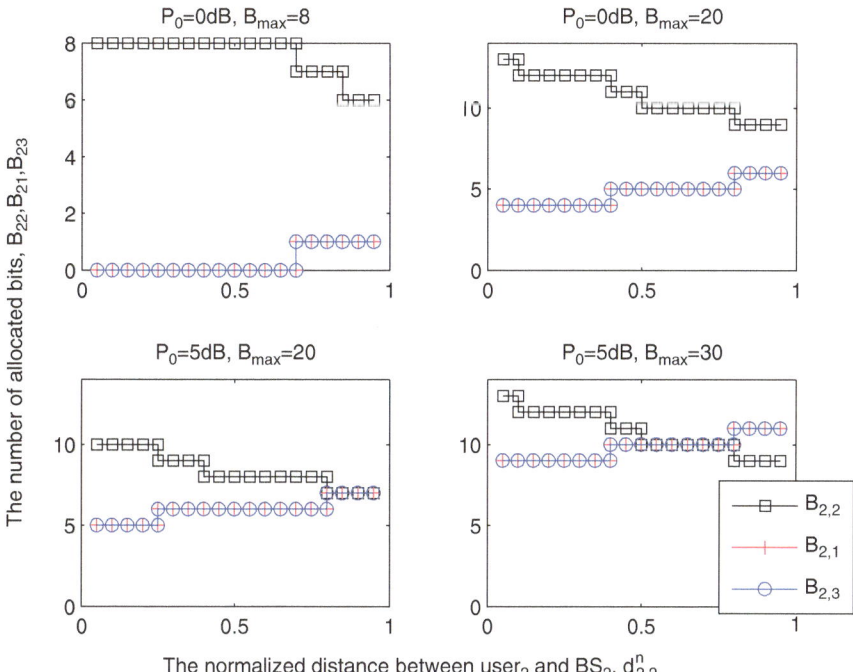

Fig. 7.16 The bit partitioning for different cell-edge power levels and different $B_{\max}$ values

the distance between the user and the interfering BSs is not the same. As shown in Figs. 7.17–7.20, the bit partitioning among the BSs changes depending on the user location in the cell, such as when the user is far from BS_3; the number of quantization bits to cancel interference caused by BS_3 is also decreased.

7.5.3.2   Adaptive Rate Limited Feedback Channel

We compare the performance results for the case where the number of feedback bits is fixed for each user. The equal sharing means that in the non-cooperative region, B_{avg} bits are used to quantize the CDI of the serving BS and in the cooperative region, $B_{avg}/2$ bits are used to quantize the CDIs of the serving and interfering BS. As it is shown in Fig. 7.21, the proposed strategy with adaptive bit partitioning increases the user rate significantly in the cell-edge region while the average feedback rate remains the same.

According to simulation results shown in Fig. 7.22, the bit partitioning scheme [54] gives better performance at the cell-edge region than restriction scheme.

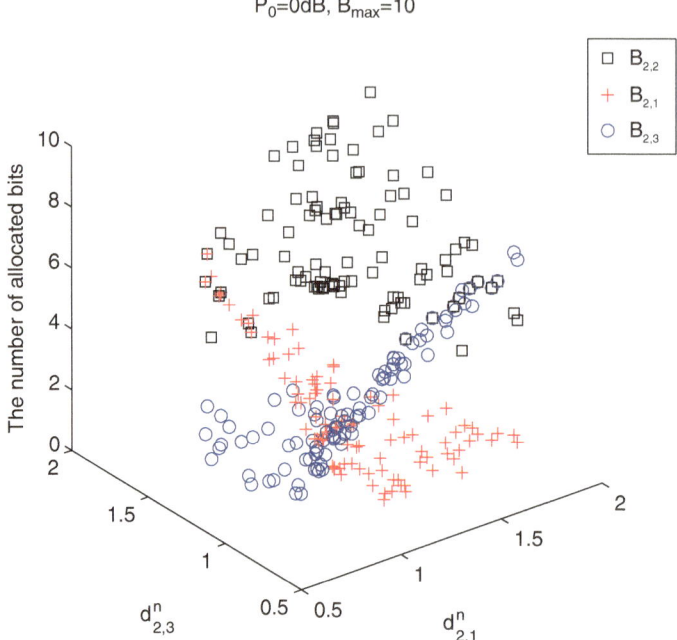

Fig. 7.17 The bit partitioning for user 2 at different positions at $P_0 = 0$dB and $B_{\max} = 10$

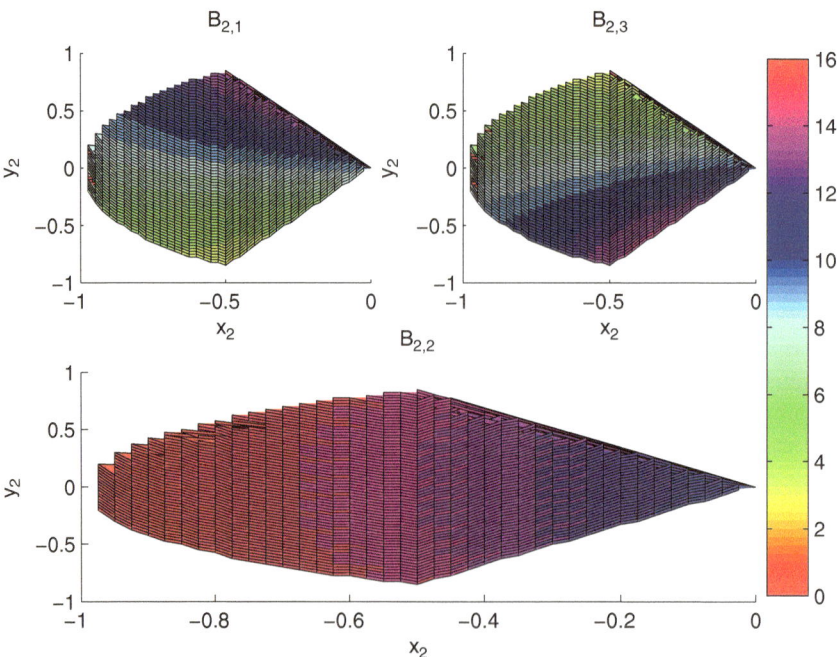

Fig. 7.18 The bit partitioning for user 2 at different positions at $P_0 = 0$dB and $B_{\max} = 30$

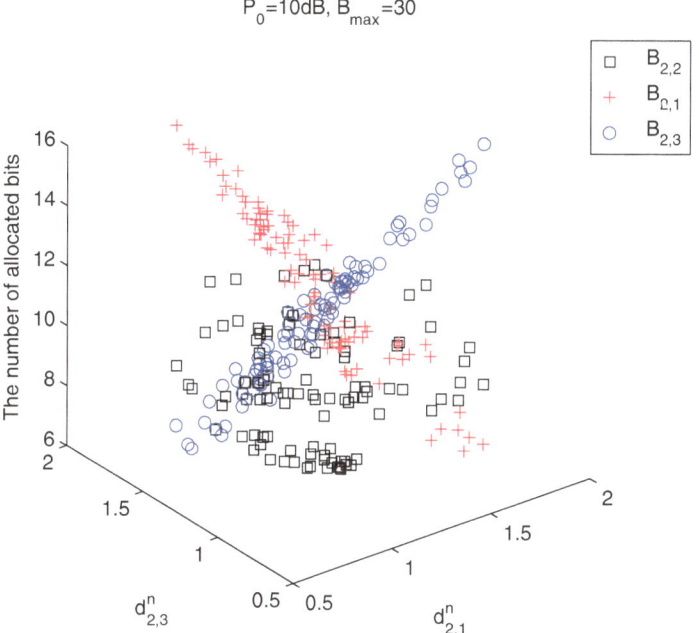

Fig. 7.19 The bit partitioning for user 2 at different positions at $P_0 = 0$dB and $B_{max} = 30$

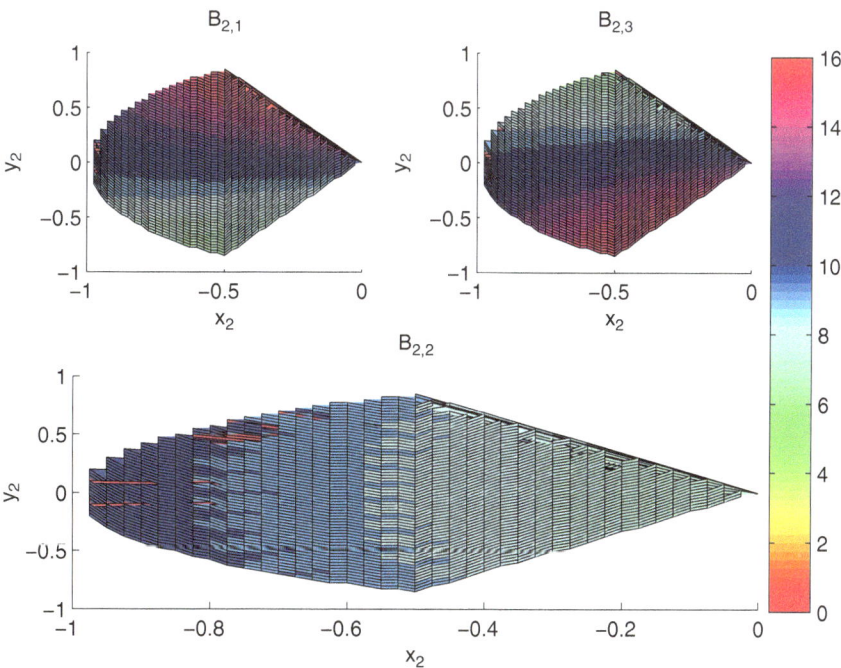

Fig. 7.20 The bit partitioning for user 2 at different positions at $P_0 = 10$dB and $B_{max} = 30$

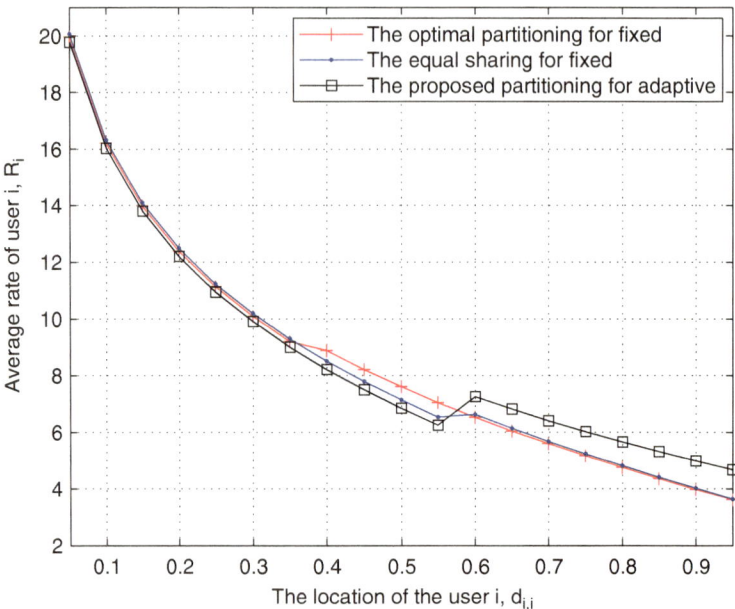

Fig. 7.21 The average rate of user i at different locations in the cell for $d_{th} = 0.6$ and $B_{avg} = 12$

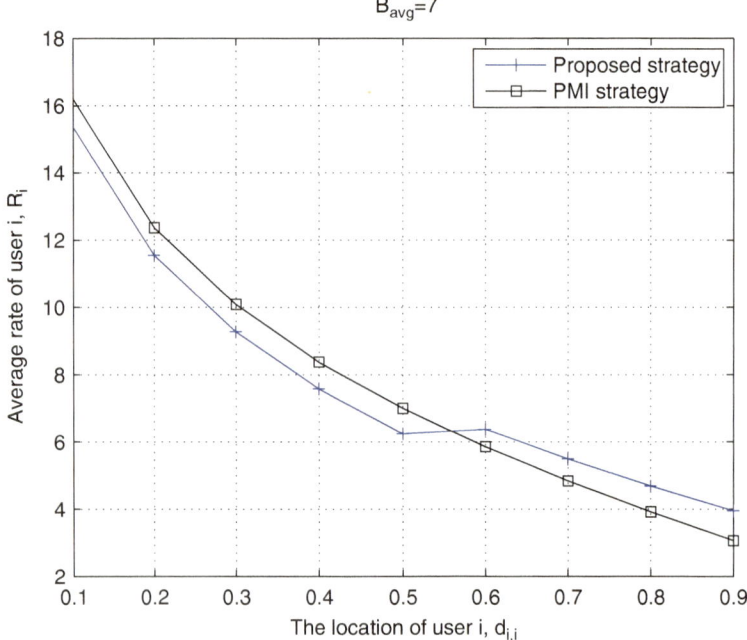

Fig. 7.22 The comparison of the average rate of user i at different locations in the cell

7.6 Interference Alignment for K-User Interference Channel

Interference channels are wireless networks consisting of multiple transmitter and receiver pairs communicating over the same radio resources and are a good model for communication in cellular networks, wireless local area networks and ad-hoc networks [48]. Recent works on the interference channel [6, 35] have shown that at high SNR sum rates can scale linearly with the number of users using a transmission strategy known as interference alignment. Interference alignment is a linear precoding technique that aligns interfering signals in time, frequency, or space. In MIMO networks, thanks to the spatial dimension offered by multiple antennas, the users coordinate their transmissions such that the interfering signals lie in a reduced dimensional subspace at each receiver.

Let's consider the K-user interference channel composed of K interfering transmitter and receiver pairs where transmitter i has a message for its intended receiver i. We assume that each transmitter has N_t antennas while each receiver is equipped with N_r antennas. Each transmitter employs a linear precoder to transmit d_k data streams and the received vector at the kth receiver can be described as follows:

$$\mathbf{y}_k = \mathbf{H}_{kk}\mathbf{F}_k\mathbf{s}_k + \sum_{j=1, j \neq k}^{K} \mathbf{H}_{kj}\mathbf{F}_j\mathbf{s}_j + \mathbf{n}_k \qquad (7.49)$$

where $\mathbf{s}_k$ is the data signal of the kth transmitter, $\mathbf{F}_k$ is the precoding matrix at the kth transmitter, $\mathbf{H}_{kj}$ is the channel matrix from transmitter j to receiver k, and $\mathbf{n}_k$ is the AWGN noise at the kth receiver. In this system model the global interference channel is assumed to be known by each transmitter.

Extending the definition of the DoF, the DoF region D of the K-user interference channel is defined as [25]

$$\mathscr{D} = \Big\{ (d_1, d_2, \ldots, d_K) \in R_+^K : \forall (w_1, w_2, \ldots, w_K) \in R_+^K,$$

$$w_1 d_1 + w_2 d_2 + \ldots + w_K d_K$$

$$\leq \lim_{SNR \to \infty} \sup \left[\sup_{R(SNR) \in C(SNR)} \frac{[w_1 R_1(SNR) + w_2 R_2(SNR) + \ldots + w_k R_k(SNR)]}{\log(SNR)} \right]$$

$$(7.50)$$

where $C(SNR)$ is the capacity region of the K-user interference channel, composed of the set of all achievable rate tuples $R(SNR) = (R_1(SNR), \ldots, R_K(SNR))$, i.e. the sets of rate tuples $R_1(SNR), \ldots, R_K(SNR)$ for which each transmitter-receiver pair is able to communicate reliability [25]. If the receiver is able to suppress all undesired interference, the kth transmitter-receiver pair will be able to achieve d_k DoF.

For the Kth user single input single output interference channel, it was demonstrated in [25] that the total number of DoF can be expressed as follows:

$$\mu = \max_{\mathbf{d} \in \mathscr{D}}(d_1 + d_2 + \ldots + d_K) = \frac{K}{2} \qquad (7.51)$$

We force each transmitter to use only half of its available signaling space and each receiver to partition its received space into two equally sized subspaces: one for the desired signal and one left for the interference. As a consequence the total number of DoF is $K/2$ irrespective of the number of interferers and the capacity of the SISO K-user interference channel scales linearly with the number of users.

DoF for the K-user interference channel with N_t antennas at each transmitter and N_r antennas at each receiver have been obtained in [19]. When the ratio $R = \lfloor \max(N_t, N_r)/ \min(N_t, N_r) \rfloor$ is an integer, the total number of DoF is given by

$$\mu = \begin{cases} \min(N_t, N_r)K & \text{if } K \leq R \\ \min(N_t, N_r)\frac{R}{R+1}K & \text{if } K > R \end{cases} \qquad (7.52)$$

There is no DoF penalty from interference when the number of users $K \leq R$, such as zero forcing suffices and everyone gets $\min(N_t, N_r)$ DoF. On the other hand, when $K > R$, the DoF per user is $\min(N_t, N_r)\frac{R}{R+1}$ because of interference alignment.

Assuming a feasible interference alignment setting, there exist precoding matrices $\mathbf{F}_k$, $k = 1, \ldots, K$ and projection matrices $\mathbf{Z}_k \in \mathbb{C}^{N_r \times d_k}$, $k = 1, \ldots, K$ such that

$$\mathbf{Z}_k^H \mathbf{H}_{kj} \mathbf{F}_j = 0 \qquad \forall k, j \in \{1, \ldots, K\}, k \neq j \qquad (7.53)$$

$$\text{rank}(\mathbf{Z}_k^H \mathbf{H}_{kk} \mathbf{F}_k) = d_k \qquad (7.54)$$

Having introduced interference alignment, an important question is how to align the signals in the matter previously described. Analytic solutions to interference alignment problem are in general difficult to obtain [6]. Gomadam et al. [18] have proposed an iterative scheme based on the minimization of the so-called interference leakage $\text{tr}(\mathbf{Z}_k^H \mathbf{H}_{kj} \mathbf{F}_j \mathbf{F}_j^H \mathbf{H}_{kj}^H \mathbf{Z}_k)$. The goal of this algorithm is to achieve interference alignment by iteratively reducing the leakage interference. This algorithm exploits the channel reciprocity assuming time-division duplex operations with synchronized time slot and only local channel knowledge and covariance matrix of its effective noise are required at each node. If interference alignment is feasible, then leakage interference will be zero. This algorithm makes no attempt to maximize the desired signal power within the desired signal subspace. Therefore, while the interference is eliminated within the desired space, no array gain for the desired signal is obtained with interference alignment.

In [38], the authors propose an algorithm based on an alternating minimization over the precoding matrices at the transmitters and the interference subspaces at the receivers. This algorithm is similar to the previous iterative algorithm but does not rely on channel reciprocity.

While this is optimal at high SNR, it is not optimal at low/moderate SNR values. Consequently, other algorithms have been designed which will perform better than the interference alignment algorithm at intermediate SNR values. In [18] the max SINR algorithm is introduced where the goal is to maximize SINR at the receivers instead of minimizing the leakage interference. An alternative approach to sum utility maximization based on weighted minimum mean squared error has been proposed in [47].

A centralized and iterative algorithm has been recently proposed to design optimal linear transmitters and receivers and select the users for weighted sum rate (WSR) maximization in [36].

Since channel reciprocity may not hold, one of the main difficulties to the practical use of interference alignment comes from the need to collect the CSI relative to the global multiuser channel at each transmitter. Consequently, the study of how CSIT requirements for interference alignment methods can somehow be alleviated has become an active research topic [2, 12, 13, 29, 45, 57]. In [57] the authors use Grassmannian codebooks to quantize and feedback the wideband channel coefficients in K-user SISO interference channel. The extension to K-user MIMO interference channel has been derived in [29]. Both references provide DoF-achieving quantization schemes and show that like for multiuser channels, the multiplexing gain is preserved by scaling the number of feedback bits with the SNR. Tresch and Guillaud [58] provide an analysis of the effect of non-perfect CSI on the mutual information of the interference alignment scheme. In [2], instead of using quantized feedback, Ayach and Heath have proposed to use interference alignment with analog feedback.

While most of the works assume node cooperation to obtain the precoding vectors such cooperation may not be possible and two practical strategies can be considered:

- Centralized processor: One source node only estimates all channels, calculates the interference alignment solution and feeds it forward to all the other sources.
- Distributed processing: Each source locally estimates all channels and calculates the complete set of precoders for interference alignment.

As a conclusion, while interference alignment for K-user interference channel is a promising technique with a large range of applications ranging from cellular networks to heterogeneous networks [49], further works on the practical issues need to be performed.

7.7 Conclusion

In this chapter, we have mainly focused on resource allocation techniques for cooperative multicell networks to mitigate ICI. The resource allocation including user scheduling and power allocation has been presented for cooperative multicell networks with and without a centralized unit. It is shown that the optimization problem is not convex and some iterative solution can be proposed with some relaxations. Joint and iterative user scheduling and power allocation algorithms have been examined and the performance analysis has been shown according to sum cell capacity. The limited feedback link for multiple antennas has been considered and the capacity analysis of quantized feedback has been presented in detail based on different criteria. In order to achieve sum cell capacity improvement on the multicell network, two classical beamforming strategies for the design of the beamforming vectors have been examined: MRC and partial ZF–BF. It has been shown that the required number of bits to quantize for transmission strategies can be chosen according only to the users' location and the degree of the received power at the edge of the cells.

References

1. Abrardo A, Alessio A, Detti P, Moretti M (2007) Centralized radio resource allocation for OFDMA cellular systems. In proc. of IEEE International Conference on Communications. 1: 5738–5743.
2. Ayach O E, Heath R W (2012) Interference alignment with analog channel state feedback. IEEE Trans. Wireless Commun. 11: 626–636.
3. Boccardi F, Huang H (2007) Limited downlink network coordination in cellular networks", In proc. of the IEEE Int. Symp. on Personal Indoor and Mobile Radio Comm. (PIMRC), Athens, Greece.
4. Bhagavatula R, Heath R W (2011) Adaptive limited feedback for sum-rate maximizing beamforming in cooperative multicell systems.IEEE Transactions on Signal Processing.59:800–811
5. Bhagavatula R, Heath R W (2011) Adaptive bit partitioning for multicell intercell interference nulling with delayed limited feedback.IEEE Transactions on Signal Processing.59:3824–3836.
6. Cadambe V R, Jafar S. A. (2008) Interference alignment and degrees of freedom of the K-user interference channel. IEEE Trans. Inf. Theory. 54: 3425–3441.
7. Chiang M, Tan C W, Palomar D P, Neill D O, Julian D (2007) Power control by geometric programming.IEEE Transactions on Wireless Communications. 6:2640–2651.
8. Clerckx B, Kim G, Choi J, Kim S J(2008) Allocation of feedback bits among users in broadcast MIMO channels.IEEE Global Telecommunications Conference, New Orleans, LO.
9. Chiang M(2005)Geometric programming for communication systems.Foundations and Trends in Communications and Information Theory.2: 1–2.
10. Charafeddine M, Paulraj A (2009) Maximum sum rates via analysis of 2-user interference channel achievable rate region. Proc. of Conf. Inf. Sci. Syst. 1: 170–174.
11. Chongxian Z, Chunguo L, Luxi Y, Zhenya H (2008)Dynamic resource allocation for the downlink of multi-cell systems with full spectral reuse.International Conference on Neural Networks and Signal Processing. 1: 173–177.

12. de Kerret P., Gesbert D. (2012) MIMO interference alignment algorithms with hierarchical CSIT. Proc. of International Symposium on Wireless Communication Systems (ISWCS). 1: 581–585.

13. de Kerret P., Gesbert D. (2013) Degrees of Freedom of Certain Interference Alignment Schemes with Distributed CSI. In Proc of SPAWC.

14. Gesbert D, Hanly S, Huang H, Shamai S, Simeone O (2010) Multi-cell MIMO Cooperative Networks: A new Look at Interference. IEEE Journal on Selected Areas in Communications. 28: 1380–1408.

15. Gesbert D, Kountouris M (2011) Rate scaling laws in multicell networks under distributed power control and user scheduling. IEEE Transactions on Information Theory. 57: 234–244.

16. Gjendemsjo A, Gesbert D, Oien G E, Kiani S G (2008) Binary power control for sum rate maximization over multiple interfering links.IEEE Transactions on Wireless Communications.7:3164–3173.

17. Goldsmith A, Chua S (1998). Adaptive coded modulation for fading channels. IEEE Trans. Commun., 46: 595–602.

18. Gomadam K., Cadambe V. R., and Jafar S. A. (2011) A distributed numerical approach to interference alignment and applications to wireless interference networks. IEEE Trans. Inf. Theo. 57: 3309–3322

19. Gou T., Jafar S. A. (2010) Degrees of Freedom of the K User $M \times N$ MIMO Interference Channel. IEEE Transactions on Information Theory. 56: 6040–6057

20. Huang J, Berry R, Honig M L (2006) Distributed interference compensation for wireless networks.IEEE Journal on Selected Areas in Communications.24:1074–1084.

21. Han Z, Ji Z, Liu K J R (2007) Non-cooperative resource competition game by virtual referee in multi-cell OFDMA networks.IEEE Journal on Selected Areas in Communications. 25:1079–1090.

22. Hammouda S, Tabbane S, Godlewski P (2011) Improved reuse partitioning and power control for downlink multi-cell OFDMA systems. In proc. of the International Workshop on Broadband Wireless Access for Ubiquitous Networking, New York, USA.

23. Jindal N, Andrews J G, Weber S (2009) Rethinking MIMO for wireless networks: Linear throughput increases with multiple receive antennas. In proc. of IEEE Intl. Conf. on Communications (ICC), Dresden, Germany.

24. Jindal N (2006) MIMO broadcast channels with finite rate feedback. IEEE Transactions on Information Theory, 52:.1455–1468

25. Jafar S. A., Shamai S. (2010) Degrees of Freedom Region of the MIMO X Channel. IEEE Transactions on Information Theory. 54: 151–170

26. Kiani S, Oien G, Gesbert D (2007) Maximizing multicell capacity using distributed power allocation and scheduling. In Proc. of Wireless Communications Networking Conference. 1: 1690–1694.

27. Kiani S, Gesbert D (2008) Optimal and distributed scheduling for multicell capacity maximization. IEEE Transactions on Wireless Communications. 7: 288–297.

28. Kim J, Kim D, Lee W, Ihm B C (2009) PMI restriction with adaptive feedback mode. IEEE C802.16m-09/0023.

29. Krishnamachari R. and Varanasi M. (2010) Interference alignment under limited feedback for MIMO interference channels. in Proc. IEEE International Symposium on Information Theory (ISIT)

30. Lengoumbi C, Godlewski P, Martins P (2006) Dynamic subcarrier reuse with rate guaranty in a downlink multicell OFDMA system. Proc. of IEEE 17th International Symposium on Personal, Indoor and Mobile Radio Communications. 1:1–5.

31. Liu L, Zhang J, Yu J C, Lee J (2010) Intercell interference coordination through limited feedback. International J. Digital Multimedia Broadcasting. doi:10.1155/2010/134919.

32. http://www.3gpp.org/article/lte.Cited2013

33. Liu H, Song Y, Li D, Cai L, Yang H, Lu D, Wu K (2011) Scalable limited channel feedback for downlink coordinated multi-cell transmission. Proc. of IEEE 73rd Vehicular Technology Conference (VTC Spring).1:1–5.

34. Lee N, Shin W (2011) Adaptive Feedback Scheme on K-cell MISO Interfering Broadcast Channel with Limited Feedback.IEEE Transactions on Wireless Communications. 10:401–406.
35. Maddah-Ali M., Motahari A., Khandani A. (2008) Communication over MIMO X channels: Interference alignment, decomposition, and performance analysis. IEEE Trans. Inf. Theory. 54: 3457–3470
36. Negro F., Shenoy S., Ghauri I., Slock D. (2010) Weighted sum rate maximization in the MIMO Interference Channel. Proc. IEEE 21st International Symposium on Personal Indoor and Mobile Radio Communications (PIMRC). 1: 684–689
37. Papadogiannis A, Bang H J, Gesbert D, Hardouin E (2011) Efficient Selective Feedback Design for Multicell Cooperative Networks. IEEE Transactions on Vehicular Technology. 60:196–205.
38. Peters S., Heath R. (2009) Interference alignment via alternating minimization. Proc. International Conference on Acoustics, Speech and Signal Processing. 1: 2445–2448
39. Pietrzyk S, Janssen G J M (2004) Radio resource allocation for cellular networks based on OFDMA with QoS guarantees. Proc. of IEEE Global Telecommunications Conference. 1: 2694–2699.
40. Pischella M, Belfiore J C (2008)Weighted sum throughput maximization in multi-cell OFDMA networks.IEEE Transactions on Vehicular Technology. 59:896–905.
41. Pischella M, Belfiore J C (2008)Distributed weighted sum throughput maximization in multi-cell wireless networks.IEEE 19th International Symposium on Personal, Indoor and Mobile Radio Communications.1:1–5.
42. Pischella M, Belfiore J C (2011) Distributed resource allocation for rate constrained users in multi-cell OFDMA networks.IEEE Communications Letters.12:250–252.
43. Papadogiannis A, Hardouin E, Gesbert D (2009) Decentralising multicell cooperative processing: A novel robust framework.EURASIP Journal on Wireless Communications and Networking. 28:1–10.
44. Rhee W, Cioffi J M (2000) Increase in capacity of multiuser OFDM system using dynamic subchannel allocation. Proc. in Vehicular Technology Conference. 2: 1085–1089.
45. Rezaee M., Guillaud M. (2013) Interference alignment with quantized Grassmannian feedback in the K-user MIMO interference channel. submitted to IEEE Trans. Inf. Theory. [Online]. Available: http://arxiv.org/abs/1207.6902
46. Su D, Hou X, Yang C (2011) Quantization Based on Per-cell Codebook in Cooperative Multi-cell Systems. Proc. of IEEE Wireless Communications and Networking Conference (WCNC).1:1753–1758
47. Schmidt D., Shi C., Berry R., Honig M. L., Utschick W. (2009) Minimum mean squared error interference alignment. Proc. IEEE Asilomar Conference on Signals, Systems and Computers (ACSSC).
48. Suh C., Tse D. (2008) Interference alignment for cellular networks. in 46th Annual Allerton Conference Communication, Control, and Computing. 1: 1037–1044
49. Suh C., Ho M., Tse D. (2011) Downlink Interference Alignment. IEEE Transactions on Communications. 59: 2616–2626
50. Shen Z, Andrews J, Evans B (2005) Adaptive resource allocation in multiuser OFDM systems with proportional rate constraints. IEEE Transactions on Wireless Communications. 4: 2726–2737.
51. Seong K, Mohseni M, Cioffi J M (2006) Optimal resource allocation for OFDMA downlink systems.Proc. of IEEE International Symposium on Information Theory.1:1394–1398.
52. Svedman P, Wilson S K, Cimini L J, Ottersten B (2007) Opportunistic Beamforming and Scheduling for OFDMA systems", IEEE Trans. on Communications, 55:941–952.

53. Ozbek B, Le Ruyet D (2010) Adaptive limited feedback for intercell interference cancelation in cooperative downlink multicell networks. In Proc. of 7th International Symposium on Wireless Communication Systems (ISWCS), York, UK.

54. Ozbek B, Le Ruyet D (2011) Adaptive bit partitioning strategy for cell-edge users in multi-antenna multicell networks. In Proc. IEEE 12th International Workshop on Signal Processing in Advanced Wireless Communications (SPAWC), San Francisco, USA.

55. Ozbek B, Le Ruyet D, Pischella M (2012) Reduced feedback links for power minimization in distributed multicell OFDMA networks. Proc. of IEEE International Conference on Communications (ICC), Ottawa, Canada.

56. Qiu X, Chawla K (1999) On the performance of adaptive modulation in cellular systems.IEEE Transactions on Communications. 47:884–895.

57. Thukral J. and Boelcskei H. (2009) Interference alignment with limited feedback. in Proc. IEEE International Symposium on Information Theory (ISIT)

58. Tresch R., Guillaud M. (2009) Cellular interference alignment with imperfect channel knowledge. Proc. IEEE International Conference on Communications (ICC).

59. Venturino L, Prasad N, Wang X (2008)A successive convex approximation algorithm for weighted sum-rate maximization in downlink OFDMA networks. Proc. of Conference on Information Sciences Systems. 1: 379–384.

60. Venturino L, Prasad N, Wang X (2009) Coordinated scheduling and power allocation in downlink multicell OFDMA networks.IEEE Transactions on Vehicular Technology. 58:2825–2834.

61. Wong C Y, Cheng R, Letaief K, Murch R (1999) Multiuser OFDM with adaptive subcarrier, bit, and power allocation. IEEE Journal on Selected Areas in Communications. 17: 1747–1758.

62. Wang T, Vandendorpe L (2011)Iterative resource allocation for maximizing weighted sum min-rate in downlink cellular OFDMA systems.IEEE Transactions on Signal Processing.59:223–234.

63. Yu W, LuiR (2006) Dual methods for nonconvex spectrum optimization of multicarrier systems.IEEE Transactions on Communications. 54:1310–1322.

64. Yoo T, Jindal N, Goldsmith A (2007) Multi-antenna broadcast channels with limited feedback and user selection.IEEE Journal Selected Areas in Communications.25: 1478–1491

65. Zakhour R, Gesbert D (2007) A two-stage approach to feedback design in MU-MIMO channels with limited channel state information. In Proc. of IEEE Personal, Indoor and Mobile Radio Communications (PIMRC), Athens, Greece.

66. Zhang Y J, Letaief K (2004) Multiuser adaptive subcarrier-and-bit allocation with adaptive cell selection for OFDM systems. IEEE Transactions on Wireless Communications. 3: 1566–1575.

67. Zhang J, Andrews J G(2010) Adaptive spatial intercell interference cancellation in multicell wireless network.IEEE Journal on Selected Areas in Communications. 28: 1455–1468.

Chapter 8
Feedback Strategies in LTE Systems

8.1 Introduction

Mobile communication technologies are divided into generations: long-term evolution (LTE) is often called 4G and starting from Release 10 it is labeled as LTE Advanced. The evolution on the standards continue with the usage of closed-loop transmission and MIMO techniques in order to increase the peak data rate and cell average or cell-edge spectral efficiency. During the last decade, MIMO techniques have been proposed for many wireless communication standards including third generation partnership project (3GPP) LTE and LTE-A, WIMAX 802.16e and 802.16m, and WIFI 802.11n [1–3, 5, 9, 12, 13]. In addition to that, the feedback information becomes more critical to perform closed-loop transmission for MIMO-based systems.

In this chapter, we will consider feedback information for LTE and LTE-Advanced-based systems. Section 8.2 will give an overview on LTE standard including frame structure, transmission modes and reference signals for the different releases. Sections 8.3 and 8.4 will introduce, respectively, link adaptation considering channel quality information (CQI) and channel measurement reporting modes including CQI, precoding matrix indicator (PMI) and RI. Sections 8.4 and 8.6 will give brief information about single user and multiuser MIMO strategies used in Releases 8,9, and 10. Section 8.7 will introduces multicell MIMO transmission including static and dynamic coordination techniques. Finally, the codebooks in downlink and uplink transmission will be examined in detail. Further works have been listed in Sect. 8.9.

8.2 Overview of LTE Standard

The 3GPP standard for LTE was one of the first wireless communication standards that have considered various MIMO technologies from the beginning. In this section, we will provide a short overview of LTE standard and discuss the different

B. Özbek and D. Le Ruyet, *Feedback Strategies for Wireless Communication*,   295
DOI 10.1007/978-1-4614-7741-9__8, © Springer Science+Business Media New York 2014

multiple antenna techniques and feedback mechanisms introduced in the different LTE releases. We will describe the main solutions introduced for both downlink and uplink transmissions.

The first release of LTE is termed LTE Release 8 and was published in 2008. In LTE Release 8, downlink transmission supports up to four antennas at the base station. It allows transmit diversity, codebook-based beamforming, and spatial multiplexing with up to four layers per user, and, moreover, codebook-based MU-MIMO. The codebook-based MU-MIMO scheme relies heavily on the accuracy of the rank indicator (RI)/CQI/PMI feedback and is optimized for SU-MIMO transmission. Consequently, the performance of MU-MIMO scheme are limited. For uplink transmissions, only one antenna is supported for transmission at the user side but is possible to perform antenna selection up to two transmit antennas. MU-MIMO up to 8 users is also supported in the uplink.

In LTE Release 9, a new transmission mode (TM8) has been added to support dual-layer transmission on each of the two virtual antenna ports with non-codebook-based MU-MIMO beamforming such as ZF beamforming. This transmission mode supports MU-MIMO beamforming transmission for up to 4 users rank 1 or up to 2 users rank 2. Accordingly, dedicated reference signal has been extended to two additional antenna ports.

LTE Release 10, the first release of LTE Advanced (LTE-A) that enhanced MIMO technologies, was finalized in June 2011. LTE-A has been designed to meet various fourth generation requirements of the International Telecommunications Union (ITU-R) including peak data rate and average and cell-edge spectrum efficiencies. The ITU-targeting peak data rate in downlink is 100 Mbps in high-mobility applications and 1 Gbps for low-mobility applications such as nomadic/local wireless access. Configurations with up to 8×8 MIMO are supported. In the downlink, new reference signals have been introduced to support both demodulation of the DM-RS and channel state information estimation (CSI-RS) reference signals. As a matter of fact, user-specific RS allows the estimation of an equivalent channel, including the precoding weights between base station and users. Compared to the previous release, the impact of multiuser interference on the performance is significantly reduced. New codebooks and feedback designs are introduced to support spatial multiplexing with up to eight independent spatial streams and enhanced MU-MIMO transmissions. Furthermore, a new transmission mode (TM9) has been defined which allows dynamic switching between SU-MIMO and MU-MIMO to enhance scheduling flexibility with respect to the channel conditions and traffic loads without the need for the UEs to be reconfigured via higher-layer signaling. In the uplink, in order to achieve the LTE Advanced requirements (peak spectrum efficiency of 15 b/s/Hz and average spectrum efficiency of 2.0 b/s/Hz/cell), SU-MIMO has been introduced with up to four transmit antennas at the user side using precoded spatial multiplexing as well as transmit diversity techniques for the UL control channel.

The standardization for Release 11 is focusing on CSI feedback and control signalling given the importance of the MU-MIMO operation to meet the increasing user demands. A particular attention is given to enhance the MU-MIMO for heterogeneous networks with small-cell deployment. Advanced multicell MIMO technique such as coordinated multipoint (CoMP) will be also supported in LTE Release 11.

8.2.1 Frame Structure and Transmission Modes

In this section, we will review the main characteristics of frame structure and transmission modes of both downlink and uplink LTE system. For a more detailed description of the LTE standard, please refer to the books of Dahlman, Parkvall and Skoldand [2] or Ghosh and Ratasuk [5, 6].

In LTE standard, the base station is referred to evolved Node B (eNB) while the user is called user equipment (UE). The multiple-access technique of LTE system in downlink is OFDMA and in uplink is SC-FDMA. The LTE radio frame of duration 10 ms is composed of 10 subframes. A subframe of duration 1 ms consists of two consecutive slots. This duration corresponds to one transmission time interval (TTI). One slot corresponds to 6 or 7 OFDM symbols depending if the cyclic prefix is normal (duration 4.7 μs) or extended (duration 16.6 μs). The slot is divided in resource block (RB) defined as 12 consecutive subcarriers in the frequency domain with 15 KHz spacing. The number of RB per slot, N_{RB}, is 6, 15, 25, 50, 75, and 100 for an overall channel bandwidth of 1.4, 3, 5, 10, 15, and 20 MHz, respectively. In LTE system, a resource element (RE) is one subcarrier in a single OFDM or SC-FDMA symbol. Consequently, a RB is composed of 72 or 84 RE.

The modulation schemes supported by the LTE standard are QPSK, 16QAM, and 64QAM. The channel coding scheme at the physical layer is parallel concatenated convolutional coding (PCCC) or turbo coding, with a coding rate of R=1/3 composed of two 8-state recursive convolutional encoders (RSC) and a quadratic permutation polynomial (QPP) interleaver. In order to adapt the coding rate, a circular-buffer rate matching scheme is used. Before the turbo coding, transport blocks are segmented into information block of fixed length. Filler bits are added at the beginning of the first information block. Depending on the size of the transport block, the length of the information block ranges from 40 bits to 6,144 bits. Error detection is performed by adding a 24-bit cyclic redundancy check (CRC) at the end of each information block.

For MIMO transmission, LTE has opted for multiple codewords (MCW) on the downlink. In MCW, there is one codeword for each individual stream that can use different modulations and coding schemes. MCW allows for efficient inter-stream interference cancellation using MMSE-SIC receiver. In closed-loop SU-MIMO, each codeword requires one CQI report and one HARQ process.

The physical channels defined in the downlink are:

- Physical Downlink Shared Channel (PDSCH)
- Physical Multicast Channel (PMCH)
- Physical Downlink Control Channel (PDCCH)

Table 8.1 The transmission modes for LTE up to Release 10

Transmission mode	Description
1	Single transmit antenna, port 0
2	Transmit diversity
3	Open-loop spatial multiplexing with cyclic delay diversity
4	Closed-loop spatial multiplexing
5	Multiuser MIMO
6	Closed-loop spatial multiplexing using a single transmission layer
7	Beamforming with single antenna port, port 5
8	Dual-layer beamforming, ports 7 and 8 (Release 9)
9	Multiuser MIMO (Release 10)

Table 8.2 Maximum number of supported streams for LTE up to Release 10

MIMO mode	Release 8	Release 9	Release 10
DL-SU-MIMO	4	4	8
DL-MU-MIMO	2	4	4
UP-SU-MIMO	1	1	4
UP-MU-MIMO	8	8	8

- Physical Broadcast Channel (PBCH)
- Physical Control Format Indicator Channel (PCFICH)
- Physical Hybrid ARQ Indicator Channel (PHICH)

The physical channels defined in the uplink are:

- Physical Random Access Channel (PRACH)
- Physical Uplink Shared Channel (PUSCH)
- Physical Uplink Control Channel (PUCCH)

In addition, signals are defined as reference signals (RS), primary synchronization signals (PSS), and secondary synchronization signals (SSS).

In Table 8.1, we list the different transmission modes for release up to 10.

In Table 8.2, we summarize the maximum number of supported streams for Releases 8, 9, and 10 in both uplink and downlink SU-MIMO and MU-MIMO.

8.2.2 Reference Signals

Downlink

A reference signal (RS), or pilot, allows measurements of the spatial channel properties and/or permits coherent demodulation at the UE. Two different forms

Fig. 8.1 Mapping of CRS
(antenna ports 1–4) and DRS
(antenna port 5) for normal
cyclic prefix

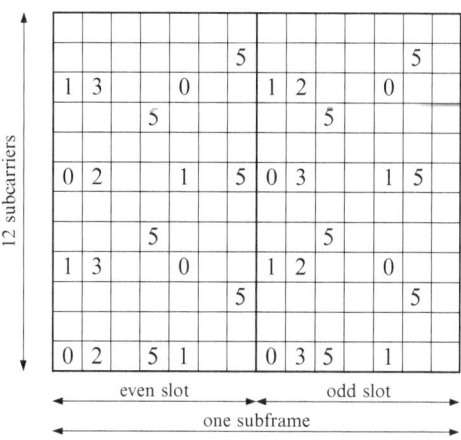

of RS were initially defined. The first form is the dedicated or demodulated RS (DRS), which is UE specific and used for demodulation. The second form of RS is the common RS (CRS), which is shared among a group of UEs and can be used for CSI measurements or for demodulation. RS can be precoded or non-precoded. DRSs are commonly precoded and are transmitted via virtual antenna ports with a spatial precoding weight to exploit beamforming gain. The overhead of DRS is proportional to the number of transmitted streams. CRSs are not precoded signals and are transmitted via physical antenna without a spatial precoder to allow for channel estimation of the MIMO channel of different UEs. The overhead of CRS is proportional to the number of physical antennas.

In LTE Release 8, non-precoded CRSs supporting up to four antenna ports have been defined for channel measurements and coherent demodulation purposes. In order to limit the signaling overhead, the CRSs for antenna ports 3 and 4 are twice as less as CRSs for antenna ports 1 and 2. With the use of non-precoded CRS, the spatial precoder used at the transmitter should be indicated to a terminal. LTE Release 8 also supports single-layer beamforming using rank-1 precoded DRS using antenna port 5. The CRS mapping for four antenna ports and of DRS mapping for port 5 is shown in Fig. 8.1.

In LTE Release 9, UE-specific demodulation reference signal (DM-RS) has been defined for the two additional antenna ports (port 7 and 8). For rank 2, code division multiplexing (CDM) is used to multiplex the DM-RS patterns. This new dual-layer transmission mode is designed for beamforming schemes and supports MU-MIMO transmission for up to 4 UEs rank 1 or up to 2 UEs rank 2. Since the antenna port and scrambling code allocations are wideband, it is not possible to ensure orthogonality when two users are multiplexed in MU-MIMO mode.

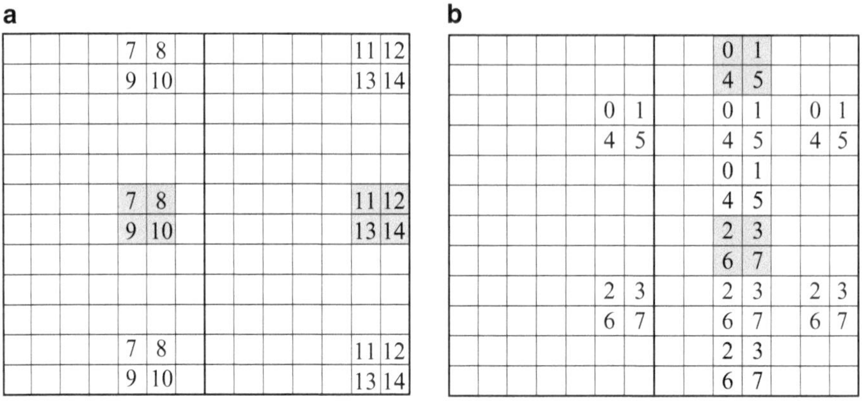

a

		7	8					11	12		
		9	10					13	14		
		7	8					11	12		
		9	10					13	14		
		7	8					11	12		
		9	10					13	14		

b

									0	1			
									4	5			
						0	1		0	1		0	1
						4	5		4	5		4	5
									0	1			
									4	5			
									2	3			
									6	7			
						2	3		2	3		2	3
						6	7		6	7		6	7
									2	3			
									6	7			

Fig. 8.2 (**a**) Mapping of DM-RS (antenna ports 7–14). (**b**) Example of 5 CSI-RS patterns (antenna ports 0–7) for normal cyclic prefix

In LTE Release 10, in order to achieve the peak data rate, average and cell-edge spectrum efficiency required by ITU-R, eight antenna port transmissions supporting a high-dimensional SU-MIMO operation (up to 8×8 MIMO) and improved MU-MIMO operation have been added. In order to support these new features, new RSs have been introduced in the downlink: the demodulation reference signal (DM-RS), user-specific, is used for demodulation, while the channel state information reference signals (CSI-RS) shared by all users in the cell are used for channel estimation measurements and determining user CSI feedback.

The DM-RS, introduced in Release 9, is extended in Release 10 to support up to rank 8 transmissions in the new transmission mode. The DM-RS patterns are optimized for each rank to minimize the pilot overhead. Rank 2 DM-RS patterns are multiplexed by CDM as in Release 9. The DM-RS patterns for ranks 3 to 8 are natural extensions. For ranks 3 and 4, The DM-RS pattern for the 3rd and 4th layer and the 1st and 2nd layers are multiplexed by frequency division multiplexing (FDM). For ranks 5 to 8, hybrid CDM/FDM DM-RS with 2 CDM groups is adopted.

The CSI-RS is only used for CSI measurements in the new transmission modes. Compared to CRS that is transmitted at every subframe, CSI-RS is transmitted in a fraction of subframes to reduce overhead. CSI-RS has also been designed for CoMP where coordination allows the cells to listen CSI-RS of the neighboring cells. While DM-RS is defined with respect to transmission layer, CSI-RS is defined with respect to transmit antennas. An example of DM-RS mapping for eight antenna ports and of CSI-RS mapping is shown in Fig. 8.2.

Uplink

In Release 8, two reference signals are defined for demodulation and channel sounding by the eNB: demodulation reference signals (DM-RS) and sounding reference signal (SRS). DM-RS is transmitted by each user to enable uplink channel estimation at the base station for the purpose of demodulating the uplink transmission and SRS is used by the eNB to measure the uplink channel quality of the transmitting user for enabling scheduling and link adaptation. Consequently, SRS is a wideband reference signal transmitted across different frequency resource blocks using distributed FDMA. The periodicity of SRS transmissions is highly adjustable from 2 to 320 ms and its overhead is significantly lower than the one of DM-RS. Since the uplink access is SC-FDMA, both DM-RS and SRS are time-multiplexed to guaranty the low peak-to-average power ratio (PAPR) or cubic metric (CM) of the SC-FDMA.

Both DM-RS and SRS are built from Zadoff–Chu sequences that have the property that cyclic shifted versions of the same sequence are orthogonal to each other.

In Release 10, DM-RS and SRS are extended to enable uplink SU-MIMO. Since codebook-based precoding are supported for uplink MIMO data transmission, DM-RS should also be precoded in the same manner as the corresponding data transmission to provide the precoding gain for channel estimation. To support multilayer transmission, one DM-RS resource is allocated for each of the scheduled transmission layers in order to allow the eNB to estimate the uplink channels associated with the different transmission layers and perform coherent demodulation. DM-RS is defined in the frequency domain by the cell-specific base sequence and consequently, the multiplexing scheme for DM-RS is time-domain cyclic shift (CS) separation. Furthermore, orthogonal cover code (OCC) separation is used to separate the DM-RS associated to each transmission layer. On the other hand, since SRS are used for rank and precoding adaptation, they are not precoded and are defined with respect to the UE transmit antennas.

8.3 Link Adaptation in LTE

For resource scheduling and link adaptation in LTE, the UEs are required to feed back their CQI to the eNB. Depending on the selected mode, this reported information can be wideband, relative to a subband, or relative to a group of subband. Depending on the total bandwidth, the size of the subband ranges from 4 to 8 RBs. For all these modes, the users will use effective mapping SINR introduced in Chap. 3 to generate the CQI values from the SNRs of the RBs that constitute the subbands. The 16 values of CQI defined in Release 8 are given in Table 8.3.

Table 8.3 Channel quality indicator (CQI) table

CQI index	Modulation	Coding rate × 1,024	Efficiency b/s/Hz
0	Out of range		
1	QPSK	78	0.1523
2	QPSK	120	0.2344
3	QPSK	193	0.3770
4	QPSK	308	0.6016
5	QPSK	449	0.8770
6	QPSK	602	1.1758
7	16QAM	378	1.4766
8	16QAM	490	1.9141
9	16QAM	616	2.4063
10	64QAM	466	2.7305
11	64QAM	567	3.3223
12	64QAM	666	3.9023
13	64QAM	772	4.5234
14	64QAM	873	5.1152
15	64QAM	948	5.5547

Each CQI is mapped to a modulation and coding scheme. The UE will send back to the eNB the index corresponding to the highest data rate. For this index, the estimated received downlink transport block BLER shall not exceed a given threshold (e.g. 10%).

In spatial multiplexing, the UE also reports the number of simultaneous streams which can be received. This number is called the rank indicator (RI). The RI is always evaluated of the whole system band, that is, the same channel rank is assumed on all the subbands.

For MIMO closed-loop transmission modes, the UE will also provide the PMI about the preferred precoding matrix in codebook-based precoding. The PMI can be evaluated over the total band or per subband. The CQI information is always calculated assuming that the selected PMI is applied.

The LTE system employs HARQ type II also called incremental redundancy based HARQ as described in Chap. 3. Regarding timing and adaptivity properties of HARQ, in LTE downlink, asynchronous adaptive HARQ is used while synchronous nonadaptive hybrid ARQ is performed in uplink. Thanks to asynchronous adaptive HARQ in downlink, the BS can shift the retransmission in time or frequency to avoid collision with system information. For downlink spatial multiplexing, the link adaptation and HARQ are performed independently for each codeword. The HARQ round-trip time is fixed to eight subframes or 8 ms taking into account a 3 ms processing time for both BS and UE [2].

8.4 Reporting Modes

Uplink

As seen in the previous section, in LTE standard, different reporting modes are defined to transmit the channel measurement reports including CQI, PMI and RI from the UE to the eNB. These information will be used by the eNB for the precoding, scheduling and resource allocation. Two of the uplink physical channels are used for the reporting, namely, the physical uplink shared channel PUSCH and the stand-alone uplink physical channel PUCCH. PUCCH has been optimized for a large number of UEs that transmit CQI, PMI, and RI with a relatively small number of control signalling bits per UE.

Two different reporting classes are considered in the standard:

- Aperiodic reporting done on the PUSCH
- Periodic reporting done on the PUCCH

The different aperiodic reporting modes done on the PUSCH have been defined as shown in Table 8.4.

Mode 1-2: This reporting mode is used for closed-loop transmission modes. In this mode, a wideband CQI of size 4 bit is reported. Furthermore, for each subband a preferred PMI is reported. Depending on the total bandwidth, the size of the subband ranges from 4 to 8 RB. The total number of bits for the reporting of N PMI is $2N$ or $4N$ when two or four antenna ports are, respectively, present where N is the number of subbands.

Mode 2-0: This mode is used for open-loop transmission modes (TM 1, TM 2 and TM 3). In this mode, the UE determines and reports the positions of the M best selected subbands, one wideband CQI value and one differentially encoded (with respect to the wideband value) CQI value reflecting transmission over the M selected subbands. As said previously, this CQI is computed using effective mapping SINR such as EESM or MI-SM. The position of the M selected subbands requires $L = \lceil \log_2 \binom{N}{M} \rceil$ bits.

Mode 2-2: This reporting mode is used for closed-loop transmission modes. The UE performs the selection of the M preferred subbands and reports the positions. The UE also reports the preferred PMI, one wideband CQI and one CQI value selected over the M subbands.

Table 8.4 Aperiodic reporting modes on PUSCH for LTE Release 8

Aperiodic reporting mode	CQI	PMI
1-2	Wideband	Multiple
2-0	UE-selected subband	None
2-2	UE-selected subband	Multiple
3-0	Higher-layer configured subband	None
3-1	Higher-layer configured subband	Single

Table 8.5 Periodic reporting modes on PUCCH for LTE Release 8

Periodic reporting mode	CQI	PMI
1-0	Wideband	None
1-1	Wideband	Wideband
2-0	UE-selected subband	None
2-1	UE-selected subband	Wideband

Mode 3-0: This mode is used for open-loop transmission modes. Compared to UE-selected subband, with higher-layer configured subband, the CQI of all subbands is reported. Again the CQI is computed using effective mapping SINR. One wideband CQI value using 4 bits and N differentially encoded (with respect to the wideband value) CQI values are reported. Since 2 bits are used per differentially encoded CQI, the total number of feedback bits is $4 + 2N$.

Mode 3-1: This reporting mode is used for closed-loop transmission modes. In addition to mode 3.0, a single PMI is reported. The subbands CQI are calculated assuming the use of the PMI in all subbands.

Three different channel coding schemes are used to protect the signaling bits on PUSCH: simplex code (3,2) for the 2-bit ACK/NACK/RI, Reed–Muller(32, K) code for the K CQI/PMI bits if $K < 11$ bits, and tail-biting convolutional code of rate 1/3 for the K CQI/PMI bits if $K \geq 11$ bits.

The different periodic reporting modes done on the PUCCH are shown in Table 8.5. Compared to PUSCH, on PUCCH, the RI, the PMI/wideband CQI and the UE-selected subband CQI are reported sequentially in different subframes.

Mode 1-0: This mode is used for open-loop transmission modes. A wideband CQI of size 4 bits is reported. In TM 3, the RI of size 1 or 2 bits is also reported.

Mode 1-1: This mode is used for closed-loop transmission modes. RI is reported in one subframe and PMI/wideband CQI is reported in the second subframe. 4 CQI bits and 4 PMI bits are reported when the rank is 1. If the rank is higher than 1, 3 additional bits for differential CQI are reported. It represents the difference between the wideband CQI of the first and the second codeword.

Mode 2-0: This mode is used for open-loop transmission modes. RI is reported in one subframe. When wideband reporting is done, the UE reports one wideband CQI value using 4 bits. When UE-selected subband reporting is done, the UE reports sequentially the CQI value associated to the selected subband among the N subbands of each of the J bandwidth parts (BP). Depending on the total bandwidth, J ranges between 1 and 4. This report consists of $L+4$ bits where $L = \lceil \log_2 N \rceil$.

Mode 2-1: This mode is used for closed-loop transmission modes. RI is reported in one subframe. When wideband reporting is done, the UE reports one wideband PMI value and one wideband CQI value for the first codeword calculated by taking into account the wideband PMI. A 3 bits differential CQI is added to report the wideband CQI of the second codeword if the rank is higher than 1.

When UE-selected subband reporting is done, the UE reports sequentially the CQI value associated to the selected subband among the N subbands. The UE also reports the position of the selected subband using $L = \lceil \log_2 N \rceil$ bits.

The channel code for periodic reporting on PUCCH is a Reed–Muller $(20, K)$ code. The choice of aperiodic versus periodic reporting depends mainly on the downlink data traffic and there is a trade-off between the accuracy and overhead.

The UE feedback definition has been also extended in LTE Release 10 to take into account the use of the dual codebook structure for the 8 Tx antennas since the feedback information the PMI corresponding to both matrix $\mathbf{W}_1$ and $\mathbf{W}_2$ has to be included. Consequently, the LTE Release 10 has added three periodic reporting modes on PUCCH [16]:

Mode 1-1 submode 2: RI reported in one subframe, $\mathbf{W}_1$, wideband $\mathbf{W}_2$ and wideband CQI reported in a second subframe

Mode 1-1 submode 1: RI and $\mathbf{W}_1$ reported in one subframe, wideband $\mathbf{W}_2$ and wideband CQI reported in a second subframe

Mode 2-1: RI reported in one subframe, $\mathbf{W}_1$, wideband $\mathbf{W}_2$ and wideband CQI reported in a second subframe, subband $\mathbf{W}_2$ and subband CQI associated to the selected subband reported in a third subframe

It should be noted that when only 2- or 4-transmit antenna are configured/used at the eNB, the feedback includes only the PMI for $\mathbf{W}_2$, and the $\mathbf{W}_1$ is the identity matrix.

Downlink

The PDCCH is the control channel that carries downlink or uplink resource assignments and power control commands defined by the downlink control information (DCI). The PDCCH is mapped in the first OFDM symbols of a subframe. The number of symbols that are used to carry the PDCCH is given in an additional PCFICH (1, 2, 3, or 4 OFDM symbols are possible). The information carried on PDCCH is referred to as DCI. The different DCI format and their purposes are given in Table 8.6.

Release 10 supports uplink transmission with up to 4 layers, 2 per codeword with codebook-based precoding. The UE precoder is defined using rank indicator and precoding matrix, which are selected by the eNB based on uplink channel measurements and conveyed to UE through the RI and PMI fields using DCI format 4.

Table 8.6 DCI format

DCI format	Release	Purpose
0	R8	Transmission of resource grants for the PUSCH
1	R8	Transmission of resource assignments for one PDSCH codeword for single antenna operation
1A	R8	Compact signalling of resource assignments for one PDSCH codeword for single antenna operation
1B	R8	Compact signalling of resource assignments for one PDSCH codeword with closed-loop single-rank transmission
1C	R8	Very compact transmission of PDSCH codeword for downlink transmission of paging
1D	R8	Compact signalling of resource assignments for one PDSCH codeword with precoding and power offset information
2	R8	Transmission of resource assignments for PDSCH in closed-loop spatial multiplexing mode
2A	R8	Transmission of resource assignments for PDSCH in open-loop spatial multiplexing mode
2B	R9	Transmission of resource assignments for PDSCH in dual-layer beamforming
2C	R10	Transmission of resource assignments for PDSCH in closed-loop single-user or multi-user MIMO
3	R8	Power control commands for PUCCH and PUSCH with 2 bits power adjustments
3A	R8	Power control commands for PUCCH and PUSCH with a single bit power adjustments
4	R10	Transmission of resource grants for the PUSCH for uplink single-user MIMO

8.5 Single User MIMO Transmission

We have seen in Chap. 4 that when the channel matrix is full rank, the capacity gain of SU-MIMO systems is scaled by $\min(N_t, N_r)$ at high SNR. Since in typical downlink cellular systems, the number of receive antennas at the UE is usually smaller than the number of transmit antennas at the eNB, the gain of SU-MIMO systems is limited by the number of receive antennas. When the SU-MIMO channel matrix is not full rank due to a strong line of sight path or correlations among antenna elements, the capacity of SU-MIMO decreases.

Release 8 allows the SU-MIMO closed-loop spatial multiplexing mode. Thanks to a codebook-based precoder exploiting the PMI feedback information, in Release 8, the eNB can transmit up to 4 layers in 4×4 SU-MIMO. New reference signals have been added in Release 10 to support a high-dimensional SU-MIMO operation up to 8×8 MIMO as we have seen previously.

8.6 Multiuser MIMO Transmission

Under the assumption that the transmitter has perfect CSIT, we have seen in Chap. 6 that the sum capacity of MU-MIMO is scaled by $\min(N_t, N_r K)$ where K is the number of users multiplexed into the MU-MIMO transmission, so that N_t-fold increase in the sum rate if $N_r K > N_t$. However, in practice, the eNB has only an unperfect CSIT knowledge and inter-user interference can significantly limit the performance of MU-MIMO.

In Release 8, it is possible to perform a rather minimal MU-MIMO transmission mode that is a direct extension of the SU-MIMO closed-loop spatial multiplexing mode. The MU-MIMO in Release 8 LTE belongs to the class of codebook-based precoding. Indeed, in Release 8 MU-MIMO mode, the UE uses the non-precoded common RS that is broadcasted from the eNB for channel estimation and data demodulation. Each UE first estimates its non-precoded MIMO channel, then selects the best rank 1 PMI and then feeds back the PMI and CQI index via the uplink channel to the eNB. Since the UE cannot estimate the SINR, the CQI is computed by the UE assuming that there is no inter-user interference. The UE also estimates the precoded MIMO channel by multiplying the estimated channel with the selected PMI. The eNB collects the PMI and CQI reports from the different UEs and performs user grouping such that the paired UEs experience minimum level of inter-user interference among themselves. The precoder is built from orthogonal PMIs.

Release 8 MU-MIMO has been defined for highly correlated channels. Since those channels are characterized by non-frequency-selective eigenbeam directions, a single wideband PMI reporting per UE is sufficient to report the PMI.

Codebook-based precoding is clearly a suboptimal precoding strategy for MU-MIMO since we have shown in Chap. 6 that all the efficient precoding techniques such as DPC or ZF are non-codebook based where the eNB chooses a precoding matrix outside the predefined codebook.

Non-codebook-based precoding is possible starting from Release 9, thanks to the two new reference signals DM-RS and CSI-RS described previously. Indeed, the UE can now separately determine the MIMO channel and precoded MIMO channel. The UE derives the PMI and CQI values from the estimated MIMO channel using CSI-RS and can estimate the precoded MIMO channel using DM-RS. Consequently, the eNB is no more constrained to choose a precoding matrix inside a predefined codebook [3, 10].

In Release 10, considering the trade-off between performance, scheduling complexity and signaling overhead, the following decisions on the dimensioning of MU-MIMO systems have been made: no more than 4 UEs are co-scheduled, no more than 2 layers are allocated per UE and no more than 4 layers are transmitted in total.

With the use of DM-RS, the eNB can dynamically switch per-subframe basis between the SU-MIMO and MU-MIMO modes to get the maximal benefit of the two modes.

8.7   Multi-cell MIMO Transmission

Today's deployed LTE networks are mostly based on homogeneous macrocells since all the BSs belong to the same type and power class. As mentioned In Release 8 of LTE, the MIMO transmission in each cell is controlled independently of that of its neighbors. ICIC is supported by means of coordination message exchanged between base stations. The goal of ICIC is to provide the scheduler at one cell with information about the current or prospective interference situation at its neighbors.

For LTE-A systems, CoMP transmission and reception has been proposed as one of the key technologies to enhance cell average and cell-edge throughput [8]. CoMP refers to a system where the transmission and/or reception at multiple, geographically separated antenna sites is dynamically coordinated in order to improve system performance.

The key technical issues in LTE/LTE-A is the efficient technology that combines OFDMA with MIMO in downlink and uplink radio transmissions. Other important aspects are the new deployment strategies using heterogenous networks which include macro-, femto- and relay BS to increase spectral efficiency especially at cell edges. The achievement of LTE-A with Rel-10 outperforms the targets of IMT-Advanced in terms of spectral and energy efficiency and the cost reduction. However, this drastic coverage and capacity achievements bring some challenging issues such as intercell interference (ICI) caused by neighboring BSs. When the user approaches its cell edge, it suffers from the high ICI from neighboring cells. In order to eliminate ICI, interference management must be performed by performing coordination/cooperation. Coordination techniques in LTE/LTE-A can be mainly grouped in two class: The first group includes static coordination techniques including fractional frequency reuse (FFR) and soft frequency reuse (SFR) which are supported since LTE Rel-8. The second group includes enhanced intercell interference coordination (eICIC) including network MIMO, distributed MIMO and multicell MIMO which requires information exchange between the cells. These schemes are conceptually the same by providing improvements of spectral efficiency and are unified into CoMP transmission and reception on LTE-A.

While macrocells constitute the basis of a mobile network's coverage, the deployment of low-power nodes (e.g., pico- or femtocells, or relay nodes) within the macrocell is foreseen as a widely used solution in the future to cope with the ever-increasing mobile traffic demands. Such networks, called heterogeneous networks, are characterized by harsh intercell interference between the macro- and the low-power nodes, due to their closer proximity and different power classes.

8.7.1   Static Coordination Techniques

FFR in LTE is a mechanism that targets the ICI mitigation by fractionally assigning frequency subbands to cell sites through static frequency planning. The basic idea

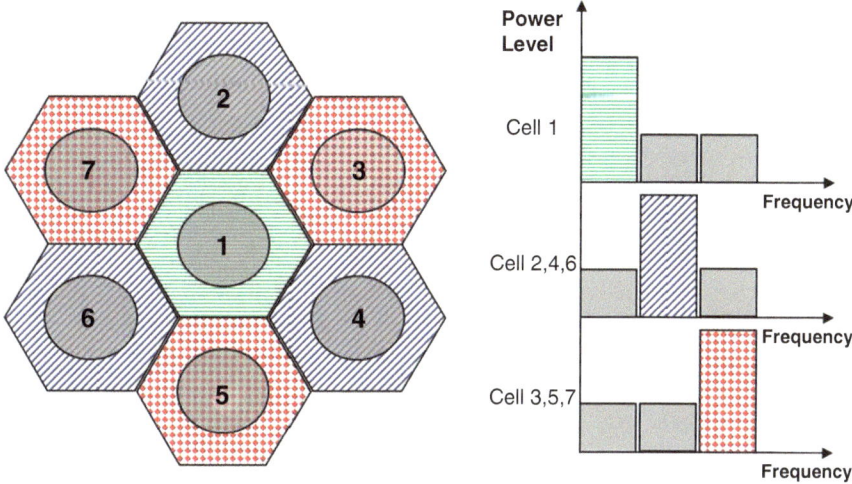

Fig. 8.3 Soft frequency reuse

of FFR is to create partitions between cell-edge and cell-center users based on SINR. Users in the cell-center area with higher signal quality and lower level of interference can use lower reuse factor to achieve better spectral efficiency while the cell-edge users with lower SINR utilize higher reuse factor to alleviate the ICI.

In the case of SFR cell-center users can utilize all resources with low-power transmission while cell-edge users can utilize only $1/F_r$ of the spectrum with frequency reuse of F_r with high-power transmission as illustrated in Fig. 8.3. SFR was first introduced for GSM and then adapted by 3GPP as an ICIC method for LTE [18, 19]. In LTE Rel-10, the time division multiplexing (TDM) was exploited to mitigate interference from macrocell to picocells in heterogenous network. The macrocell transmits an almost blank subframe or multimedia broadcast over single frequency network subframe with a certain periodicity, but picocells transmit normal subframes all the time. The macrocell is not allowed to schedule any UE in these subframes. The UE connected to picocells in the expanded region can be scheduled during transmission of these subframes from the macrocells. When UE is closed to picocells, it may be scheduled in any subframe.

Static coordination techniques developed in LTE Rel-8/9/10 can provide benefits in managing ICI with relatively little backhaul overhead and low implementation complexity. However, the achievable benefit is limited compared to dynamic coordination techniques which provide further performance improvement.

In heterogeneous networks, UEs associated with picocell may experience strong interference from the macrocell, especially when they are located in the range expansion region to balance the traffic load between the macro- and the low-power node cell. The ICI effect is even more detrimental when the user is located in this region. Therefore, the UE may be associated with a cell that does not provide the strongest received signal power. In addition to that the concept of resource partitioning and almost blank subframes is an effective way of mitigating this interference.

8.7.2 Dynamic Coordination Techniques

Coordinated multipoint transmission and reception actually refers to a wide range of techniques that enable dynamic coordination or transmission and reception with multiple geographically separated BS. Its aims are to enhance the overall system performance, utilize the resources more effectively, and improve the end-user service quality. One of the key parameters for LTE as a whole, and in particular LTE-A, is the high data rates that are achievable. These data rates are relatively easy to maintain close to the base station, but as distances increase they become more difficult to maintain. Obviously the cell edges are the most challenging. Not only is the signal lower in strength because of the distance from the BS, but also interference levels from neighboring eNBs are likely to be higher as the UE will be closer to them. LTE-A CoMP, CoMP requires close coordination between a number of geographically separated BSs. They dynamically coordinate to provide joint scheduling and transmissions as well as providing joint processing of the received signals. In this way, a UE at the edge of a cell is able to be served by two or more BSs to improve signal reception/transmission and increase throughput particularly under cell-edge conditions.

For downlink CoMP techniques, specification efforts mainly focus on transmission schemes, CSI reporting, interference management, reference signal design, and control signaling [20]. CSI reporting facilitates link adaptation in the network while interference measurement and reference signal are essential in deriving a CSI report. In the framework of 3GPP downlink CoMP transmission is categorized into three different types of techniques depending on the required constraints on the backhaul link:

- Joint transmission (JT)
- Dynamic cell selection (DCS) including dynamic point selection (DPS) and dynamic point blanking (DPB)
- Coordinated scheduling/beamforming (CS/CB)

8.7.2.1 Joint Transmission

JT can be seen as an advanced downlink CoMP technique to achieve spectral efficiency requirement for LTE-A. In JT, a CoMP set consists of a number of cell sites that coordinate to optimize the cell-edge performance by jointly processing cell-edge users data as a unique entity [21]. In basic case of JT-CoMP which can be implemented as closed loop and open loop, it is assumed that UE's data can be available in all BSs of the CoMP set.

The closed-loop JT can be performed as follows:

- The CSI feedback from UE consists of per-subband PMI, CQI and wideband RI.
- Multicell transmission is performed between cells by providing a unique pre-coder for all transmission points of the CoMP set.
- The cells jointly process the PMI to resource allocation based on the preferred UEs' PMIs.
- The data transmitted in downlink are coherently combined at the receiver.

The open-loop JT can be performed as follows:

- The CSI feedback from UE consists of per sun-band CQI and wideband RI.
- Single-cell precoding is used by each site and only CQI information is exchanged between the cells.
- The transmitted data in downlink are not coherently combined at the receiver.

The preferable implementation in downlink of LTE-A is the closed-loop MIMO.

8.7.2.2 Dynamic Cell Selection

According to the instantaneous channel condition on per-subframe basis, the BS is dynamically selected for DPS. One common way is to select the BS with the best link quality to exploit channel variations opportunistically (e.g. selection diversity gain is achieved). Therefore, the users located at cell edges have the opportunity to reselect serving cell through the best measured SINRs or the least path loss. Then, the cell-edge UEs select the best link for the next frame. For this frame period other cooperating cells are muted for the resources this UE uses. In order to obtain further gain, DPS can be combined with DPB in which the dominant interferers are identified and muted dynamically (no signal is transmitted from the identified BS). By muting the dominant interferer, SINR of the UE may be enhanced significantly and a few interferers are taken into account to calculate interference power. A BS is muted only when there is gain on the whole coordination area.

8.7.2.3 Coordinated Scheduling/Beamforming

In coordinated scheduling/beamforming (CS/CB) coordination strategy, it is not necessary to share the UE's data across multiple eNBs. In order to achieve the coordination, the UE needs to feed back information about the CSIs of the serving cell and the other cells in the CoMP set.

UE-side: A CSI reporting scheme that allows the network to have an estimate of the channel quality experienced by a UE as a function of the transmission beam employed by one, suitably selected, strong interfering cell.

Network-side: By assuming an ideal backhaul among the nodes forming a multicell cluster, namely, one macro-BS along with its interfering BSs, one single centralized scheduling algorithm can jointly perform an optimal scheduling for all BSs in the clusters. Then the centralized unit then informs all the nodes about the decision.

The idea of CB emerged in the mid-1990s, mainly targeting a so-called SINR leveling problem, in which the power levels and the beamforming coefficients are calculated to achieve some common SINRs in the system or to maximize the minimum SINR.

In CS/CB, the scheduling and beamforming across different BSs are aligned to reduce interference. In ideal, the scheduling and beam selection should be performed jointly to optimize the performance. However, it is impractical to make joint decisions on UE and beams on different BSs. A simple and practical solution is iterative scheduling where each cell revisits the choice of UE and underlying transmit beams based on scheduling decision not only accounts.

There are numerous proposals for LTE/LTE-A that face the problem of intercell beam collision and can be performed by using coordinated beam (CB) switching and coordinated scheduling (CS)-based CoMP. Different approaches jointly combining CS and CB have been studied in LTE-Advanced, which can be classified by increasing order of complexity and requirements in terms of CSI feedback and CSI sharing. Coordinated beam selection is a low feedback overhead approach targeting highly loaded cells that consists of coordinating the precoders (beamforming matrices) in the cooperating transmission points (TP)s in a predefined manner in order to reduce interference variation and enable accurate link adaptation.

In CB-CoMP, each cell determines a sequence of beams over a serving cell in a distributed way. In CS-CoMP, multiple BSs collaborate to mitigate ICI using multicell PMI between cooperating BSs. Multicell PMI coordination is different from CBS-CoMP in the sense that the coordination can be done using preferred UE's PMIs. This is accomplished by recommending a set of PMIs to the other cells of the CoMP set rather than restricting the use of beams to resources.

There are two different strategies of PMI coordination based on how feedback PMI sets are selected by BSs [22]:

- Best PMI reporting improves cell-edge performance due to ICI reduction by selecting a set of least-interfering PMIs to the BS.
- Worst PMI reporting allows a cell to mitigate ICI by restricting strongly interfering PMIs.

Best PMI set reporting causes degradation in neighbor cell average throughput more severely than worst PMI set reporting.

In [23], further modifications have been done to cope ICI by combining PMI coordination and ICIC mechanism allowing CoMP to be compatible with the previous interference coordination schemes such as FFR.

The mechanism for PMI coordination in a simplified case of 2 BSs and 2 cell-edge users is in the following:

- The UE sends the feedback information based on the collaborating BSs' reference signals. This feedback consists of the reference PMIs for the interfering BSs, worst or best PMI set. As a result of this feedback, the cell-edge UE is fed back its preferred PMI indices for each frequency subband to its serving BS. With this information, the UE informs its serving BS about its preferred or restricted precoding index that corresponds to neighboring cells.
- The BSs exchange signaling information that indicates the PMI restrictions. As a result, the neighboring cell is obtained to either use the preferred precoder or not to use the restricted precoder through this multicell coordination.
- Each BS decides the precoding vector for its own UE and transmits data to its own UE in the downlink.

PMI recommendation is more effective than PMI restriction in suppressing the interference. In order to avoid excessive feedback overhead, the PMI information can be limited to one or two strongly interfering cells.

8.8 Codebooks in the LTE Standard

In Chap. 4, we have discussed the design of codebook for quantized feedback. In practice, the design is a trade-off between performance, complexity and feedback load.

8.8.1 Codebooks for Two Transmit Antennas

In Release 8, two codebooks are defined for the SU-MIMO precoding in downlink with two antennas as shown in Table 8.7.

Table 8.7 Codebook for 2 transmit antennas

Index	Rank 1	Rank 2
0	$\frac{1}{\sqrt{2}}\begin{bmatrix} 1 \\ 1 \end{bmatrix}$	$\frac{1}{2}\begin{bmatrix} 1 & 0 \\ 0 & 1 \end{bmatrix}$
1	$\frac{1}{\sqrt{2}}\begin{bmatrix} 1 \\ -1 \end{bmatrix}$	$\frac{1}{2}\begin{bmatrix} 1 & 1 \\ 1 & -1 \end{bmatrix}$
2	$\frac{1}{\sqrt{2}}\begin{bmatrix} 1 \\ j \end{bmatrix}$	$\frac{1}{2}\begin{bmatrix} 1 & 1 \\ j & -j \end{bmatrix}$
3	$\frac{1}{\sqrt{2}}\begin{bmatrix} 1 \\ -j \end{bmatrix}$	

Table 8.8 Set of generating vectors

s_0	1	-1	-1	-1	s_8	1	-1	1	1
s_1	1	$-j$	1	j	s_9	1	$-j$	-1	$-j$
s_2	1	1	-1	1	s_{10}	1	1	1	-1
s_3	1	j	1	$-j$	s_{11}	1	j	-1	j
s_4	1	$\frac{(-1-j)}{\sqrt{(2)}}$	$-j$	$\frac{(1-j)}{\sqrt{(2)}}$	s_{12}	1	-1	-1	1
s_5	1	$\frac{(1-j)}{\sqrt{(2)}}$	j	$\frac{(-1-j)}{\sqrt{(2)}}$	s_{13}	1	-1	1	-1
s_6	1	$\frac{(1+j)}{\sqrt{(2)}}$	$-j$	$\frac{(-1+j)}{\sqrt{(2)}}$	s_{14}	1	1	-1	-1
s_7	1	$\frac{(-1+j)}{\sqrt{(2)}}$	j	$\frac{(1+j)}{\sqrt{(2)}}$	s_{15}	1	1	1	1

- A beamforming codebook with six vectors including two vectors corresponding to antenna selection
- A linear precoding codebook with three matrices

8.8.2 Codebooks for Four Transmit Antennas

The codebook design for the SU-MIMO precoding for 4 transmit antennas in downlink has been designed based on the following performance and complexity criteria:

- Low computational complexity: the Householder-based codebook design has been selected since it offers competitive performance compared to the DFT-based design and low complexity.
- Elements of each precoding matrix or vector selected from a small set such as QPSK $\pm 1, \pm j$ and 8PSK alphabet sets in order to limit the searching complexity.
- Constant modulus property for avoiding unnecessary increase in PAPR.
- Quantization accuracy using a metric such as the maximization of the minimum chordal distance.
- Nested property across ranks to accommodate rank override. For each precoder matrix of rank 2, 3, or 4, there should exist at least one corresponding column subset in all the codebooks of the lower ranks.

Using these criteria a four-bit codebook is specified for beamforming and linear precoding with 1, 2, 3, and 4 streams.

Let $H(\mathbf{u})$ be the $4 \times N$ submatrix of the 4×4 Householder matrix given by

$$H(\mathbf{u}) = \mathbf{I}_4 - \frac{2}{||\mathbf{u}||^2} \mathbf{u}\mathbf{u}^H \tag{8.1}$$

The generating vectors and the precoding matrix of the codebook defined in Release 8 for 4 transmit antennas are, respectively, given in Tables 8.8 and 8.9.

Table 8.9 Set of precoding matrices of the codebook for 4 transmit antennas

Index	Rank 1	Rank 2	Rank 3	Rank 4
0	$H(s_0,1)$	$\dfrac{H(s_0,[14])}{\sqrt{2}}$	$\dfrac{H(s_0,[124])}{\sqrt{3}}$	$\dfrac{H(s_0,[1234])}{2}$
1	$H(s_1,1)$	$\dfrac{H(s_1,[12])}{\sqrt{2}}$	$\dfrac{H(s_1,[123])}{\sqrt{3}}$	$\dfrac{H(s_1,[1234])}{2}$
2	$H(s_2,1)$	$\dfrac{H(s_2,[12])}{\sqrt{2}}$	$\dfrac{H(s_2,[123])}{\sqrt{3}}$	$\dfrac{H(s_2,[3214])}{2}$
3	$H(s_3,1)$	$\dfrac{H(s_3,[12])}{\sqrt{2}}$	$\dfrac{H(s_3,[123])}{\sqrt{3}}$	$\dfrac{H(s_3,[3214])}{2}$
4	$H(s_4,1)$	$\dfrac{H(s_4,[14])}{\sqrt{2}}$	$\dfrac{H(s_4,[124])}{\sqrt{3}}$	$\dfrac{H(s_4,[1234])}{2}$
5	$H(s_5,1)$	$\dfrac{H(s_5,[14])}{\sqrt{2}}$	$\dfrac{H(s_5,[124])}{\sqrt{3}}$	$\dfrac{H(s_5,[1234])}{2}$
6	$H(s_6,1)$	$\dfrac{H(s_6,[13])}{\sqrt{2}}$	$\dfrac{H(s_6,[134])}{\sqrt{3}}$	$\dfrac{H(s_6,[1324])}{2}$
7	$H(s_7,1)$	$\dfrac{H(s_7,[13])}{\sqrt{2}}$	$\dfrac{H(s_7,[134])}{\sqrt{3}}$	$\dfrac{H(s_7,[1324])}{2}$
8	$H(s_8,1)$	$\dfrac{H(s_8,[12])}{\sqrt{2}}$	$\dfrac{H(s_8,[124])}{\sqrt{3}}$	$\dfrac{H(s_8,[1234])}{2}$
9	$H(s_9,1)$	$\dfrac{H(s_9,[14])}{\sqrt{2}}$	$\dfrac{H(s_9,[134])}{\sqrt{3}}$	$\dfrac{H(s_9,[1234])}{2}$
10	$H(s_{10},1)$	$\dfrac{H(s_{10},[13])}{\sqrt{2}}$	$\dfrac{H(s_{10},[123])}{\sqrt{3}}$	$\dfrac{H(s_{10},[1324])}{2}$
11	$H(s_{11},1)$	$\dfrac{H(s_{11},[13])}{\sqrt{2}}$	$\dfrac{H(s_{11},[134])}{\sqrt{3}}$	$\dfrac{H(s_{11},[1324])}{2}$
12	$H(s_{12},1)$	$\dfrac{H(s_{12},[12])}{\sqrt{2}}$	$\dfrac{H(s_{12},[123])}{\sqrt{3}}$	$\dfrac{H(s_{12},[1234])}{2}$
13	$H(s_{13},1)$	$\dfrac{H(s_{13},[13])}{\sqrt{2}}$	$\dfrac{H(s_{13},[123])}{\sqrt{3}}$	$\dfrac{H(s_{13},[1324])}{2}$
14	$H(s_{14},1)$	$\dfrac{H(s_{14},[13])}{\sqrt{2}}$	$\dfrac{H(s_{14},[123])}{\sqrt{3}}$	$\dfrac{H(s_{14},[3214])}{2}$
15	$H(s_{15},1)$	$\dfrac{H(s_{15},[12])}{\sqrt{2}}$	$\dfrac{H(s_{15},[123])}{\sqrt{3}}$	$\dfrac{H(s_{15},[1234])}{2}$

8.8.3 Dual Codebook for Eight Transmit Antennas

For configurations with 8 transmit antennas in downlink, a dual-codebook approach is introduced in Release 10. Since an antenna array composed of 8 transmit antennas means in practice closely spaced antennas, two antenna setups have been considered in Release 10:

- An uniformly linear array of eight closely spaced $((\lambda/2))$ uni-polarized antenna elements
- An uniformly linear array of four closely spaced $((\lambda/2))$ dual-polarized antenna elements

For these two antenna setups, LTE Release 10 defines a new feedback framework based on the so-called dual codebook. This dual codebook has been designed in order to cope with both high azimuth angular spread channels (suitable for SU-MIMO) and low azimuth angular spread channels (suitable for MU-MIMO). Consequently, the dual codebook defined in Release 10 is a double 4-bit codebook obtained via the multiplication of two precoding matrices $\mathbf{W} = \mathbf{W}_1 \times \mathbf{W}_2$, where $\mathbf{W}_1$ is targeting long-term and wideband channel information and $\mathbf{W}_2$ is targeting short-term and subband channel information [14, 15]. The ideas behind the dual codebook are related to the codebook structure for spatial correlated channel introduced in

Chap. 4 [11]. The dual codebook achieves a good trade-off between performance and overhead in both antenna scenarios. The matrix $\mathbf{W}_1$ has a block diagonal structure matching the spatial covariance matrix of dual-polarized antennas:

$$\mathbf{W}_1 = \begin{bmatrix} \mathbf{X}_k & \mathbf{0} \\ \mathbf{0} & \mathbf{X}_k \end{bmatrix} \tag{8.2}$$

In case of dual-polarized antennas, the two matrices $\mathbf{X}_k$ perform beamforming in each polarization independently.

In spatially correlated channels with $\lambda/2$ spaced antenna arrays and low azimuth spread, the spatial covariance matrix can be well approximated using its eigenvectors. In that case we have seen in Chap. 4 that the maximum array gain can be obtained by using a grid of beams obtained from oversampled DFT vectors. On the other hand, with higher azimuth spread, the channels are less spatially correlated and the beamforming vector should include more than one DFT vectors. As a consequence, for rank 1 to 4 dual codebook, the matrices $\mathbf{X}_k$ are $4 \times N_b$ matrices consisting of N_b adjacent column vectors extracted from the oversampled DFT matrix $\mathbf{V}$ of size $4 \times N$ given as

$$\mathbf{V} = \begin{bmatrix} 1 & 1 & 1 & \cdots & 1 \\ 1 & e^{j2\pi/N} & e^{j4\pi/N} & \cdots & e^{j2(N-1)\pi/N} \\ 1 & e^{j4\pi/N} & e^{j8\pi/N} & \cdots & e^{j3(N-1)\pi/N} \\ 1 & e^{j6\pi/N} & e^{j12\pi/N} & \cdots & e^{j4(N-1)\pi/N} \end{bmatrix}$$

$$= \begin{bmatrix} \mathbf{v}_0 & \mathbf{v}_1 & \cdots & \mathbf{v}_{N-1} \end{bmatrix} \tag{8.3}$$

The oversampling factor is $N/4$.

For rank 1 and 2, we have $N = 32$ and $N_b = 4$. The 16 different matrices $\mathbf{X}_k$ are defined as follows:

$$\mathbf{X}_k \in \left\{ \begin{bmatrix} \mathbf{v}_{(2k)\text{mod}32} & \mathbf{v}_{(2k+1)\text{mod}32} & \mathbf{v}_{(2k+2)\text{mod}32} & \mathbf{v}_{(2k+3)\text{mod}32} \end{bmatrix} \right\} \quad 0 \le k \le 15 \tag{8.4}$$

For rank 3 and 4, $N = 16$ and $N_b = 8$ are chosen. The four different matrices $\mathbf{X}_k$ are defined as follows:

$$\mathbf{X}_k \in \left\{ \begin{bmatrix} \mathbf{v}_{(4k)\text{mod}16} & \mathbf{v}_{(4k+1)\text{mod}16} & \cdots & \mathbf{v}_{(4k+7)\text{mod}16} \end{bmatrix} \right\} \quad 0 \le k \le 3 \tag{8.5}$$

For rank 5 to 7, the four different matrices $\mathbf{X}_k$ are obtained by rotation from the 4DFT matrix as follows:

$$\mathbf{X}_0 = \frac{1}{2} \begin{bmatrix} 1 & 1 & 1 & 1 \\ 1 & j & -1 & -j \\ 1 & -1 & 1 & -1 \\ 1 & -j & -1 & j \end{bmatrix} \tag{8.6}$$

$$\mathbf{X}_1 = \mathrm{diag}\{1, e^{j\pi/4}, j, e^{j3\pi/4}\}\mathbf{X}_0$$

$$\mathbf{X}_2 = \mathrm{diag}\{1, e^{j\pi/8}, e^{j2\pi/8}, e^{j3\pi/8}\}\mathbf{X}_0$$

$$\mathbf{X}_3 = \mathrm{diag}\{1, e^{j3\pi/8}, e^{j6\pi/8}, e^{j9\pi/8}\}\mathbf{X}_0$$

For rank 8, only the matrix $\mathbf{X}_0$ is allowed and consequently $\mathbf{W}_1$ is fixed.

$\mathbf{W}_2$ is the matrix targeting short-term and computed per subband channel. The structure provides good performance in both high and low spatial correlation channels.

For rank 1, the 16 matrices $\mathbf{W}_2$ of size $2N_b \times 1$ with $N_b = 4$ are of the form

$$\mathbf{W}_2 = \frac{1}{\sqrt{2}}\begin{bmatrix} \mathbf{e}_k \\ \alpha\mathbf{e}_k \end{bmatrix} \tag{8.7}$$

where $\mathbf{e}_k$ is a $N_b \times 1$ beam selection vector that has 1 on the kth row and zeros elsewhere and α is the QPSK co-phasing factor such that $\alpha \in \{1, -1, j, -j\}$.

For rank 2, the 16 matrices $\mathbf{W}_2$ of size $2N_b \times 2$ with $N_b = 4$ are of the form

$$\mathbf{W}_2 = \frac{1}{\sqrt{2}}\begin{bmatrix} \mathbf{e}_k & \mathbf{e}_l \\ \alpha\mathbf{e}_k & \beta\mathbf{e}_l \end{bmatrix} \tag{8.8}$$

where $(k, l) \in \{(1, 1), (2, 2), (3, 3), (4, 4), (1, 2), (2, 3), (1, 4), (2, 4)\}$ and $(\alpha, \beta) \in \{(1, -1), (j, -j)\}$.

For rank 3, the 16 matrices $\mathbf{W}_2$ of size $2N_b \times 3$ with $N_b = 8$ are of the form

$$\mathbf{W}_2 = \frac{1}{\sqrt{2}}\begin{bmatrix} \mathbf{Y}_1 & \mathbf{Y}_2 \\ \mathbf{Y}_1 & -\mathbf{Y}_2 \end{bmatrix} \tag{8.9}$$

where

$$\begin{aligned}
(\mathbf{Y}_1, \mathbf{Y}_2) \in \{&(\mathbf{e}_1, [\mathbf{e}_1 \quad \mathbf{e}_5]), (\mathbf{e}_2, [\mathbf{e}_2 \quad \mathbf{e}_6]), (\mathbf{e}_3, [\mathbf{e}_3 \quad \mathbf{e}_7]), (\mathbf{e}_4, [\mathbf{e}_4 \quad \mathbf{e}_8]), \\
&(\mathbf{e}_5, [\mathbf{e}_1 \quad \mathbf{e}_5]), (\mathbf{e}_6, [\mathbf{e}_2 \quad \mathbf{e}_6]), (\mathbf{e}_7, [\mathbf{e}_3 \quad \mathbf{e}_7]), (\mathbf{e}_8, [\mathbf{e}_4 \quad \mathbf{e}_8]), \\
&([\mathbf{e}_1 \quad \mathbf{e}_5], \mathbf{e}_5), ([\mathbf{e}_2 \quad \mathbf{e}_6], \mathbf{e}_6), ([\mathbf{e}_3 \quad \mathbf{e}_7], \mathbf{e}_7), ([\mathbf{e}_4 \quad \mathbf{e}_8], \mathbf{e}_8), \\
&([\mathbf{e}_5 \quad \mathbf{e}_1], \mathbf{e}_1), ([\mathbf{e}_6 \quad \mathbf{e}_2], \mathbf{e}_2), ([\mathbf{e}_7 \quad \mathbf{e}_3], \mathbf{e}_3), ([\mathbf{e}_8 \quad \mathbf{e}_4], \mathbf{e}_4)\} \quad (8.10)
\end{aligned}$$

For rank 4, the 16 matrices $\mathbf{W}_2$ of size $2N_b \times 4$ with $N_b = 8$ are of the form

$$\mathbf{W}_2 = \frac{1}{\sqrt{2}}\begin{bmatrix} \mathbf{Y} & \mathbf{Y} \\ \mathbf{Y} & -\mathbf{Y} \end{bmatrix} \quad \text{or} \quad \mathbf{W}_2 = \frac{1}{\sqrt{2}}\begin{bmatrix} \mathbf{Y} & \mathbf{Y} \\ j\mathbf{Y} & -j\mathbf{Y} \end{bmatrix} \tag{8.11}$$

where $\mathbf{Y} \in \{[\mathbf{e}_1 \quad \mathbf{e}_5], [\mathbf{e}_2 \quad \mathbf{e}_6], [\mathbf{e}_3 \quad \mathbf{e}_7], [\mathbf{e}_4 \quad \mathbf{e}_8]\}$

Table 8.10 Size of dual codebooks for 8 transmit antennas

Rank	Nbr beams	$\mathbf{W}_1$	$\mathbf{W}_2$
1–2	32	16	16
3	16	4	16
4	16	4	8
5–7	1	4	1
8	1	1	1

For rank $r = 5, 6, 7$, and 8 the matrices $\mathbf{W}_2$ of size $8 \times r$ are, respectively, given by

$$\mathbf{W}_2 = \frac{1}{\sqrt{2}} \begin{bmatrix} \mathbf{e}_1 & \mathbf{e}_1 & \mathbf{e}_2 & \mathbf{e}_2 & \mathbf{e}_3 \\ \mathbf{e}_1 & -\mathbf{e}_1 & \mathbf{e}_2 & -\mathbf{e}_2 & \mathbf{e}_3 \end{bmatrix} \quad \text{for rank 5} \tag{8.12}$$

$$\mathbf{W}_2 = \frac{1}{\sqrt{2}} \begin{bmatrix} \mathbf{e}_1 & \mathbf{e}_1 & \mathbf{e}_2 & \mathbf{e}_2 & \mathbf{e}_3 & \mathbf{e}_3 \\ \mathbf{e}_1 & -\mathbf{e}_1 & \mathbf{e}_2 & -\mathbf{e}_2 & \mathbf{e}_3 & -\mathbf{e}_3 \end{bmatrix} \quad \text{for rank 6} \tag{8.13}$$

$$\mathbf{W}_2 = \frac{1}{\sqrt{2}} \begin{bmatrix} \mathbf{e}_1 & \mathbf{e}_1 & \mathbf{e}_2 & \mathbf{e}_2 & \mathbf{e}_3 & \mathbf{e}_3 & \mathbf{e}_4 \\ \mathbf{e}_1 & -\mathbf{e}_1 & \mathbf{e}_2 & -\mathbf{e}_2 & \mathbf{e}_3 & -\mathbf{e}_3 & \mathbf{e}_4 \end{bmatrix} \quad \text{for rank 7} \tag{8.14}$$

$$\mathbf{W}_2 = \frac{1}{\sqrt{2}} \begin{bmatrix} \mathbf{e}_1 & \mathbf{e}_1 & \mathbf{e}_2 & \mathbf{e}_2 & \mathbf{e}_3 & \mathbf{e}_3 & \mathbf{e}_4 & \mathbf{e}_4 \\ \mathbf{e}_1 & -\mathbf{e}_1 & \mathbf{e}_2 & -\mathbf{e}_2 & \mathbf{e}_3 & -\mathbf{e}_3 & \mathbf{e}_4 & -\mathbf{e}_4 \end{bmatrix} \quad \text{for rank 8} \tag{8.15}$$

where $\mathbf{e}_k$ is a 8×1 beam selection vector that has 1 on the kth row and zeros elsewhere.

In Table 8.10, we summarize the size of the different dual codebooks.

8.8.4 Codebooks for Uplink

The UL multiple access in LTE is SC-FDMA, where a DFT is followed by the OFDM transmission scheme. Since DFT spreads each modulation symbol over all the subcarriers assigned, it enables single-carrier transmission. Like for the design of downlink codebook, the PMI searching complexity and the trad-off between quantization accuracy and efficiency are important criteria. However, in order to limit the power backoff, the uplink codebook has been designed in order not to increase the cubic metric with respect to single-layer transmissions. Compared to the PAPR, the cubic metric (CM) can be easily computed with an empirical formula and has been agreed as a more reliable predictor of the power derating needed at the transmitter to avoid the degradation due to the nonlinearities of the UE power amplifier [17]. The CM refers to the third power term of the signal, which is the main cause of intermodulation distortions. A low CM property translates to higher power efficiency and to longer operation time. Consequently, all the nonzero elements of the precoding codewords are constrained to be QPSK and are thus constant modulus.

Table 8.11 Codebook for uplink 2 transmit antennas

Index	Rank 1	Rank 2
0	$\frac{1}{\sqrt{2}}\begin{bmatrix} 1 \\ 1 \end{bmatrix}$	$\frac{1}{\sqrt{2}}\begin{bmatrix} 1 & 0 \\ 0 & 1 \end{bmatrix}$
1	$\frac{1}{\sqrt{2}}\begin{bmatrix} 1 \\ -1 \end{bmatrix}$	
2	$\frac{1}{\sqrt{2}}\begin{bmatrix} 1 \\ j \end{bmatrix}$	
3	$\frac{1}{\sqrt{2}}\begin{bmatrix} 1 \\ -j \end{bmatrix}$	
4	$\frac{1}{\sqrt{2}}\begin{bmatrix} 1 \\ 0 \end{bmatrix}$	
5	$\frac{1}{\sqrt{2}}\begin{bmatrix} 0 \\ 1 \end{bmatrix}$	

This choice enables the power amplifiers to transmit with the same power level and reduces the computational complexity required for the PMI searching complexity. Since the chordal distance criterion alone does not yield to a codebook with constant modulus, the uplink codebooks in Release 10 have been designed as a trade-off between chordal distance minimization and constant modulus.

The rank 1 for 2 transmit antennas codebook Table 8.11 is composed of four constant modulus vectors and two antennas selection vectors as shown in Table 8.11. The inclusion of constant modulus vectors enables the signals from multiple UE antennas to be combined constructively at the receiving eNB by introducing a relative phase shift of 0°, 90°, 180°, or 270° between the UE antennas. The two antenna selection vectors reduce the transmit power and allow UE power saving.

As shown in Table 8.12, the rank 1 for 4 transmit antennas codebook consists of 16 constant modulus vectors for constructive combining and 8 antenna-pair turn-off vectors for UE power saving. For constant modulus vectors, four relative phase shifts (0°, 90°, 180°, or 270°) are allowed between the first and second antennas. For each of these relative phase shifts, only two relative phase shifts are considered between the third and fourth antennas. Additional relative phase shifts are applied between the first and third antennas and between the second and fourth antennas in order to take into account the antenna array composed of dual-polarized antennas (Table 8.12–8.14).

In order to preserve the CM constraint, each row of the rank 2 and rank 3 codebook matrices has at most one nonzero element as shown in Tables 8.13 and 8.14. As a consequence, each transmit antenna is not allowed to convey more than one layer. This CM-preserving constraint reduces the precoding gain compared to Grassmannian codebook.

The rank 2 codebook is composed of 16 matrices: 8 matrices involved with antenna grouping (1,2) and (3,4) corresponding to the lower intergroup correlation for dual-polarized antennas, 4 matrices with antenna grouping (1,3) and (2,4), and 4 matrices with antenna grouping (1,4) and (2,3).

Table 8.12 Codebook for uplink 4 transmit antennas rank 1

Index 0 to 7
$$\frac{1}{2}\begin{bmatrix}1\\1\\1\\-1\end{bmatrix}\ \frac{1}{2}\begin{bmatrix}1\\1\\j\\j\end{bmatrix}\ \frac{1}{2}\begin{bmatrix}1\\1\\-1\\1\end{bmatrix}\ \frac{1}{2}\begin{bmatrix}1\\1\\-j\\-j\end{bmatrix}\ \frac{1}{2}\begin{bmatrix}1\\j\\1\\j\end{bmatrix}\ \frac{1}{2}\begin{bmatrix}1\\j\\j\\1\end{bmatrix}\ \frac{1}{2}\begin{bmatrix}1\\j\\-1\\-j\end{bmatrix}\ \frac{1}{2}\begin{bmatrix}1\\j\\-j\\-1\end{bmatrix}$$

Index 8 to 15
$$\frac{1}{2}\begin{bmatrix}1\\-1\\1\\1\end{bmatrix}\ \frac{1}{2}\begin{bmatrix}1\\-1\\j\\-j\end{bmatrix}\ \frac{1}{2}\begin{bmatrix}1\\-1\\-1\\-1\end{bmatrix}\ \frac{1}{2}\begin{bmatrix}1\\-1\\-j\\j\end{bmatrix}\ \frac{1}{2}\begin{bmatrix}1\\-j\\1\\-j\end{bmatrix}\ \frac{1}{2}\begin{bmatrix}1\\-j\\j\\-1\end{bmatrix}\ \frac{1}{2}\begin{bmatrix}1\\-j\\-1\\j\end{bmatrix}\ \frac{1}{2}\begin{bmatrix}1\\-j\\-j\\1\end{bmatrix}$$

Index 16 to 23
$$\frac{1}{2}\begin{bmatrix}1\\0\\1\\0\end{bmatrix}\ \frac{1}{2}\begin{bmatrix}1\\0\\-1\\0\end{bmatrix}\ \frac{1}{2}\begin{bmatrix}1\\0\\j\\0\end{bmatrix}\ \frac{1}{2}\begin{bmatrix}1\\0\\-j\\0\end{bmatrix}\ \frac{1}{2}\begin{bmatrix}0\\1\\0\\1\end{bmatrix}\ \frac{1}{2}\begin{bmatrix}0\\1\\0\\-1\end{bmatrix}\ \frac{1}{2}\begin{bmatrix}0\\1\\0\\j\end{bmatrix}\ \frac{1}{2}\begin{bmatrix}0\\1\\0\\-j\end{bmatrix}$$

For the rank 3 codebook, the first layer is assigned to two transmit antennas, whereas layers 2 and 3 are each assigned to one antenna. Each of six antenna groupings for the first layer is covered.

8.9 Further Works

Up to now, only implicit feedback employing PMI, RI, and CQI has been considered in LTE mainly due to backward compatibility. However explicit feedback where channel matrix $\mathbf{H}$ and/or covariance matrix $\mathbb{E}\{\mathbf{H}^H\mathbf{H}\}$ is quantized and reported from the UE to the eNB is a potential alternative since explicit feedback provides better scheduling flexibility compared to implicit feedback.

In Releases 10 and 11, the adaptation of the antennas at the eNB is only performed in azimuth direction. A promising evolution considered in Release 12 consists to use a two-dimensional antenna array that provides control over both the azimuth and elevation directions. A two-dimensional antenna array enables new strategies such as sector-specific elevation beamforming (e.g., adaptive control over the vertical pattern beamwidth and/or downtilt), advanced sectorization in the vertical domain, and user-specific elevation beamforming. UE-specific elevation beamforming promises to increase the SINR statistics seen by the UEs by pointing the vertical antenna pattern in the direction of the UE while sending less interference to the adjacent sectors due to the ability to steer the transmitted energy in elevation. In order to validate these techniques, new channel models that will modelize both vertical and horizontal dimension of the environment will have to be defined. Feedback strategies in this context are also very challenging [4, 7].

Table 8.13 Codebook for uplink 4 transmit antennas rank 2

Index 0 to 7

$$\frac{1}{2}\begin{bmatrix}1&0\\1&0\\0&1\\0&-j\end{bmatrix} \quad \frac{1}{2}\begin{bmatrix}1&0\\1&0\\0&1\\0&j\end{bmatrix} \quad \frac{1}{2}\begin{bmatrix}1&0\\-j&0\\0&1\\0&1\end{bmatrix} \quad \frac{1}{2}\begin{bmatrix}1&0\\-j&0\\0&-1\\0&-j\end{bmatrix}$$

$$\frac{1}{2}\begin{bmatrix}1&0\\-1&0\\0&1\\0&-j\end{bmatrix} \quad \frac{1}{2}\begin{bmatrix}1&0\\-1&0\\0&1\\0&j\end{bmatrix} \quad \frac{1}{2}\begin{bmatrix}1&0\\j&0\\0&1\\0&1\end{bmatrix} \quad \frac{1}{2}\begin{bmatrix}1&0\\j&0\\0&1\\0&-1\end{bmatrix}$$

Index 8 to 15

$$\frac{1}{2}\begin{bmatrix}1&0\\0&1\\1&0\\0&1\end{bmatrix} \quad \frac{1}{2}\begin{bmatrix}1&0\\0&1\\-1&0\\0&1\end{bmatrix} \quad \frac{1}{2}\begin{bmatrix}1&0\\0&1\\0&1\\0&1\end{bmatrix} \quad \frac{1}{2}\begin{bmatrix}1&0\\0&-1\\0&1\\1&0\end{bmatrix}$$

$$\frac{1}{2}\begin{bmatrix}1&0\\0&1\\1&0\\0&-1\end{bmatrix} \quad \frac{1}{2}\begin{bmatrix}1&0\\0&1\\-1&0\\0&1\end{bmatrix} \quad \frac{1}{2}\begin{bmatrix}1&0\\0&1\\0&1\\-1&0\end{bmatrix} \quad \frac{1}{2}\begin{bmatrix}1&0\\0&-1\\0&1\\-1&0\end{bmatrix}$$

Table 8.14 Codebook for uplink 4 transmit antennas rank 3

Index 0 to 5
$$\frac{1}{2}\begin{bmatrix}1&0&0\\1&0&0\\0&1&0\\0&0&1\end{bmatrix}\quad \frac{1}{2}\begin{bmatrix}1&0&0\\-1&0&0\\0&1&0\\0&0&1\end{bmatrix}\quad \frac{1}{2}\begin{bmatrix}1&0&0\\0&1&0\\1&0&0\\0&0&1\end{bmatrix}\quad \frac{1}{2}\begin{bmatrix}1&0&0\\0&1&0\\-1&0&0\\0&0&1\end{bmatrix}\quad \frac{1}{2}\begin{bmatrix}1&0&0\\0&1&0\\0&0&1\\1&0&0\end{bmatrix}\quad \frac{1}{2}\begin{bmatrix}1&0&0\\0&1&0\\0&0&1\\-1&0&0\end{bmatrix}$$

Index 6 to 11
$$\frac{1}{2}\begin{bmatrix}0&1&0\\1&0&0\\1&0&0\\0&0&1\end{bmatrix}\quad \frac{1}{2}\begin{bmatrix}0&1&0\\1&0&0\\-1&0&0\\0&0&1\end{bmatrix}\quad \frac{1}{2}\begin{bmatrix}0&1&0\\1&0&0\\0&0&1\\1&0&0\end{bmatrix}\quad \frac{1}{2}\begin{bmatrix}0&1&0\\1&0&0\\0&0&1\\-1&0&0\end{bmatrix}\quad \frac{1}{2}\begin{bmatrix}0&1&0\\0&0&1\\1&0&0\\1&0&0\end{bmatrix}\quad \frac{1}{2}\begin{bmatrix}0&1&0\\0&0&1\\1&0&0\\-1&0&0\end{bmatrix}$$

References

1. Boccardi F, Clerckx B, Ghosh A, Hardouin E, Jongren G, Kusume K, Onggosanusi E, Tang Y (2012) Multiple-Antenna Techniques in LTE-Advanced. IEEE Comm. Mag. 50: 114–121
2. Dahlman E, Parkvall S, Skold J (2011) 4G LTE/LTE-Advanced for Mobile Broadband. Academic Press.
3. Duplicy J, Badic B, Balarj R, Ghaffar R, Horvath P, Kaltenberger F, Knopp R, Kovacs I, Nguyen H T, Tandur D, Vivier G (2011) MU-MIMO in LTE Systems. EURASIP Journal On Wireless Commun. and Networking. vol. 2011, ID 496763, 1–13
4. Halbauer H, Saur S, Koppenborg J, Hoek C (2012) Interference Avoidance with Dynamic Vertical Beamsteering in Real Deployments. Wireless Communications and Networking Conference Workshops (WCNCW). 294–299
5. Ghosh A, Ratasuk R, Mondal B, Mangalvedhe N, Thomas T (2010) LTE-Advanced: Next-Generation Wireless Broadband Technology. IEEE Wireless Commun. 17: 10–22
6. Ghosh A, Ratasuk R (2011) Essentials of LTE and LTE-A. Cambridge University Press
7. Koppenborg J, Halbauer H, Saur S, Hoek C (2012) 3D Beamforming Trials with an Active Antenna Array. Workshop on Smart Antennas (WSA), Dresden, Germany. 110–114
8. Irmer R, Droste H, Marsch P, Grieger M, Fettweis G, Brueck S, Mayer H, L Thiele, V Jungnickel (2011) Coordinated Multipoint: Concepts, Performance, Field Trial Results. IEEE Communications Magazine 49: 102–111
9. Li Q, Furuskar A, Li G, Lee W, Lee M, Mazzarese D, Clerckx B, Li Z (2010) MIMO Techniques in WiMAX and LTE: A Feature Overview. IEEE Commun. Mag. 48: 86–92
10. Lim C, Yoo T, Clerckx B, Lee B, Shim B (2013) Recent Trend of Multiuser MIMO in LTE-Advanced. IEEE Communications Magazine. 51: 127–135
11. Love D J, Heath R W (2006) Limited feedback diversity techniques for correlated channels. IEEE Trans. on Veh. Technol. 55: 718–722
12. Park C S, Wang Y P E, Jongren G, Hammarwall D (2011) Evolution of Uplink MIMO for LTE-Advanced. IEEE Communications Magazine. 49: 112–121
13. Parkvall S, Furuskar A, Dahlman E (2011) Evolution of LTE toward IMT-Advanced. IEEE Communication Magazine. 49: 84–91
14. Shuang T, Koivisto T, Maattanen H, Pietikainen K, Roman T, Enescu M (2011) Design and evaluation of LTE-Advanced double codebook. Proc. IEEE Vehicular Technology Conference Spring (VTC). 1–5
15. 3GPP TSG RAN WG1 #62 (2010) Way forward on 8 Tx Codebook for Release 10 DL MIMO. R1-105011, Madrid, Spain
16. 3GPP TSG RAN WG1 #62 (2010) Way Forward on Aperiodic PUSCH CQI modes in Release 10. R1-105058, Madrid, Spain
17. 3GPP TSG RAN WG1 #44 (2006) Cubic Metric in 3GPP-LTE. R1-060385

18. 3GPP (2005) Soft Frequency Reuse Scheme for UTRAN LTE. 3rd Generation Partnership Project (3GPP), Project Document R1-050507
19. 3GPP, Downlink inter-cell interference coordination /avoidance evaluation of frequency reuse, 3rd Generation Partnership Project (3GPP), Project Document, Ericsson R1-061374
20. 3GPP TR 36.819 v11.1.0 (2006) Evolved Universal Terrestrial Radio Access (E-UTRA) and Evolved Universal Terrestrial Radio Access Network (E-UTRAN); Coordinated Multi-Point Operation for LTE Physical Layer Aspects (Release 11). TSG RAN.
21. 3GPP (2010) Mobile Broadband Innovation path to 4G: Release 9,10 and Beyond. 3rd Generation Partnership Project (3GPP)
22. 3GPP (2009) CoMP Configurations and UE/eNB Behaviors in LTE Advanced. 3rd Generation Partnership Project (3GPP), LG Electronics R1-090782
23. 3GPP (2009) Multi-cell PMI coordination for downlink CoMP. 3rd Generation Partnership Project (3GPP), ETRI R1-091490

Chapter 9
Conclusions

We conclude this book by summarizing and highlighting its contribution in the field of feedback strategies for wireless communication. We will also discuss some open research topics that were identified during the writing of this book.

We first studied in Chap. 3 the case of a single user wireless communication system where both the transmitter and the receiver are equipped with a single antenna. After a review of the capacity of this channel, we have studied the adaptive transmission over time and frequency where rate and power are adapted in order to maximize the spectral efficiency. The adaptive modulation and coding where a joint optimization of the coding rate and modulation is a practical scheme to approach capacity. We have then studied channel prediction at the transmitter to compensate the delay due to feedback link. While average spectral efficiency is not affected by the time delay, there is a significant degradation of the average bit error depending on the length of the prediction filter. For wideband channel, due to correlation of the channel in frequency, the amount of feedback can be reduced by performing data compression. Another classical feedback strategy is the ARQ schemes that allow to build a reliable data transmission using ACK/NACK feedback message and data retransmission.

Chapter 4 was dedicated to single user MIMO wireless communication system. We have studied beamforming and linear precoding where the transmitted data are premultiplied, respectively, by a vector and a matrix. We have seen that the codebook can also be seen as a finite set of subspaces in the Grassmannian manifold. We have derived the codebook criterion and have studied the different methods of construction of codebooks. We have studied the different techniques to exploit the spatial and time and frequency correlation of the channel. We have shown that the codebook design should be adapted to spatial correlation by performing local packing. For time correlation, differential feedback strategies allow to reduce the quantization errors.

In Chap. 5, we have considered multiuser wireless systems composed of single antenna transmitter and receiver. In addition to limited feedback, user scheduling and reduced feedback strategies have to be studied. The user scheduling has been developed by considering different criteria such as the maximization of sum rate or

B. Özbek and D. Le Ruyet, *Feedback Strategies for Wireless Communication*,
DOI 10.1007/978-1-4614-7741-9_9, © Springer Science+Business Media New York 2014

the fairness. We have demonstrate that it is important to perform a selection at the user side to reduce the feedback load. Finally, the user selection strategies based on threshold criterion have been examined for single-carrier and multicarrier systems.

Chapter 6 has been dedicated to multiple antenna for multiuser systems. We have mainly focused on the downlink transmission by considering both users with one and multiple receive antennas. The precoding and user selection algorithms are examined and the performance results for multiuser schemes which allocate more than one user have been illustrated with full and quantized feedback. In order to further reduce feedback load, the user selection at the receiver side has been performed for OFDMA-based multiuser MIMO systems. Both the quantization and delay issues have been examined by giving MAT algorithm and reducing the quantization error on the CDI by designing the CDI codebooks using a local packing by taking into account the users' direction. In the last part, we have considered multiuser MIMO systems with multiple receive antennas with precoding and user scheduling algorithms and the performance results have been illustrated for quantized feedback. The performance results have indicated that multiuser MIMO systems with multiple receive antennas improves the performance compared to multiuser MISO systems.

In Chap. 7 we consider resource allocation techniques and quantized feedback for cooperative multicell networks to mitigate ICI. The resource allocation including user scheduling and power allocation has been presented for cooperative multicell networks with and without a centralized unit. It is shown that the optimization problem is not convex and some iterative solution can be proposed with some relaxations. Joint and iterative user scheduling and power allocation algorithms have been examined and the performance analysis has been shown according to sum cell capacity. The limited feedback link for multiple antennas has been considered and the capacity analysis of quantized feedback has been presented in detail based on different criteria. In order to achieve sum cell capacity improvement on the multicell network, maximum ratio combining and partial zero forcing beamforming strategies for the design of the beamforming vectors have been examined. It has been shown that the required number of bits to quantize for transmission strategies can be chosen according only to the users' location and the degree of the received power at the edge of the cells.

In Chap. 8, we have describe some practical issues such as codebook designs and algorithms used in wireless communication system standards including LTE and LTE-advanced.

Some promising research directions in the field of feedback strategies have been also pointed out in this book such as:

- 3D beamforming that is generalization of the concept of 2D beamforming.
- Improvement of the prediction of the time/frequency channel to extend the range of application of closed-loop MIMO and the usage of outdated channel state information in MIMO systems.

- Feedback of CSI in the context of interference channels. While interference alignment is a multiplexing gain optimal transmission scheme for interference channel, it requires network CSI at the transmitters.
- Efficient limited feedback designs for massive MIMO systems that include large number of transmit antennas.

Index

B. Özbek and D. Le Ruyet, *Feedback Strategies for Wireless Communication*,
DOI 10.1007/978-1-4614-7741-9, © Springer Science+Business Media New York 2014